POLYMERIC DELIVERY SYSTEMS

MIDLAND MACROMOLECULAR MONOGRAPHS

A series of monographs based on special symposia held at the Midland Macromolecular Institute

Editor: Hans-Georg Elias

International Standard Serial Number: 0141-0342

Volume 1 Trends in Macromolecular Science
Edited by Hans-Georg Elias

Volume 2 Order in Polymer Solutions
Edited by Karel Šolc

Volume 3 Polymerization in Organized Systems
Edited by Hans-Georg Elias

Volume 4 Molecular Basis of Transitions and Relaxations
Edited by Dale J. Meier

Volume 5 Polymeric Delivery Systems
Edited by Robert J. Kostelnik

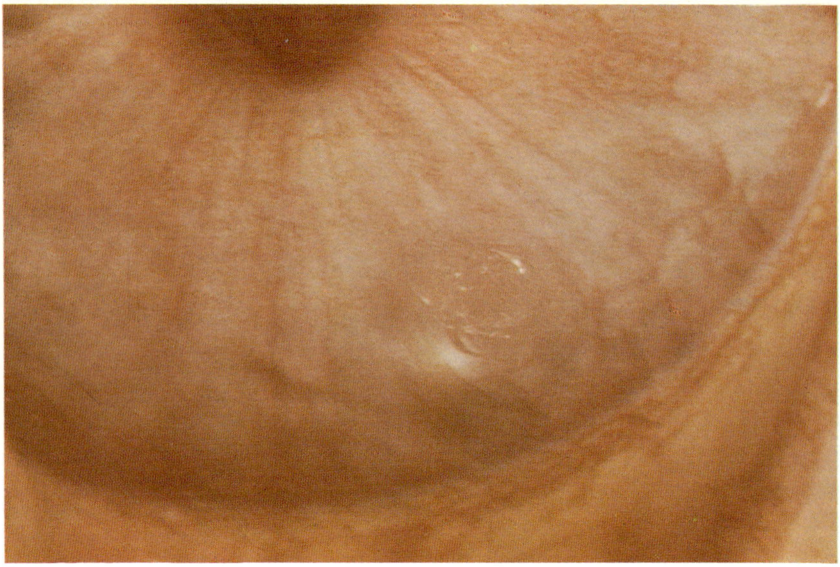

(a)

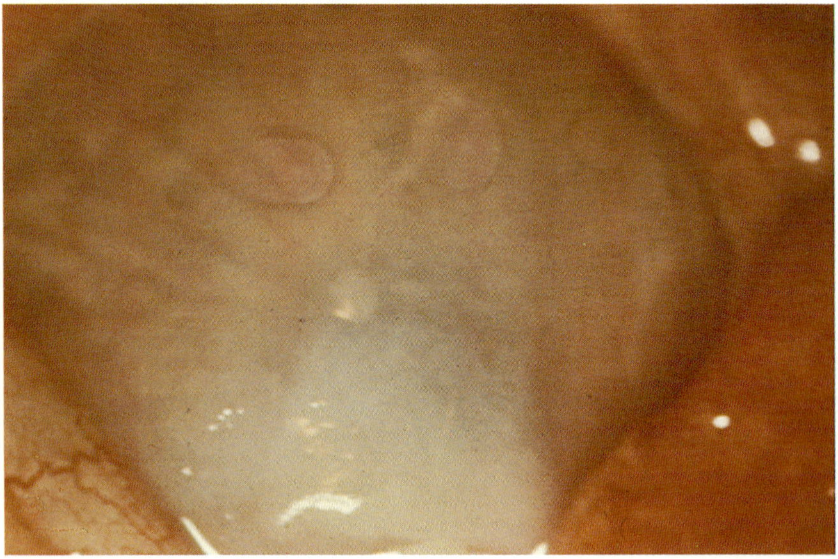

(b)

FIGURE 3. (a) Rabbit cornea 3 days after implantation of washed ethylene-vinyl acetate copolymer pellet. (b) Rabbit cornea 3 days after implantation of poly(vinylpyrrolidone).

POLYMERIC DELIVERY SYSTEMS

Edited by
ROBERT J. KOSTELNIK
Midland Macromolecular Institute

GORDON AND BREACH SCIENCE PUBLISHERS
New York London Paris

Gordon and Breach Science Publishers, Inc.
One Park Avenue
New York, NY10016

Gordon and Breach Science Publishers Ltd.
42 William IV Street
London WC2N 4DE

Gordon & Breach
7–9 rue Emile Dubois
75014 Paris

Library of Congress Cataloging in Publication Data
Main entry under title:
Polymeric delivery systems.
(Midland macromolecular monographs; v. 5)
Includes index.
1. Delayed-action preparations—Congresses.
2. Drugs—Dosage forms—Congresses. I. Kostelnik, Robert J. II. Series. [DNLM: 1. Polymers—Congresses. 2. Macromolecular systems—Congresses. QD281.P6 P783 1976]
RS201.D4P64 615′.191 77-19089
ISBN 0-677-15940-4

Library of Congress catalog card number 77-19089. ISBN 0 677 15940 4. ISSN 0141-0342.

Printed in Great Britain.

Preface

This volume comprises 18 papers presented at the Midland Macromolecular Symposium, *Polymeric Delivery Systems*, held at the Midland Macromolecular Institute, Midland, Michigan on August 23–27, 1976.

This symposium, the fifth in MMI's continuing program of international symposia on special topics in macromolecular science, attracted over 90 participants and truly exemplified the interdisciplinary nature of macromolecular science. The attendees included polymer chemists, polymer physicists, chemists (organic and physical), physicists, biochemists, biologists, pharmacologists, chemical engineers and medical doctors.

The program featured seven invited papers reviewing various topics concerning polymeric delivery systems and twelve contributed papers reporting current research results. The invited papers were given by H. Hopfenberg, G. Banker, A. Trouet, H. Ringsdorf, W. Leonard, A. Casola, R. Langer and J. Folkman. A contributed paper on "The Use of Polymers in Long Acting Anti-Fertility Drug Delivery Systems. A Review of World Health Organization Activities" presented by M. Goodman is not included in this monograph at the request of the WHO.

In many cases the discussion following a lecture is as important as the lecture itself. For this reason the discussions of the papers as edited by the participants are also included in these proceedings.

I wish to take this opportunity to express my sincere thanks to all speakers, chairpersons and others who contributed to the success of this meeting. My special appreciation is extended to Dr. D. Bennett who originally suggested the topic for this fifth Midland Macromolecular Meeting.

Midland, Michigan ROBERT J. KOSTELNIK

Contents

Formation and Structure of Synthetic Membranes, Films, and Microcapsules

HAROLD B. HOPFENBERG

Department of Chemical Engineering, North Carolina State University, Raleigh, NC27607, U.S.A.

The various techniques for fabrication of films, synthetic membranes, and microcapsules are compared and contrasted. The discussion of film, membrane, and microcapsule formation is complemented by presentation of those structural features and macroscopic properties which distinguish films from semi-permeable membranes. The historical development of film and membrane synthesis forms the basis of the more recent development of microencapsulation.

Transport properties of the various structures are discussed in detail since, in most applications of these polymeric forms, and especially in polymeric delivery systems, permselectivity or barrier properties are of paramount importance. In this regard, steady-state and unsteady-state transport behaviour are discussed in terms of the specific requirements of the various polymeric delivery forms.

1 FORMATION OF FILMS, MEMBRANES, AND MICROCAPSULES

1.1 Films

In the late 1920's and through the 1930's Carothers and other chemists directed their polymer-related research toward the synthesis of "super polymers" suitable for formation of high strength fibers.[1] Much of the early effort was directed toward the development of a synthetic replacement for silk. In the course of these early investigations, it became apparent that many of the polymers which were suitable for fiber-forming applications were also appropriate selections for the preparation of transparent films.

For approximately 20 years, however, regenerated cellulose dominated the transparent film market. At the time, cellulose esters including cellulose acetate were found too expensive to make a significant inroad into the burgeoning cellophane market. The discovery of the aromatic polyester, polyethylene terephthalate, led to the concomitant characterization and recognition that polyethylene terephthalate provided significant mechanical advantages compared with cellophane.[2]

Polyethylene terephthalate was followed in the late 1940's and early 1950's by large-volume applications of polyethylene. By 1960 non-cellulosic based polymers had captured 40% of the film market. In addition to polyesters and polyamides (nylons), various vinyl compounds including poly(vinyl chloride) and polyvinylidene chloride copolymers had entered the marketplace. Biaxial orientation of polystyrene has rendered that otherwise brittle commodity polymer into a mechanically tough film material. Polyolefin films are typically blown during the extrusion process, introducing biaxial orientation and, in turn, improved mechanical properties. Poly(vinyl fluoride) and rubber hydrochloride are examples of halogenated film-forming fibers. Poly(phenylene oxide) and polycarbonate provide glassy polymer films with remarkable toughness. Polyvinyalcohol, poly(vinylpyrollidone), and poly-(ethylene oxide) are water soluble film formers.

In addition to conventional homopolymers, polymer blends, graft copolymers, and block copolymers have been used successfully as film forming materials. The various combinations of individual homopolymers, blending techniques, conventional copolymerization, and block and graft copolymerization provides a virtually endless array of possibilities to the film formulator. There are many subtle molecular variables which affect film formation and, in turn, ultimate properties. Specifically, for a given copolymer with a known composition the comonomer can be distributed along the polymer chain in a variety of ways. The comonomer groups might alternate or conversely be distributed randomly along the polymer chain. The comonomer might be grouped as blocks along the polymer chain and, furthermore, hydrolysis or other side reactions might have occurred either inadvertently or as an additional processing variation. Tacticity of the pendant comonomer groups is an additional structural variable, related to film formation and, in turn, ultimate properties.

In addition to the myriad compositional variables that affect ultimate film properties, it has become increasingly common to laminate polymer films to achieve desired properties. Lamination may be achieved by a process as complex as coextrusion[3] or, more classically, by a simple coating of an already formed and consolidated polymeric substrate. Cellophane is typically coated to provide suitable barrier properties.

An interesting example of composite film manufacture is related to the sandwich construction of polyethylene–poly(ethylene-co-vinylalcohol)–polyethylene.[3] Whereas poly(vinylalcohol) is an excellent barrier to oxygen or nitrogen transport at low relative humidities, poly(vinylalcohol) provides very little resistance to gas transport when the poly(vinylalcohol) film is swollen on contact with humid air. Transient barrier performance can be provided by laminating the poly(vinylalcohol) between films of polyethylene. Although at suitably long times, the sandwiched poly(vinylalcohol) will, in fact, become swollen when contacted with humid air, transient barrier properties can be achieved by this lamination technique. In any event, the permeability of the laminate to oxygen and/or nitrogen under any humidity condition will be lower than the permeability of the unprotected poly(vinylalcohol).

Polymer films can be formed either by melt extrusion or by casting from an appropriate solvent or solvent mixture. In either case, some degree of asymmetry is involved in the film forming process since the heat or mass (solvent or solvents) are removed unidirectionally. Typically, a polymer solution is cast onto a suitable substrate and the solvent is removed from only one surface. Alternatively, in the case of melt extrusion the polymer may, in fact, be cooled on both surfaces; however, the cooling is rarely identically balanced and some degree of asymmetry is typically imposed upon the resulting film structure.

In general, polymeric materials derive mechanical properties of commercial interest by introducing one or more of the three basic architectural features; crystallinity, crosslinking, or immobilization of chain segments resulting in glassy behavior are required to provide stable film structures. Although polymeric films are typically either crystalline and/or glassy, they are not usually crosslinked. In contrast, coatings are typically glassy and, moreover, in many cases crosslinked.

The detailed nature of the crystalline morphology of a polymer film is related in a complex, albeit understandable, manner to the detailed thermal and mechanical history imposed during film preparation. The gross mechanical properties as well as important film properties including transparency and permeability are, of course, significantly affected by the detailed nature of the resulting polymer morphology.

Some films are prepared by dissolving a polymer in a solvent or solvent mixture followed by the application of the resulting liquid film onto a suitable substrate and ultimately by complete evaporation of the solvent. Complete solvent removal is achieved by maintaining the formed-film at temperatures in excess of the glass transition temperature consequent to film consolidation, since solvent transport in the resulting rubbery polymer is markedly more rapid than small molecule transport in polymers below their respective glass transition temperatures. The presence of even small quantities of residual solvent markedly affects most of the important film properties. The rate of crystallization in many cases is strongly affected by the rate of evaporation for solvent cast film systems. The crystallization rate is essentially zero until an indication period, corresponding to early evaporation, is completed.[4] Consequent to initial evaporation, the rate of crystallization becomes increasingly rapid. Thermal annealing at temperatures in excess of the glass transition temperature can alter the development of the crystallinity and facilitate complete removal of the casting solvent. The presence of non-volatile, permanent plasticizers is quite akin to the use of multicomponent solvent mixtures including high boiling, albeit removable, solvents. Plasticizers and high boiling solvents tend to improve long-term film consolidation during the post-forming processing.[5] Moreover, film properties are affected by the relative humidity[6] and/or the partial pressure of solvating components in the environment contacting the film during consolidation.

In summary, film formation can be achieved by either extrusion or casting of polymer melts or by the casting of polymer solutions. In addition to the wide variety of polymers available as film formers, a vast array of processing variables are effective in altering film properties.

1.2 Membranes

Membranes are usually distinguished from films in that membranes are either highly permeable or permselective to the transport of one or more of the solute or solvent species contacting the upstream membrane surface. In contrast, films are typically barriers to the transport of most materials. In general the world "film" is used to describe a two-dimensional polymeric form whose ultimate application does not necessarily depend upon transport behavior. In most cases, properties such as transparency, tear strength, impact strength, abrasion resistance, stiffness, dielectric properties, tenacity, and economic considerations override the requirements for specific barrier properties or permeability to specific components. When in fact transport properties are of interest, barrier properties, or impermeability, is usually desirable.

In contrast, the term "membrane" is assigned to a two-dimensional polymeric structure which offers specific permeability to a component or components in a process feed to the membrane.

Since economic considerations usually dictate that the membrane productivity or flux be maximized, there is a very large incentive for synthesizing membranes with extremely thin, rate-determining transport barrier layers. This requirement would appear to be at odds with the fundamental requirement that the membrane exhibit reasonable mechanical integrity. This seeming dichotomy between the requirements for ultrathin, selective barrier layers and mechanical integrity, usually attributed to microscopic film structures, has been resolved in several rather ingenious ways.

Ultrathin membranes have been developed wherein the barrier layer is less than 1 micron thick.[7,8] These ultrathin layers can be placed on top of porous supports with suitably small pore diameters. Alternatively, ultrathin films can be cast directly onto porous supports.[9] The resulting composite can be treated as a mechanically stable entity; however, the transport properties of the overall composite are dictated rather completely by the geometry and physico-chemical description of the ultrathin barrier layer.

Alternatively, essentially the same level of asymmetry can be imparted to a membrane by a direct casting procedure pioneered by Loeb and Sourirajan.[10] The Loeb–Sourirajan technique involves phase-inversion casting of cellulose acetate from a ternary solvent system including water, magnesium perchlorate, and acetone. Rather surprisingly, the magnesium perchlorate–water solution, in sufficiently high concentrations of the perchlorate salt, serves as a solvent for the cellulose acetate. Pure acetone is also a solvent for the cellulose acetate. Whereas the magnesium perchlorate is completely non-volatile, the acetone is exceedingly volatile and, consequent to casting, the acetone readily volatilizes reducing the "solvent power" of the remaining solution, resulting in concomitant gelling of a surface layer. Clearly, the compositional details of the casting solution are extremely critical variables which, in turn, affect ultimate membrane properties. Specifically, the acetyl content of the cellulose acetate[11] and the concentration of cellulose acetate affect membrane performance. Data describing the quantitative effect of acetyl content on the permeability of cellulose acetate to sodium chloride and water is presented in Figure 1. Cellulose acetate is available in a variety of degrees of acetylation and in various molecular weight (viscosity) ranges.

Consequent to selection of a suitable cellulose acetate fraction, the weight concentration of cellulose acetate in the solvent system is specified. The solvent is formulated in such a way that the polymer is precariously dissolved in the complex solvent and consequent to initial evaporation of acetone, the surface layer gels. The gelled membrane is then immersed in water at a prescribed temperature for prescribed time periods. The counter-current transport of magnesium perchlorate and water, consequent to water-immersion, results in virtually complete gelling of the membrane structure. The resulting structure is best characterized as a thin, rather dense surface layer overlaying a porous, albeit continuous, reticulum of cellulose acetate supporting material. The consolidation of the surface layer can be improved by annealing in water at temperatures in the range of 70–85°C.

The effect of annealing temperature on the water flux and salt rejection of cellulose acetate membranes is presented in Figure 2. As annealing temperatures increase, the salt rejection increases and the permeability to water decreases.[12] This phenomenon involving increased rejection at the expense of product throughput appears to be a fundamental limitation impeding the optimization of membrane systems. This general relationship is presented quantitatively in Figure 3. The inherent asymmetry of membrane systems permits the apparent increase in overall permeability, calculated on the basis

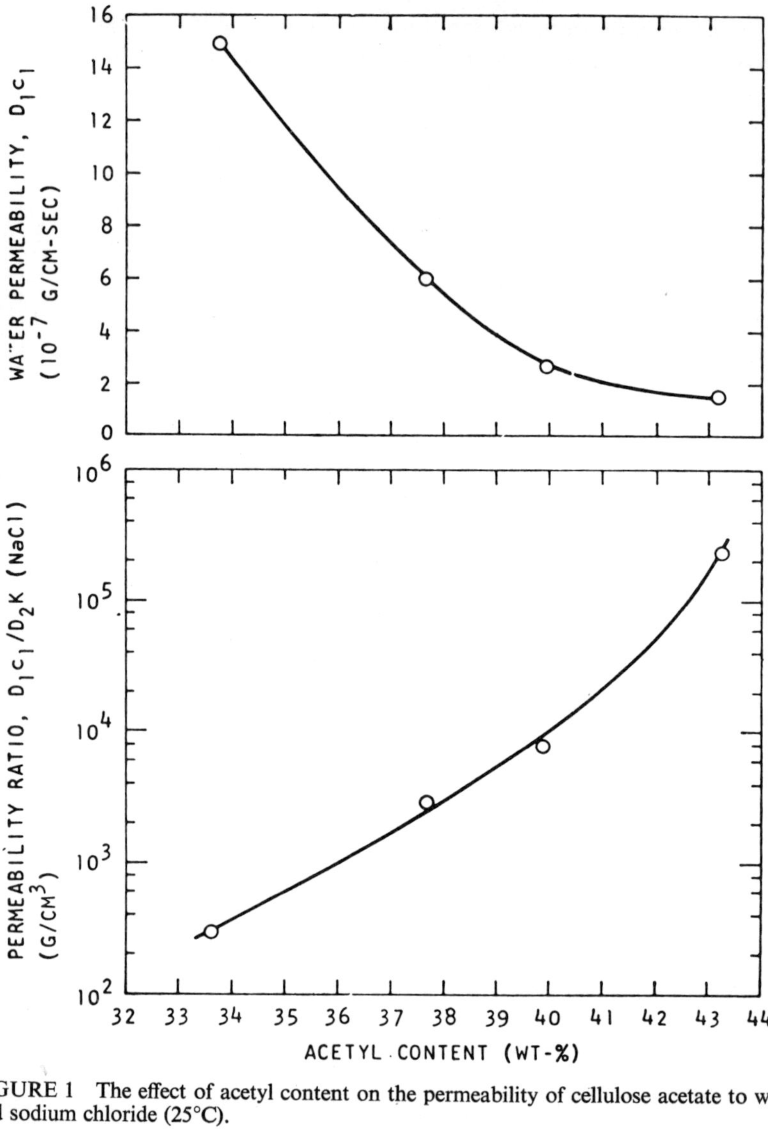

FIGURE 1 The effect of acetyl content on the permeability of cellulose acetate to water and sodium chloride (25°C).

of the macroscopic membrane thickness. The actual permeability, however, of the thin (approximately 0.2 micrometer) dense layer is essentially identical to the permeability of a dense consolidated, symmetrical cellulose acetate film. The striking advantage of the Loeb–Sourirajan technique (later refined by Manjikian,[13] eliminating the magnesium perchlorate) resides in the ability to prepare membranes which are at once highly permeable, owing to the thin barrier layer, and mechanically stable as a consequence of the macroscopic dimensions of the supporting sublayer of the composite membrane system.

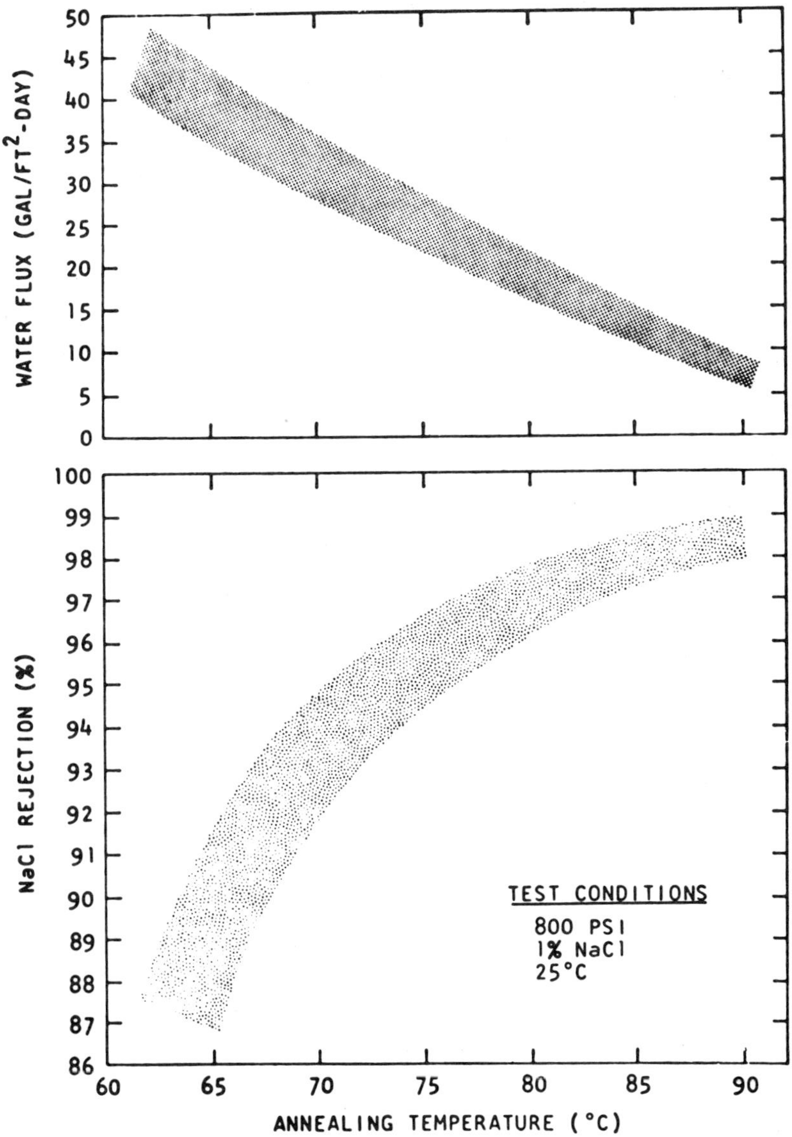

FIGURE 2 Effect of annealing temperature on the product flux and salt rejection of asymmetric cellulose acetate membranes.

The "fine tuning" of the transport behavior of the controlling dense layer, by various post-forming procedures, is directed toward improving solute rejection without excessive sacrifice of product throughout.

The explosion of interest in membrane processes is in large measure a consequence of the large-scale funding provided by the Office of Saline Water

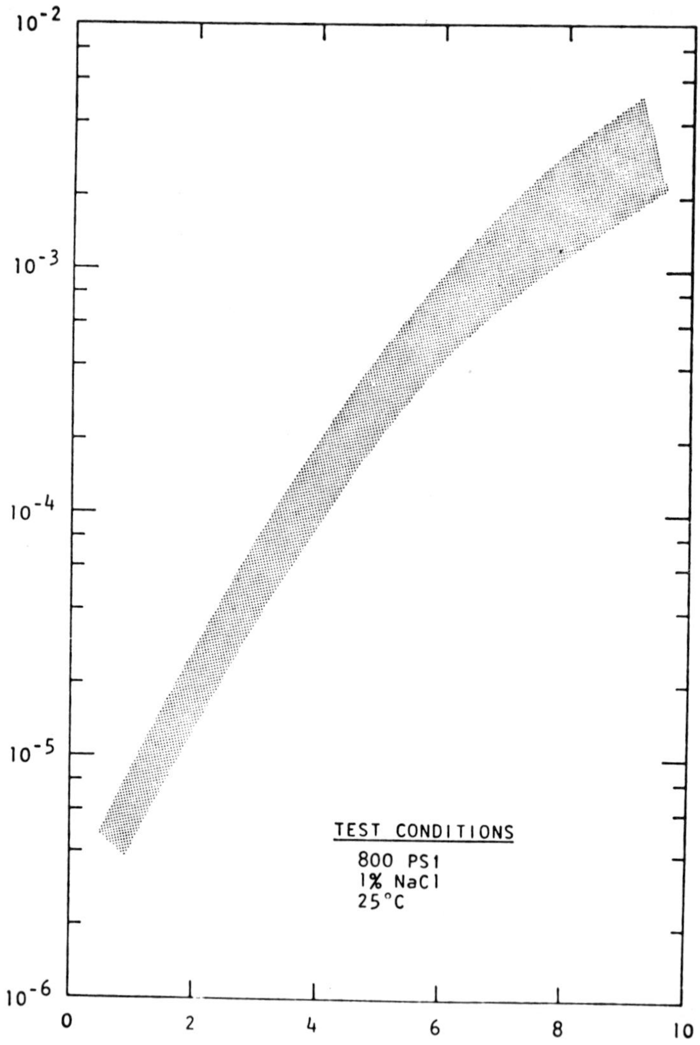

FIGURE 3 Relationship between salt permeability and water permeability for asymmetric cellulose acetate membranes.[12,46]

during the 1960's.[14] During this period, a wide variety of polymer types and polymer combinations were evaluated as candidates for electrolyte separations from water. In addition to cellulose acetate, the polyamides and aromatic polyamides were observed to show suitable flux-rejection behavior to merit commercialization. Moreover, a variety of cellulose ester membranes was developed. Each of these various types of membranes appears to be suitable for a specific salt separation. These membrane systems include a blend of cellulose diacetate and cellulose triacetate,[15,16] cellulose acetate butyrate,[17] as well as thin film composites involving dense ultrathin cellulose acetate membranes.[18–20]

The systematic study of formation, structure, and properties of the cellulose acetate membrane system was complimented by a variety of surveys of alternate membrane materials. None of these studies was performed in nearly the detail of the complimentary investigations of cellulose acetate. However, a rather large body of literature exists which describes probing experiments to evaluate each of these alternate non-cellulosic membrane materials including dynamically-formed membranes prepared by deposition of colloidal material on suitable porous supports.[21–27] In general, these materials were studied in the form of dense, consolidated, virtually symmetrical films, and transport across these membranes was, in most cases, governed by simple diffusion kinetics uncomplicated by parallel pore transport.

It is quite common to describe porous materials as membranes since these structures are synthesized to effect selective transport. In some cases, rejection of colloidal-sized particles is desired. Porous cellulose-derivative membranes, prepared commercially, are examples of membranes which exclude particles larger than a fraction of a micrometer, while passing low molecular weight electrolytes and solvents. These membranes appear to be a cohesive system of open cell foams comprised of the cellular network bound together by chainlike ribs which extend in three dimensions.[28,29]

In contrast, membranes recently offered by the Nucleopore Corporation are characterized by a regular array of cylindrical capillaries formed by controlled irradiation of, for example, polycarbonate substrates. Similarly regular structures were prepared by Reimschuessel in membranes termed ionotropic gel structures.[30]

In summary, membranes permit transport of molecular species at rates which are consistent with product or process requirements. Moreover, in many cases, membranes provide selective transport of the desired component to the exclusion of all other species contacting the membrane surface. Membranes may be dense, asymmetric, microporous, or various combinations of these basic morphologies. Typically, asymmetry is introduced by preparing laminated or coated composites of ultrathin barriers overlaying porous supports or alternatively by a complex, phase-inversion casting process.[31] Membranes can be prepared as flat sheets, cylindrical tubes, spiral-wound modules or fine hollow fibers. All of these independent configurations are asymmetric, transverse to the principle planar dimensions. Photomicrographs of cross-sections of flat membranes and hollow fine fibers are presented as Figures 4 and 5, respectively.

Membrane transport may be controlled by solution-diffusion through

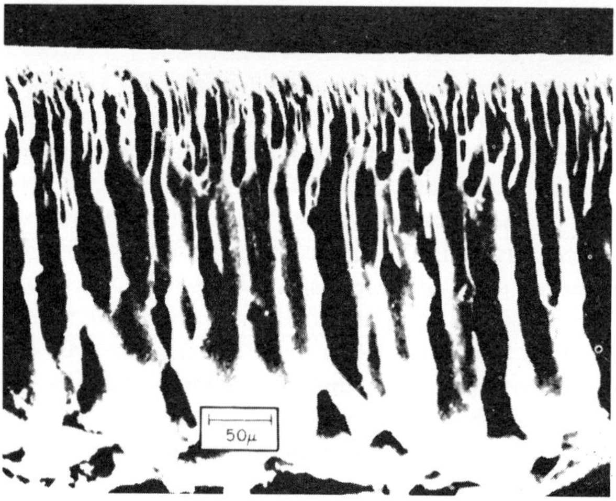

FIGURE 4 Photomicrograph of cross-section of a typical asymmetric membrane cross-section.[45,47]

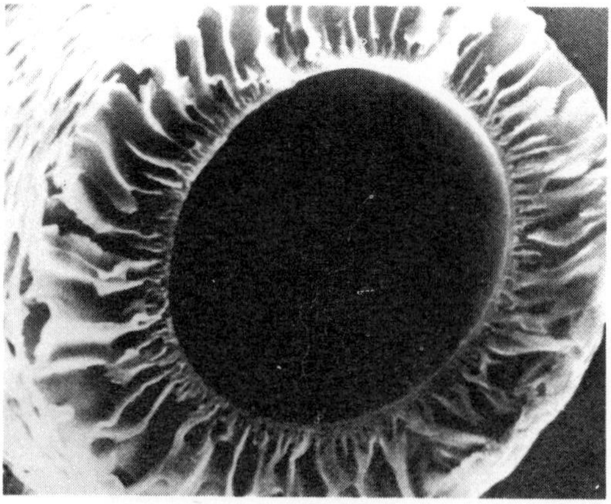

FIGURE 5 Photomicrograph of cross-section of an asymmetric fine hollow fiber.

dense surface layers or through consolidated symmetrical films. In contrast, the overall transport may be controlled by convective or diffusive flow through well defined, liquid-filled pores.

1.3 Microcapsules

Microcapsules may be formulated to provide barrier properties usually associated with film structures or selective transport behavior usually associated with membranes. Microcapsules are formulated typically to protect the interior of a capsule from environmental attack and to control the transport of the internal core material across the microcapsule wall. These requirements are in many respects similar to the requirements of living plant and animal cells. The present implications of synthetic construction of structures which function even crudely in the manner of the basic building blocks of living forms, suggest a truly endless array of applications of this rather recent technology.

Microcapsules, ranging in size from several tenths of a micrometer to 5000 micrometers, can be prepared by a wide variety of techniques. The early development of microcapsules centered upon coacervation of gelatin and gum arabic around an oil droplet containing a dye-precursor.[32,33] Photomicrographs of microcapsules prepared in this manner are shown in Figure 6. These microcapsules were coated on the reverse side of paper. A second layer of paper coated with an acidic clay receptor surface is placed directly under the microcapsule coated paper. In this manner, consequent to rupture of the microcapsules by writing or typing, the dye-precursor is transferred to the clay receptor, producing a color-forming reaction.

FIGURE 6 Photomicrograph of a gelatin microcapsule containing oil core.

Coacervation involves the concentration of a colloid consequent to salt addition, dilution, temperature change, or variation of pH. As opposed to complete precipitation in organic systems, coacervation is characterized by continuity of water between the supernatant phase and the concentrated coacervate phase. Microcapsule formation *via* coacervation usually proceeds by preparing a coacervate-precursor phase composed typically of hydrocolloids dissolved in water. The oil phase is then dispersed as microdroplets within the hydrocolloid-containing phase. The hydrocolloid typically aggregates at the oil–water interface and, consequent to a change in concentration, temperature, pH, or chemical composition, the coacervation occurs at the interface between the oil droplet and the supernatant water. This liquid-like coacervate phase can be "hardened" by crosslinking. Crosslinking in the case of gelatin–gum arabic systems is effected by formaldehyde or gluteraldehyde.

Alternatively, polar solvents including aqueous solutions in water can be encapsulated by organic materials.[34] Timed-release encapsulated analgesics, encapsulated liquid crystals, and adhesive capsules employ this alternative technique termed "phase separation." Recent innovations in this regard involve combinations of phase separation technology with coacervation.

Microcapsules can be formed by dehydration of emulsions containing a dispersed oil phase. Early work in this regard centered upon the encapsulation of vitamins to impart long-term stability to the resulting encapsulated vitamin concentrate.

The discovery of a process for synthesis of water-soluble dextrin permitted the use of this water-soluble polymer as an encapsulating agent for surrounding flavor or fragrance oils into microcapsules.[35] Typically, these microcapsules are formed as a consequence of spray drying subsequent to the encapsulation of the dispersed oil phase. In contrast to capsules prepared by coacervation and phase separation the spray-dried microcapsules are comprised of a myriad of tiny dispersed oil droplets within an encapsulating matrix.

The use of fluidized bed technology to perform microencapsulation was pioneered at the Battelle Memorial Institute. Solid core particles can be fluidized into an environment containing a liquid coating solution or a molten encapsulating medium to form microcapsules.[36] This technology can be extended directly to the encapsulation of materials normally used in the liquid state if these liquids are first frozen into solid particles prior to fluidization. Fluid bed technology is limited to the preparation of capsules which are relatively large, typically in the range of 200 micrometers and larger. This limitation accrues from the difficulty associated with effecting stable fluidization of powders composed of smaller particles.

Chemical vapor deposition is an elegant extension of the basic techniques for fluidized bed microcapsule formation. In one example process, nuclear fuel particles are coated with carbon by condensing the products of thermally decomposed hydrocarbons on suspended, fluidized core particles.[37] Alternatively, vacuum metallization techniques have been applied to microcapsule formation.[38] These techniques are, of course, bound by the same limitations inherent to the fluidized bed techniques for microencapsulation.

In situ polymerization is, essentially, a variation of chemical vapor deposition technology. The polymerization reactions can be carried out at much lower temperatures and, therefore, this technology promises much wider commercial acceptance. Typically, core particles are suspended in a liquid medium containing a polymerizable monomer. Consequent to polymerization on a core particle, microcapsules are formed. This technique has been reduced to practice by the National Lead Company wherein particles are suspended in toluene containing a Ziegler-type catalyst; ethylene gas is bubbled through this mixture and polymerized forming the requisite microcapsules.[39]

The classic experiment, involving contact of an organic solution containing an acid chloride with an aqueous solution of a diamine to produce an interfacial film between the two liquids, forms the basis of some intriguing chemistry directly applicable to the synthesis of microcapsules. Specifically, if droplets of one solution are added to the other with appropriate stirring, microcapsules can be obtained. The resulting interfacial polymer can be suitably crosslinked to achieve the desired level of permeability or barrier properties.[40] The controlled permeability of these nylon microcapsules form the basis of investigations concerning the suitability of these membrane microcapsules as components of artificial organs. Alternatively, these membrane-microcapsules can be tailored to provide controlled and constant release of drugs.

Li of the Exxon Corporation[41] has pioneered the technology of liquid membranes. Liquid membranes are essentially microcapsules wherein the surrounding barrier layer is comprised largely of surfactants stabilized mechanically by appropriate additives. The early investigation of these particular microcapsules has been directed toward using these microcapsules as scavengers for trace components from aqueous feed streams.

In summary, microcapsules can be formed in a wide variety of ways encompassing techniques which have been developed over several decades in the combined fields of surface, colloid, and polymer science. Moreover, the resulting microcapsule wall can either be microporous and freely permeable, semi-permeable, or can be a consolidated impermeable barrier.

2 TRANSPORT BEHAVIOR COMMON TO POLYMERIC FILMS, MEMBRANES, AND MICROCAPSULES

Virtually all applications related to the use of membranes, films, and microcapsules as polymeric delivery systems are related to the diffusion-controlled transport of some biologically active component across a polymeric layer. Even for the case of porous materials, diffusion of active material through a stagnant layer of liquid, confined within the pores of the membrane, controls the solute transport of interest. It is worthwhile, therefore, to review in some detail the basis and implications of Fickian diffusion in and through two-dimensional polymeric materials.

The permeability of a flat polymeric membrane, film, or capsule wall is defined by the expression:[42]

$$\bar{P} = -J(\Delta p/l). \tag{1}$$

The minus sign derives from the definition that a positive flux, J, is driven by a negative pressure gradient, $\Delta p/l$. This equation is written for the arbitrary but instructive case of gas transport driven by the partial pressure difference of the diffusing gas species.

For the case of transport controlled completely by Fickian diffusion, the flux, J, is given by:[43]

$$J = -D\partial C/\partial x. \tag{2}$$

In the steady state, the partial derivative can be replaced by $\Delta C/l$ and Fick's law, therefore, reduces to

$$J = -D\Delta C/l. \tag{3}$$

The concentration gradient, ΔC, relates to the concentration difference within the polymeric membrane. Typically, the concentration at the film/gas boundary is given by a simple equilibrium relationship such as Henry's law wherein:

$$C = kp. \tag{4}$$

Combining the Henry's law equation with the empirically derived Fick's law relationship, the following equation is obtained:

$$J = -Dk\Delta p/l. \tag{5}$$

By combining this relationship with the defining equation for the permeability, $\bar{P}$, the simple relationship between $\bar{P}$, D, and k is obtained:

$$\bar{P} = Dk. \tag{6}$$

The permeability, therefore, is given by the simple product of the diffusivity (which may be considered as a kinetic parameter) and an equilibrium solubility, k (which is a true thermodynamic or equilibrium parameter). Although molecular size and shape of the diffusing species affect the diffusion coefficient and, in turn, the permeability in a regular, predictable, and understandable way, the equilibrium solubility is determined by the more abstract, albeit fundamental, laws governing intermolecular interactions between molecular entities.

For the case of gas transport, solubilities generally increase with the boiling point of the gas and, in turn, the molecular size of the diffusing gas molecule. In effect, this increase in solubility with molecular size, accompanying a decrease in diffusivity with molecular size compromises the simple predictions regarding the effect of size and shape of the diffusing species on the observed steady state permeabilities. Clearly, considerations of size and shape tell only half the story for a "solution-diffusion" transport process.

Equations quite analogous to Eqs. (5) and (6) can be derived for the case of liquid transport through polymer membranes or solute transport from gaseous or liquid solutions contacting polymer membranes. In the case of mixture permeation, the equilibrium parameter that appropriately described the distribution of permeating species between the polymer phase and the contacting fluid phase is typically a conventional partition (or distribution) coefficient.

These simple relationships describing steady state transport are complicated somewhat for the case of transport in glassy polymeric systems. Recent investigations have revealed quite clearly that transport in polymeric glasses is complicated by immobilization of the diffusion species on specific sites which are distributed throughout the glassy polymer matrix.

In many practical cases, the approach to the steady state represents an important portion of the overall lifetime of the transport process. For example, for the case of barrier plastics, transport is typically confined to the relatively slow unsteady state behavior throughout the projected shelflife of the packaged materials. A typical pressure *vs.* time response for CO_2 transport through Mylar[44] is presented in Figure 7.

Conversely, for many reservoir based controlled delivery systems, the polymeric barrier layer is presaturated with active ingredients during storage prior to use and, consequent to implantation, a so-called "burst effect" results from the approach to the steady state form of a previously saturated polymer film. This phenomenon is the acknowledged cause of the initially high release rates apparent in Figure 8 for pilocarpene released from an Ocusert™ device.[45]

Whereas the concentration gradient is lowest at the downstream exiting surface, consequent to initial contact of the polymer with a permeating feed,

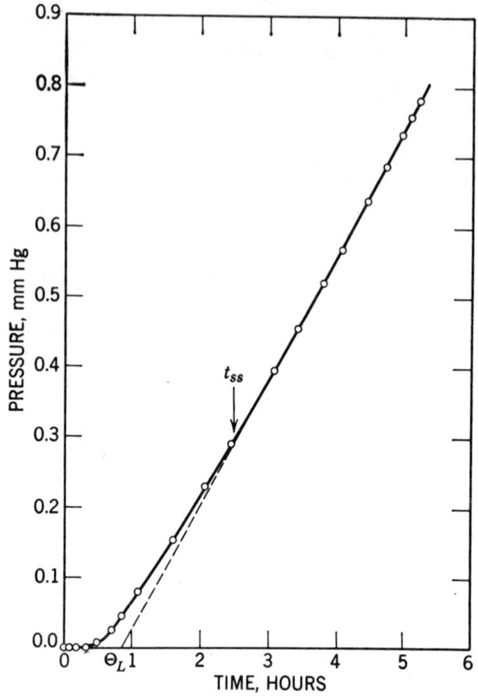

FIGURE 7 Transient response of downstream pressure *vs.* time for CO_2 permeation through Mylar.

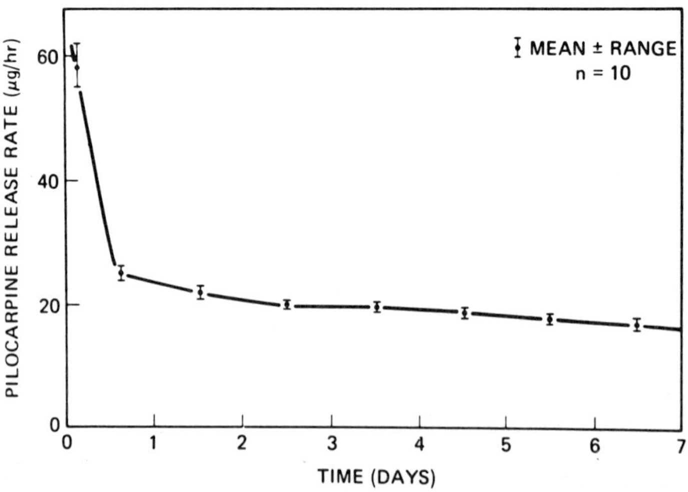

FIGURE 8 Time dependence of Pilocarpine released from Ocusert™ delivery system.

the concentration gradient is highest at the existing surface for the case of a previously saturated membrane in continuous contact with a permeating feed, consequent to initial implantation. Slow approach to a steady state results from the former condition and the "burst effect" results from the latter.

The characteristic time describing the approach to the steady state is typically proportional to the square of the membrane thickness and inversely proportional to the penetrant diffusivity. Significant variations in barrier thickness, ranging from fractions of a micrometer for the dense layer of asymmetric membranes to hundreds of micrometers for consolidated and symmetrical films alter significantly the steady state flux through polymeric systems, and even more dramatically, the time required to achieve the steady state transport behavior.

Whereas transport coefficients for solvent and solute diffusion in the liquid phase are bounded within rather narrowly defined limits, the observed solubilities, diffusivities and, in turn, steady state permeabilities vary dramatically from polymer to polymer and from penetrant to penetrant. The data presented in Table I are presented as an example of this wide range of

TABLE I

The permeability of various polymers to oxygen, water and carbon dioxide at 30°C[48,50]

Material	P_O	P_{CO_2}	Ratio P_{CO_2}/P_{O_2}	P_{H_2O}
Polyacrylonitrile	0.0002	0.0008	4.0	300
Polymethacrylonitrile	0.0012	0.0032	2.7	410
Lopac (Monsanto Co.)[a]	0.0035	0.0108	3.1	340
Polyvinylidene chloride	0.0053	0.029	5.5	1.0
Barex (Sohio Co.)[a]	0.0054	0.061	3.4	660
Polyethylene terephthalate	0.035	0.017	4.9	175
Nylon 6	0.038	0.016	3.6	275
Polyvinyl chloride (unplast.)	0.045	0.016	3.6	275
Polyethylene (dens. 0.964)	0.40	1.80	4.5	12
Cellulose acetate (unplast.)	0.080	2.40	3.0	6800
Butyl rubber	1.30	5.18	4.0	120
Polycarbonate	1.40	8.0	5.7	1400
Polypropylene (dens. 0.907)	2.20	9.2	4.2	65
Polystyrene	2.63	10.5	3.8	1200
Polyethylene (dens. 0.922)	6.90	28.0	4.0	90
Neoprene	4.0	25.8	6.5	910
Teflon	4.9	12.7	2.6	33.0
Poly(2,6-dimethyl 1,4-phenylene oxide)	15.8	75.7	4.8	4060
Natural rubber	23.3	153	6.6	2600
Poly(4-methyl pentene-1)	32.3	92.6	2.9	—
Polydimethyl siloxane	605	3240	5.3	40,000

[a] Nitrile copolymers
Units: CCS. (S.T.P.)/sq cm/cm/sec/cm Hg × 10^{10}.

variability afforded by readily available polymer systems. In this regard, the permeabilities of various polymers to oxygen, water, and carbon dioxide are tabulated. The permeability of these polymers to either oxygen or CO_2 varies by a factor of 10^6. Typically, glassy, crystalline polymers are relatively impermeable barriers and the more rubbery polymers are highly transporting.

Most material properties vary within much narrower limits than the widely ranging permeability values typically encountered amongst polymers of relatively similar molecular structure. This wide variation in normalized transport behavior, coupled with the similarly large variation in available barrier thicknesses, provide the designer of polymeric delivery systems with a virtually unique range of design possibilities.

Consequent to selection of the appropriate device geometry (typically dictated by rather straightforward physiological constraints) materials selection proceeds based largely upon considerations related to biophysical and biochemical compatibility. The seemingly endless array of quantitative design possibilities then become narrowed quite significantly. In fact, relatively few polymer systems are acknowledged to be suitably biocompatible by the scientific community and appropriate regulatory agencies.

In addition, interactions of environmental and core components can alter significantly the properties of the rate-limiting barrier layer of a delivery device. Long-term relaxations must be accommodated in device design and in accelerated test protocols which provide reliable long-term predictions.

In summary, the state-of-the-art of polymer science, and especially the sub-science relating to the transport of small molecules in and through polymeric materials, is sufficiently well developed to permit accommodation of virtually any delivery requirement by suitable tailoring of polymeric materials, coupled with specification of the appropriate physical configurations. The design challenge becomes much more realistic, however, when one couples the physico-chemical constraints required of polymeric delivery systems with the imposed biocompatibility requirements fundamental to long-term therapeutic implants.

References

1. W. H. Carothers, *Linear Condensation Polymers*, U.S. Pat. 2,071,250, February 1, 1937.
2. J. R. Whinfield and J. T. Dickson, Brit. Pat. 578,079 (June 14, 1946) and U.S. Pat. 2,465,319 (March 22, 1949).
3. Anon, *Modern Plastics International*, **8**, 7 (1975).
4. A. Backter and M. S. Nerurkar, *Eur. Polymer J.*, **4**, 685 (1968).
5. L. Boltenbruck, H. Schnell, and A. Prietzschk, Belgian Pat. 589,858 (1960).
6. K. Maier and E. Scheuermann, *Kolloid Z.*, **171**, 122 (1960).
7. P. Carnell and H. Cassidy, *J. Polymer Sci.*, **55**, 233 (1961).
8. P. Francis *et al.*, *OSW Res. Develop. Rept.* 177 (1964).

9. M. Cohen and B. Riggelman, *OSW Res. Develop. Rept.* 400 (1969).
10. S. Loeb and Sourirajan, *UCLA Rept.* 60-60 (1959).
11. H. K. Lonsdale, U. Merten, and R. L. Riley, *J. Appl. Polymer Sci.*, **9**, 1341 (1965).
12. H. K. Lonsdale, in *Industrial Processing with Membranes*, ed. R. E. Lacey and S. Loeb, Wiley-Interscience, New York, p. 138 (1972).
13. S. Manjikian, *Ind. Eng. Chem. Prod. Res. Dev.*, **6**, 23 (1967).
14. Saline Water Conversion Report for 1969, Office of Saline Water, U.S. Dept. of the Interior, U.S. Gov't Printing Office, Washington, D.C.
15. C. W. Saltonstall, Jr., "Development and Testing of High-Retention Reverse Osmosis Membranes," International Conference PURAQUA, Rome, Italy, February 17–22, 1969.
16. D. L. Hoernschemeyer, C. W. Saltonstall, Jr., O. S. Schaeffler, L. W. Schoellenbach, A. J. Secchi, and A. L. Vincent, Research and development of new and improved cellulose ester membranes, *U.S. Off. Saline Water Res. Dev. Prog. Rept.* **556**, Aerojet-General Corp.
17. S. Manjikian, S. Liu, M. Foley, C. Allen, and B. Fabrick, Development of reverse osmosis membranes, *U.S. Off. Saline Water Res. Dev. Prog. Rep.* **534**, Universal Water Corp. (1970).
18. H. K. Lonsdale, R. L. Riley, C. E. Milstead, L. D. LaGrange, A. S. Douglas, and S. B. Sachs, Research on improved reverse osmosis membranes, *U.S. Off. Saline Water Res. Dev. Prog. Rept.* **577**, Gulf General Atomic, Inc. (1970).
19. R. L. Riley, H. K. Lonsdale, C. R. Lyons, and U. Merten, *J. Appl. Polymer Sci.*, **11**, 2143 (1967).
20. R. L. Riley, H. K. Lonsdale, L. D. LaGrange, and C. R. Lyons, Development of ultrathin membranes, *U.S. Off. Saline Water Res. Dev. Prog. Rep.* **386**, Gulf General Atomic, Inc. (1968).
21. A. E. Marcinkowsky, K. A. Kraus, H. O. Phillips, J. S. Johnson, Jr., and A. J. Shor, *J. Am. Chem. Soc.*, **88**, 5744 (1966).
22. K. A. Kraus, H. O. Phillips, A. E. Marcinkowsky, and J. S. Johnson, Jr., *Desalination*, **1**, 225 (1966).
23. K. A. Kraus, A. J. Shor, and J. S. Johnson, Jr., *Desalination*, **2**, 243 (1967).
24. A. J. Shor, K. A. Kraus, W. T. Smith, Jr., and J. S. Johnson, Jr., *J. Phys. Chem.*, **72**, 2200 (1968).
25. J. R. Kuppers, A. E. Marcinkowsky, K. A. Kraus, and J. S. Johnson, Jr., *Separation Science*, **2**, 617 (1967).
26. L. C. Flowers, D. E. Sestrick, and D. A. Berg, Reverse osmosis membranes containing graphite oxide, *U.S. Off. Saline Water Res. Dev. Prog. Rept.* **224**, Westinghouse Electric Corp. (1966).
27. F. E. Litman, and G. A. Guter, Research on porous glass membranes for reverse osmosis, *U.S. Off. Saline Water Res. Dev. Prog. Rept.* **379**, McDonnell Douglas Corp, (1968).
28. J.-G. Helmcke, *Zentr. Bakteriol.*, **159**, 308 (1953).
29. J.-G. Helmcke, *Optik*, **10**, 147 (1953).
30. A. Reimschuessel, in *Synthetic Polymer Membranes*, ed. R. E. Kesting, McGraw-Hill Book Co., New York, p. 105 (1971).
31. R. E. Kesting, *Synthetic Polymer Membranes*, McGraw-Hill Book Co., New York (1975).
32. B. K. Green, and Schleicher, U.S. Pat. 2,800,457 and U.S. Pat. 2,800,458 (July 23. 1957).
33. J. A. Bakan, U.S. Pat. 3,436,355 (April 1, 1969).
34. A. Dobry, and F. Boyer-Kawenoki, *J. Polymer Sci.*, **2**, 90 (1947).
35. O. Wurzburg and W. Herbst, U.S. Pat. 3,091,567 (Feb. 17, 1961).
36. G. F. Sachsel and H. Nack, U.S. Pat. 3,202,533 (August 24, 1965).
37. C. F. Powell, J. H. Oxley, and J. M. Blocker, Jr., eds., *Vapor Deposition*, John Wiley and Sons, New York (1966).
38. C. A. Baer, U.S. Pat. 2,846,971.

39. D. F. Herman, U. Kruse, and J. J. Brancato, *J. Polymer Sci.*, **11**, 75 (1965).
40. J. E. Vandegaer and F. G. Meyer, U.S. Pat. 3,464,926 (September 2, 1969).
41. N. N. Ki, U.S. Pat. 3,410,794 (1966) and U.S. Pat. 3,410,078 (1966).
42. H. J. Bixler and O. J. Sweeting, in *The Science and Technology of Plastic Films*, Vol. **II**, ed. O. J. Sweeting, Wiley-Interscience, New York (1971).
43. A. Fick, *Ann. Phys. Leipzig*, **170**, 59 (1855).
44. C. E. Rogers, Permeability and chemical resistance, in *Engineering Design for Plastics*, ed. E. Baer, Reinhold, New York (1964).
45. A. S. Michaels, in *Permeability of Plastic Films and Coatings to Gases, Vapors, and Liquids*, ed. H. B. Hopfenberg, Plenum Press, New York, p. 418 (1975).
46. S. Hwang and K. Kammermeyer, in *Membranes in Separations*, Techniques of Chemistry Vol. 7, ed. A. Weissberger, John Wiley and Sons, New York, p. 310 (1975).
47. *Microencapsulation*, ed. J. E. Vandegaer, Plenum Press, New York, p. ii (1974).
48. H. B. Hopfenberg, in *Polymers in Medicine and Surgery*, Polymer Science and Technology Vol. 8, eds., R. L. Kronenthal, Z. Oser and E. Martin, Plenum Press, New York, p. 101 (1975).
49. Reprinted with the permission of John Wiley and Sons, Inc.
50. Reprinted with the permission of the Plenum Publishing Corp.

DISCUSSION

Dr. I. Parikh (*Burroughs Wellcome Company, Research Triangle Park, NC*): Would you briefly describe the experimental methods that you used to measure transport properties in small spheres?

Dr. H. Hopfenberg: All of the gravimetric experiments, either with small spheres or with films, were performed on a very sensitive quartz helical spring which deflected 200 mm per 100 mg of load. The deflection was detected with a cathetometer. The sample, in the case of beads, is placed on a quartz pan which is in a vacuum system. A vapor is admitted to a fixed pressure, and the increase in the sample weight is monitored to yield the change in weight with time.

Typically, the experiment takes about an hour. This procedure is good for any vapor including water.

Dr. T. Roseman (*The Upjohn Company, Kalamazoo, MI*): In the pilocarpine system that you described, there may exist a negative lag time which in effect results from an initial surge, or burst, of drug. This initial surge can be explained mathematically as you did. In the delivery of many drugs, there is a minimum effective dose level and a higher toxic dose level. My question is how do you propose to get away from a potentially toxic initial drug surge in systems where such a surge is certainly not beneficial?

Dr. H. Hopfenberg: In many cases, a burst effect could be eliminated by dipping the device in water *in vitro* prior to *in vivo* placement, thus permitting the steady state to develop prior to *in vivo* insertion. What this boils down to is getting the medical community to accept these devices as useful and getting them to handle them as they do their more conventional medical and surgical instruments and make them part of their practice, paying attention to the specific requirements of each device.

Dr. D. Meier (*Midland Macromolecular Institute, Midland, MI*): In the experiments that you described with the mixed isotactic–atactic polystyrene in which you ascribed the peak to the extraction of residual mesitylene, have you actually determined that it is indeed mesitylene that is responsible for this peak? Is it possible that the peak is due to solvent-induced crystallization of the isotactic polystyrene?

Dr. H. Hopfenberg: We don't see that if the samples are subjected to an annealing cycle prior to determining the sorption kinetics.

Dr. D. Meier: The annealing cycle may have enabled the isotactic polystyrene to have already crystallized.

Dr. H. Hopfenberg: In highly isotactic polymers, where we show very low levels of sorption, we are invoking hexane induced crystallization of the isotactic polymer to explain our results. I do think that isotactic polymer crystallization plays a part in the kinetics. We chose the experiments using the chlorinated solvent to show that casting solvent is extracted by the subsequently sorbing vapor.

Dr. D. Meier: With regard to the experiments with very small spheres, have you taken into account the possibility that surface tension or interfacial effects may also provide a contributing factor in the uptake of material?

Dr. H. Hopfenberg: We have. You can use the Kelvin equation to describe the effect of curvature on the chemical potential. This does not, however, account for the differences we see in the equilibrium sorption between the large and small spheres.

The big particles absorb less than the small particles on a mass basis. The diffusion coefficient for desorption from the larger sphere is about 100 times larger than that for desorption from the small sphere.

However, if we consider the hollow sphere model and say that it's all absorbing in the outer shell and then use the hollow sphere model for calculating diffusion coefficients for large spheres, the diffusion coefficients are identical to those calculated for the small spheres.

Dr. R. Phares (*Alza Corporation, Palo Alto, CA*): It seems that it may be possible to decrease the size of a sphere with time to the point where one has essentially a large solvated sphere with the equivalent of a small non-solvated sphere on the inside. This interior sphere may be expected to behave in a fashion similar to starting with a small sphere. Is this the case or can one investigate this possibility at this time?

Dr. H. Hopfenberg: All we have is 184 μm spheres and 0.534 μm spheres. These samples were supplied to us after careful characterization revealing a narrow particle size distribution.

We do have some real problems with these spheres. When you work at a higher temperature, you have to worry about the particles sticking together. We really like to work in a temperature-activity region where Case II transport would predominate, but when we get that high in temperature, the small spheres become a bit sticky. That's another problem.

Dr. H. Ringsdorf (*University of Mainz, Mainz, Germany*): You discussed the uptake of compounds into membranes and films in the presence of chelating agents. Can you comment on how this work may relate to the uptake of drugs into the cells of living organisms?

Dr. H. Hopfenberg: I wouldn't know how to begin to translate this work on synthetic polymers to cellular systems.

Dr. H. Ringsdorf: It may be that the uptake by a cell may be mediated by a peptide in the lipid bilayer of the membrane, and the results may be completely different than those you reported.

Dr. H. Hopfenberg: We've never looked into this.

Dr. W. Good (*Ciba–Geigy Corporation, Ardsley, NY*): I would like to comment on our experiences with two different drugs, both completely soluble in water, in the same polymer release system. We found that one system released drug according to Case II diffusion, while the other system was apparently Fickian with an extra rate process (solvent absorption). It is interesting that one can obtain different release mechanisms for different but similar drugs in the same polymer matrix. Also, your derived equation for Case II diffusion indicates that the effect on the amount released should be much more visible with spheres than with sheets, and it is interesting that we have indeed observed this effect experimentally with these drug-release systems.

Dr. E. Goldberg (*University of Florida, Gainesville, FL*): Are there analogies between your swollen-state facilitated transport picture and plasticizer effects on T_g and modulus? If so, then you can of course invoke anti-plasticizer situations where you can presumably change the transport as a direct function of the swollen state.

Dr. H. Hopfenberg: Perfect analogies. In fact that would be a fitting conclusion to my presentation.

The basis of the analysis is to understand anti-plasticizer systems which in many cases are polymer blends, and those are the type of effects that may ultimately be exploited in practical applications.

Controlled Release of Effectors from Polymers and Microcapsules

GILBERT S. BANKER

School of Pharmacy and Pharmacal Sciences, Purdue University, West Lafayette, IN47907, U.S.A.

The optimization of the action of effectors (drugs, insecticides, herbicides, etc.) typically involves the simultaneous maximization of agent safety, effectiveness and reliability. Achieving these goals through product design often involves effector entrapment or association in polymeric systems and subsequent controlled release. Principles and approaches which should be followed in the design of effective controlled release effector systems are presented. Physicochemical methods of entrapping and controlling effector release are preferred to mechanical methods, from the standpoint of effector dispersion levels attainable, reliability and predictability of entrapment or encapsulation, and precision of control of release rates that may be attained. Physicochemical methods of accomplishing effector controlled release from polymers are described and reviewed with emphasis placed on molecular scale entrapment of drugs by colloidal latices.

SIGNIFICANCE OF CONTROLLED RELEASE IN EFFECTOR PRODUCT DESIGN

Effectors are herein defined and treated as any type of chemical agent capable of affecting plant or animal life processes. Accordingly, effectors include all human and veterinary drugs, insecticides, herbicides, plant growth stimulators and regulators, etc. There are a number of reasons for wishing to control the release characteristics of effectors in such systems as shown in Table I. Specific examples, citing cases in which these objectives were actually achieved utilizing polymers, microcapsules and various methods in colloid chemistry, for controlled release of drugs and other effectors will be enumerated throughout this paper. The optimization of the action and total quality of effectors involves simultaneously maximizing the effectiveness, safety and reliability of these agents. Mathematical and statistical techniques of optimizing effectors through product design have been described.[1–3] Such optimization frequently involves some form of effector entrapment or association in a polymeric system as well as control of the release of the agent over a time period.

TABLE I

Objectives of effector product design utilizing polymeric controlled delivery systems

Objective	Example or application
Maximize duration of effectiveness	All effectors
Control specific duration of effectiveness	Controlled short acting insecticides
Control site of release in the organism	Enteric release drug systems, controlled gastric retention
Target effector in the organism	Liposomes, drug targeting to specific tissues
Facilitate dispersion, spreading and uniformity of dose or application	All potent, low dose effectors
Reduce toxicity and improve safety	Drugs and insecticides in particular
Reduce undesirable side effects of effectors	Drugs, herbicides and insecticides
Improve effector reliability	All effectors
Increase chemical and physical stability of effectors in exposure environment	All effectors with chemical lability, high volatility, etc.
Increase effector selectivity	Desirable for all effectors
Prevention or reduction of effector mismanagement or abuse	Desirable for all effectors
Optimization of total effector action and performance	Ultimate goal with all effectors

Table I suggests the importance of current and future work to better control effector release to achieve the objectives shown. The significance of the work in effector control is also apparent from the following facts. Between one-third and one-half of the world's population is still underfed. Improved animal and plant growth stimulators and regulators will continue to play a role in combatting this problem. Perhaps more important is the fact that there is strong evidence that insects have been making gains in recent years in their natural quest to inherit planet Earth. This is due to at least three factors; governmental controls such as those of the U.S. Environmental Protection Agency which have banned or restricted the use of some of the most effective insecticides, development of resistance to existing insecticides by some insect species, and misuse of insecticides and other insect control methods. Not only do insects continue to threaten our food supply, they are a leading cause of disease, especially in underdeveloped and tropical countries. The fact that the least developed countries tend to be in the tropics is probably not totally coincidental and may be related in part to insect problems. Malaria, which was coming under world-wide control, has returned with a vengeance. It now strikes 100 million people a year in Africa alone; killing 800,000, mostly children under five.[4] In the Volta River basin alone, a species of blackfly afflicts a million Africans annually with river blindness, leaving the majority sightless. The tsetse fly, carrier of sleeping sickness dominates much of the continent. In the U.S., where control of agricultural pests is among the world's best, insects still destroy about 10% of all crops grown, representing a loss of 5 to 6 billion dollars a year.[4] In the least developed countries, the average crop loss due to insects is probably closer to 50%. The need for more effective agents and systems in the war on insects is abundantly clear.

Drugs are the number one method for treating diseases, whether measured from the standpoint of frequency of use of the chemotherapy approach, or whether considering the treatment of choice for particular disease states. Great advancements have been made in chemotherapy, particularly over the last three decades. The fact remains, however, that many drug products currently marketed are not optimum systems as far as drug delivery is concerned. The optimized drug system, in addition to providing a maximum level of effectiveness, must be maximally reliable in a wide spectrum of people varying in disease state and considering the many physiologic variables that may affect drug absorption and response, as well as being maximally safe under normal or various abuse situations. The fact that not all drug products are now optimal drug delivery systems is perhaps of little consequence for some drugs or therapeutic classes, but is of much more concern for others. Optimization of the drug delivery system by precise control of its release profile is a worthy goal for drugs which are extremely potent, which treat life-threatening ailments, which require careful dose titration to produce the desired effect, which have a narrow margin of safety between the effective dose and toxic dosage levels or which are frequently abused. Drug products as a class are not a routine consumer product since they directly deal with the well-being and even the life of the consumer. In this case for a product to be good, may not be adequate. It is often desirable for these products to be optimized so as to be as simultaneously safe, effective and reliable as is humanly possible with mathematical certainty. The additional concept of maximizing safety of use and minimizing environmental impacts of other effectors, notably insecticides, while retaining effectiveness, convenience of use and reliability, regardless of environmental variables, is a fairly recent dimension.

Figure 1 illustrates the relationship between the level of use of an effector agent and its duration of action for products which provide conventional release of effectors (no system controlled agent release).

The curves shown are typical for a drug, an insecticide or an herbicide. The left hand vertical and bottom axes and the solid lines present data for an herbicide[5] while the right hand and top axes and dashed line illustrates the loss rate from blood for a drug following one dosage level.[6] In both situations, some minimal amount of material must be applied or dosed to reach an effective level. The optimum level range shown is for the herbicide. As the dosage is increased to produce peak values above this effective level, an increase in dose does produce an increase in duration of effectiveness (as shown

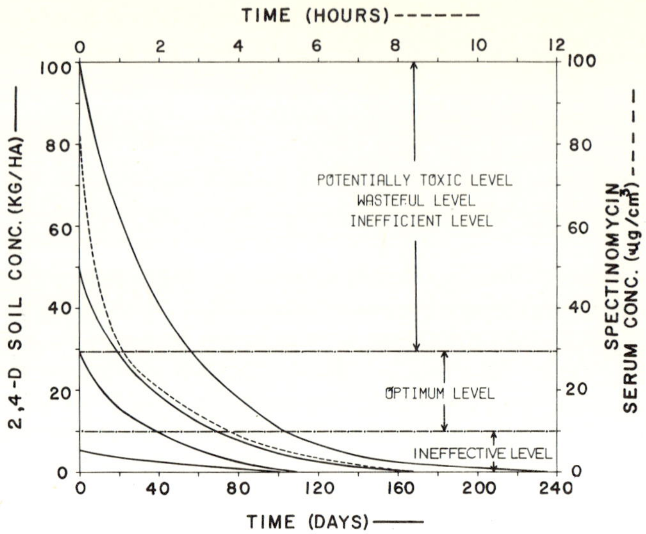

FIGURE 1 Relationship between the level of use of effectors, duration of effector action, and ineffective, optimum or toxic levels following a single application or dose.

for the herbicide). The material has a half-life of about 30 days and must maintain a surface soil concentration above about 0.5 kg/ha to be effective and above 10 kg/ha for optimum action. Using a conventional system, this would require the application at a level of at least 50 kg or over 100 lbs per hectare (about $2\frac{1}{2}$ acres) to provide protection over a 100 day growing season from a single application. Even this high level of application would be inadequate to cover the growing season for most weeds and it represents an application of 100 times the minimum effective level. Another difficulty with such massive applications is, of course, that desirable plants will be destroyed. In the case of drugs it is even less feasible to increase dosage levels in order to increase duration of activity. In this situation it becomes quite clear that controlled drug release *must* be employed if prolonged drug activity from a single dose or administration is to be achieved.

Figure 2 illustrates the type of blood level profile observed on repetitive dosing of a drug from a conventional delivery system by oral, parenteral or other routes.

The single 100 mg or other unit dose produces a high blood or tissue level which may well reach toxic levels (curve A). A series of peaks and valleys are produced after the drug is given in three divided doses of 33 mg each (curve B). This "wave-effect" is usually more pronounced in actual practice depending on how much the patient departs from the prescribed dosage regimen, and few patients take their medication exactly as ordered (the case illustrated

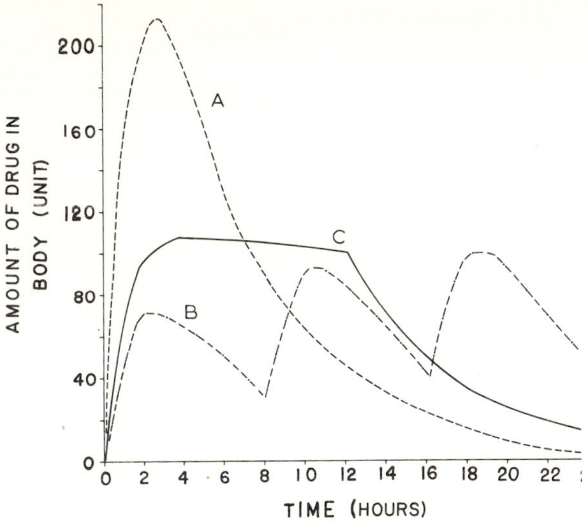

FIGURE 2 Typical body level—time profiles for a drug following a single dose (A), repetitive divided doses (B) and an ideal controlled delivery system (C).

in the figure). The problem with repetitive dosing as shown with curve B is that the valley levels of the profile may drop below the effective concentrations, which in the case of some antibiotics, anticoagulants, antihyperglycemic and many other drug classes, could be very serious, or that the peak levels rise above toxic or even life-threatening ranges, which is of greatest concern for drugs with a narrow therapeutic index (ratio of toxic to effective dose). Furthermore, on repetitive dosing the peak levels tend to climb as shown, leading to possible side effects. Curves A and B both represent conventional drug products, wherein physiological factors and not the dosage form are controlling the drug absorption. Curve C typifies the administration of a single 100 mg or unit dose in a well designed controlled and sustained release system. The dosage form is now the rate controlling factor over drug absorption. As long as the system can release drug for absorption at a rate equal to the drug's elimination and inactivation rates, while being maintained at or traversing the region of the gut where absorption can occur, a constant blood and tissue level is possible. For current marketed controlled release oral drug delivery systems, the duration over which a single dose of such a product can maintain constant levels is dictated by the gastrointestinal transit rate past the absorption site (usually the small intestine).

To summarize, the delivery of effectors by conventional techniques can be grossly inefficient, may be unreliable, provides less than optimum safety and is frequently uneconomical. Whether measured in gallons or kilograms or milliters or micrograms, for herbicides, insecticides or drugs in man, vast quantities of agents are frequently employed when much smaller amounts would be theoretically adequate for the purpose. The use of large amounts of effectors from the "dump-type" uncontrolled release system in a single application or in applications designed to be minimal in number, greatly increases the incidence of undesirable side effects because non-target organisms or organs are swamped with potentially toxic agents to insure that the target will receive an effective dose for the desired period of time. When agent applications or dosages are divided in very small increments or are controlled to provide a constant infusion rate, lesser amounts of agent are effective and overall effectiveness may be improved with less danger of harm to non-target organs or species.

PRINCIPLES OF EFFECTOR CONTROLLED RELEASE DESIGN

To optimize effector agent delivery, typically requires very precise control of release rates with the system being designed to be as insensitive as possible to biological or environmental variables which could act to alter delivery rates. Until recently most controlled and sustained release pharmaceutical systems were based on mechanical approaches and employed pharmaceutical art such as coating skills. The same is true for other effectors. Barrier coatings were the first controlled and sustained release principle. The originally used barrier coating materials for drugs lacked programmable release characteristics since they variably disintegrated or dissolved upon exposure to GI fluids and were strongly influenced by environmental pH and/or enzymes. A later approach was to embed drugs in fats and waxes which were intended to slowly erode and liberate drug. These systems suffered the same disadvantages as the barrier coatings. Other pharmaceutical controlled release mechanisms involved the use of hydrophilic matrices which were intended to slowly hydrate and release drug through a viscous gel barrier, plastic insoluble matrices which provided slow drug liberation through channels and by leaching processes, drug inclusion in insoluble beads with or without subsequent coating, and similar mechanical methods. These methods were all deficient from two standpoints; they were based on mechanical factors such as mixing, dispersion and coating, frequently involving sophisticated skills and were accordingly difficult to reproduce, and secondly the methods lacked precision in ability to control release rates with sufficient "fine-tuning." Most of the methods relied on physiological variables to provide the mechanisms of release. These are the same physiological variables that may differ greatly from subject to subject. The key to developing highly effective, near optimum or optimum effector controlled delivery systems lies not only in the selection of better polymers which are less sensitive to alterations of release rate as influenced by specific physiological or environmental factors, but to the employment of physicochemical approaches for the preparation and manufacture of effector controlled delivery systems. Using an appropriate physicochemical method such as various forms of microencapsulation or molecular entrapment and dispersion to replace a mechanical art such as coating, has provided for the first time, sufficient control of the factors of the system influencing both effector entrapment and subsequent release, to permit some of the goals listed in Table I to be effectively approached. Indeed, an effector controlled delivery system which is predicated on physicochemical principles for design and manufacture should provide for the preparation of a product whose effector concentration is stoichiometric in predictability as is the resultant release rate, under variable conditions, covering the range of variables to be encountered in practice. The ideal attributes of an effector controlled delivery system are shown in Table II. Insecticides are influenced by both physiological factors such as those noted for drugs and by environmental factors such as those shown for herbicides, as well as by the nature of the substrate to which the insecticide is applied.

TABLE II

The ideal attributes of an effector controlled delivery system

1) *Not Highly Sensitive to Environmental Factors*

A. *Drug effectors—(physiological factors)*	B. *Herbicides—(environmental factors)*
Gastric motility and emptying	Temperature
pH of gut	Temperature cycling
Fluid volume	Sunlight exposure
Presence/absence and concentration of enzymes	Water and moisture exposure
State of fasting and type of food present	Physiochemical properties of the soil substrate
Physical position and activity of subject	
Individual variability and/or disease state	

2) *Predicated on Physicochemical principles*
3) *Functions by several mechanisms not keyed to any single item no. 1 above*
4) *High order of effector dispersion possible*
5) *Flexible; capable of handling a wide variety of agents*
   Applicable to many agents, regardless of ionic form, physical properties such as solubility, and capable of achieving a wide range of latitude in controllable delivery rates.
6) *Maintenance or enhancement of physical and chemical stability*
   Capable of controlling volatility or other physical stability problems and of maintaining or enhancing chemical integrity.

NEWER EFFECTOR ENTRAPMENT AND CONTROLLED RELEASE SYSTEMS

Table III presents some of the newer effector entrapment and controlled release systems. Most of these systems in their application to effectors are less than 25 years old. Ion exchange is perhaps the oldest of the methods listed and it has been used to commercially control drug release for at least three decades. The majority of all drugs are basic amines, and many of these including alkaloids, antihistamines, amphetamines, etc. have been bound to various cation exchange resins.[7] Amphoteric compounds such as oxy and chlortetracycline have been applied to sulfonated polystyrene ion-exchange resins.[8] Acidic compounds such as aspirin have been bound to anion exchange resins and studied for benefitted effects.[9] Even weakly ionic effector compounds of very limited aqueous solubility, such as coumarin, have been successfully prepared as ion exchange systems.[10,11] Ion exchange was the first physicochemical method to be employed in practice for controlling and prolonging drug release. Since the effector is released from an ion-exchange resin according to the ionic environment, it is difficult to regulate the rate of release under physiological or environmental conditions. For example, basic drugs on cation exchange resins tend to release much faster in the acid stomach than in the more neutral to alkaline small and lower intestines. The ion exchange approach, however, has many unexplored applications for non-drug ionic effectors.

As noted in Table III, complexes and salts of effectors (notably drugs) and

TABLE III

Newer effector entrapment and controlled release systems utilizing polymers

Method	Polymer examples	Reference
Ion exchange	Cation exchange resins	7–11
Complexes and salts	Cellulosics, dextrins, PVP, acrylics, anhydride copolymers	12–31
Esters and prodrugs	—	32
Microencapsulation and encapsulation	All thermoplastic resins	33–39
Membrane controlled	Thermoplastic and lightly crosslinked polymers	5, 37, 40
Matrix entrapment	Polyvinyl chloride, ethyl cellulose, acrylic–acrylate copolymers, crosslinked polymers and copolymers	41–54
Osmotic control	Osmotic pump Insoluble–impermeable polymer coatings	55
Molecular dispersions Solid solutions Glass solutions Amorphous precipitates	Polyethylene glycols polyvinylpyrrolidone	56–60
Sorption on colloidal surfaces with or without coagulation	Colloidal latices	
Misc. new systems and recent reviews	—	61–68

polymers have been used as a controlled release approach. The ability of macromolecules to bind effectors was first noted in pharmaceutical systems as various viscosity imparting polymeric agents were seen to bind antimicrobial agents (preservatives).[12,13] This undesirable phenomena which partially or totally inactivated the preservative, was later recognized as a potential problem in drug absorption, where certain polymers which were unabsorbable, by strongly binding with drugs, would prevent the free transport of the drugs across membranes and inhibit their absorption.[14,15] It is now recognized that the complexation of a drug to a polymer is not the only mechanism by which the absorption and activity of an effector may be influenced by polymeric formulation adjuvants. Other mechanisms can include surface activity effects, viscosity effects, adhesion and localization to tissues and membranes, alteration of diffusion coefficients and other modifications.

More recently the formation of polymer–effector complexes and salts have been seen as a desirable and useful physicochemical approach of not only controlling effector release, but of achieving other beneficial effects such as chemical stabilization. Stability enhancement has been shown for benzocaine, vitamin A and a number of fatty acids complexed as inclusion compounds.[16,17] Complexation of a wide range of pharmaceuticals with Schardinger dextrins has been reported,[18–20] advantages cited and mechanisms described. Cyclodextrin complexes with esters such as aspirin, atropine and aminobenzoic acid derivatives were found to affect hydrolysis.[21] Stereochemistry of binding and ionization effects were attributed to either the catalytic or inhibitory effects on hydrolysis. Complexes between natural polymers and drugs have been investigated for their ability to increase durational effects.[22–24] A promazine–carrageenan complex, for example, produced detectable blood levels in test animals for six days versus one day for free drug. A phenobarbital–vinyl copolymer complex has been reported which greatly increases drug stability in a total water vehicle while retaining full activity.[25] Cylic iminocarbonates of dextran with coenzyme B_{12}, insulin, levarterenol, amphetamine and procaine exhibited both prolonged activity and increased stability.[26]

Hydrogen bonded complexes and complexes of basic drug moities often suffer the same disadvantage as ion-exchange products, though to a lesser extent; rapid drug release in the stomach and declining release rates thereafter. To overcome this problem in part, various vehicle substances and design approaches are used, such as buffers, use of hydrophilic gums, compression, etc., to reduce drug dissolution in the stomach.

The preparation of methylcellulose salts of basic amines has also been reported as have polymer drug salts of basic amines and carboxylic acid polymers and copolymers.[27,28] Drug release from polymer-effector salts may be a diffusionally controlled process and will be quite unlike the drug dissolution observed for a typical monomeric salt. As a high molecular weight polymer dissolves a viscous hydrated layer typically forms on the surface which impedes both polymer dissolution and the passage of effector solute molecules originally associated with the polymer, possibly as an ionic salt. Polymer drug salts may provide a mechanism of chemically stabilizing some effectors as noted as well as controlling their rate of volatility. A knowledge of the mechanisms of polymer dissolution is fundamental to many controlled delivery applications of effectors.[29,30]

Esters of effectors may also be prepared to control release, with prodrugs being one such example (Table III). Many prodrugs do not involve interaction with polymers. The principal of preparing a prodrug is to convert the drug into a form which may be more readily absorbed or which will overcome chemical degradation in the gastrointestinal tract or other environment. After absorption the prodrug ester is hydrolyzed enzymatically to release the parent drug. Many penicillins and cephalosporins are either not absorbed or are absorbed incompletely from the gastrointestinal tract. This necessitates parenteral administration or results in poor utilization of such drugs. Acyloxyalkyl esters of ampicillin such as the pivaloyloxymethyl ester, pivampicillin, or the ethoxycarbonyloxyethyl ester as well as the related lactonyl esters are all efficiently absorbed after oral administration, are subsequently hydrolyzed enzymatically, with the result that much higher blood tissue levels are obtained than after a corresponding dose of ampicillin. Similar advantages can be obtained with some of the penicillins. Erythromycin has been prepared as the estolate and as the stearate to overcome the problem of rapid decomposition in the stomach which occurs with the parent compound. Preparation of effector–polymer esters which are subsequently hydrolyzed to release the active compound provides an interesting potential method of both stabilizing certain effectors as well as controlling the delivery rate. Such esters would have limited application to overcome absorption problems in animals or man if the polymer molecular dimension precluded passage through a membrane and, hence, absorption.

Microencapsulation is not a well defined term as such, and is furthermore often loosely used when a more appropriate expression would be encapsulation or entrapment. A true microcapsule or microsphere is microscopic in size or requires microscopic examination for size and shape visualization. The most common size characterization for a microcapsule is "any size up to about 1/125 of an inch in diameter." While the selection of the upper size range of a microcapsule is somewhat arbitrary, a microcapsule may be further defined as a system in which a liquid or solid center is enclosed by a barrier coating which provides the primary release mechanism. The encapsulated material may be a liquid, suspension, a solid or semi-solid, and possibly even an emulsion. Coacervation in its several forms is a major method of producing true microcapsules. Coacervation may also be employed to produce particle sizes much larger than that indicated for the general upper range of microcapsule size. Table IV summarizes some of the many methods which have been employed to encapsulate and microencapsulate effectors and other agents. Some of the techniques indicated in Part B of Table IV are unable to produce coated particles in the true microcapsule range. Other techniques listed are quite limited in their capability to produce capsules in other than a very narrow range of sizes. As indicated, some of the methods are based on physicochemical approaches and others are purely mechanical. Time will not permit discussion of the various methods; several recent reviews describe all of the methods of microencapsulation shown in Table IV, and others.[33–36]

TABLE IV

Encapsulation and microencapsulation methods

A) *Chemical approaches to microencapsulation*
Simple coacervation
Complex coacervation
Interfacial polymerization
Organic phase separation
Other dispersion processes
B) *Mechanical methods of encapsulation and microencapsulation*
Fluidized bed
Electrostatic methods
Vacuum encapsulation
Spray drying and congealing
Centrifugal encapsulation

Equation 1 describes the release of an effector from a coated particle which has the dimension of a sphere, capsule or microcapsule.[37]

$$\mathrm{d}M_t/\mathrm{d}t = 4\,\pi\, DK\,\Delta\, C\, r_o r_i/(r_o - r_i). \tag{1}$$

As can be seen the release rate at any time is related to the radius of the coated material, r_o, the radius of the coated particle, r_i, the difference in concentration across the membrane, ΔC, the diffusion coefficient of the permeant in the membrane, D, and the distribution coefficient, K. In a typical system, the dimensions of the particles do not change appreciably during release, D and K are constant, and so long as ΔC is substantially constant with a much higher concentration gradient within the membrane than at the exterior, a constant release rate will obtain. Thus, the advantage of this type of system, which is the advantage of any membrane controlled system whether a capsule or microcapsule is that the release rate may be maintained constant and unchanging as long as the concentration of effector in solution within the particle is saturated with respect to the ability of the coating barrier membrane to release it. The kinetics of effector release from membrane controlled, constant activity sources of various geometrics (cylinder, slab or sandwich, etc.), as well as from non-constant sources and monolithic devices (effector dispersed or dissolved in a film for example), have been established and recently summarized.[37]

Coacervation is probably unmatched as a process of producing very small spherical uniformally coated particles. So-called simple coacervation, (Table IV), involves insolublizing a macromolecular substance in the presence of the agent to be entrapped. Complex coacervation takes place when two different kinds of colloidal macromolecular materials unite to form the insoluble substance. The processes of coacervation are well described by Ranney[33] and others[34–36] The use of coacervation as a technique in microencapsulation or encapsulation offers a number of advantages over the coatings of powders and beads and other mechanical methods of encapsulation. Membrane diffusional control is dependent upon membrane thickness. Unless a particle is spherical, it is difficult to coat the particle uniformly by any technique. Typical powder particles are not spherical. The preparation of spherical particles for coating is a variable and costly mechanical process. In coating small particles below 30 to 60 mesh for example, to effect satisfactory coverage may require half as much coating by weight as the mass of material to be coated. As the particle size of the material to be coated decreases and its surface increases, the amount of coating agent required increases dramatically and non-linearly.[38] The mechanical problems of coating are also compounded. The increasing cost of the polymeric coating required to effect coverage, prohibitive as it may be, is not usually as critical as that of the increasing cost of the coating solvent required to carry the polymer to the product. The handling of solvent vapors (control, disposal or recovery) according to State and Federal EPA Standards is an additional troublesome and costly matter. Finally, coating while it may be successfully controlled as a mechanical process in encapsulation, typically requires special techniques for microencapsulation. New methods of membrane coating for encapsulation of materials, involving for example direct particle immersion[39] in a fully hydrated or solvated coating solution, are being developed as means of better controlling membrane density and diffusional properties, improving coat thickness control and coverage, and development of special systems (half coated, dual and other systems).

Coacervation as a microencapsulation process has had two limitations which have probably restricted commercial reduction to practice for many effector controlled delivery systems. One major factor is that the coacervation process has been extensively patented as a controlled release and entrapment process with the patent position being largely dominated initially by one company, with this position still being controlled by a handful of companies. Commercialization of coacervated systems on a broad basis in the drug, insecticide and related fields has required licensing, and some companies have chosen other routes. Secondly, polymer selection is perhaps more restricted and less flexible for coacervated systems, especially when applied to a particular chemical agent. It is anticipated, however, that as patents become dated, as technology advances, and as the capabilities of the method become more apparent, the inherent advantages of coacervation will lead to a rapidly increasing number of coacervated effector systems on the market.

The use of porous, inert, slowly eroding or indestructable (as employed) plastic matrices (Table III) represents another form of effector entrapment and control. The kinetics of effector release from these systems has been well described, especially by Higuchi,[42,47–51] including release from planar systems having a homogeneous matrix, planar systems with granular matrices, spheres with a homogeneous matrix, or leaching release from granular spheres, as well as other matrix systems. Factors which effect the release rate from such systems include the diffusivity of the drug in the permeating fluid, the tortuosity of the capillary system if any, the total amount of drug remaining at any time, the solubility of the drug in the permeating fluid and the porosity of the matrix. Matrix systems typically demonstrate release profiles which continuously decline with time, and follow a first-order pattern with the rate of release at any time being proportional to the amount of agent yet remaining to be released. Special formulation and design techniques, such as incorporation of soluble polymers to increase porosity with time, use of laminates, etc., are often required with such systems to attempt to compensate at least in part for the natural tendency of these systems to continually decline in their release rates with time. The membrane diffusional controlled system on the other hand, is a natural zero order releasing system.

Some of the pharmaceutical uses of drugs dispersed in inert polymers matrices and films include the following:

1) Conventional sustained release tablets and other forms.
2) Catheters impregnated with antibiotics, anti-infectives, anaesthetics and other drugs.
3) Prolonged drug releasing soft contact lens or "Ocusert" systems.
4) Long acting parenteral implants.
5) Steroid release from IUD devices for fertility control.
6) Long term buccal absorption.
7) Drug releasing plastic topical bandages.
8) Topically applied plastic patches intended to provide drug absorption and systemic effects.
9) Long acting insecticide strips.
10) Long acting herbicide strips.
11) Long acting flea collars.

An interesting new concept in controlled effector release is the osmotic pump.[55] This patented system involves placing a precise small hole (laser cut) in a substantially impermeable coated particle. As long as a constant osmotic pressure is retained within the resultant sac, a constant release rate will reportedly obtain.

Over 150 articles have been published on some of the classes of solid dispersion systems, applied to drug effectors, as reported in Table III. These were comprehensively reviewed in 1971.[56] The primary application of solid dispersions in the drug field has been to enhance the solubility properties of poorly soluble drugs and improve their absorption. An effector which is molecularly dispersed need not overcome an energy of crystallization in dissolving. Mechanisms of increased dissolution rates of solid drug dispersions have been discussed.[57,58] Dispersions of the insoluble systemic antifungal agent, griseofulvin, in polyethylene glycol (6000 M.W.), have been shown to greatly increase absorption from the 30–60% commercial micronized products.[59,60] Recently just such a product has been marketed at about half the dose (twice the potency) of other existing products. While solid dispersions have not been widely employed to increase duration of effector activity, such systems have potential for increasing the potency and reliability of many drugs and insecticides. The solid dispersion technique also has obvious applications in providing homogeneous distributions of potent effectors. The solid state may stabilize unstable agents, permits dispersion of liquid or gaseous effectors, may facilitate formulation of a fast priming dose in a long acting system, or could produce long acting systems of soluble effectors by dispersion in poorly soluble or insoluble carriers. Solid dispersion techniques do have limitations, including the ability to produce a polymer melt or appropriate cosolution with the effector, thermal degradation of the effector, insolubility of the effector in desired polymers as well as general problems with the preparative methods.

New controlled delivery methods are continuously being reported. Immobilized enzymes combined with microcapsules reportedly provide continuous zero order drug release.[61] Recent reviews of controlled release in the drug field include: oral sustained release medications in general,[62] dry sustained release oral drugs,[63] matrices in prolonged release oral products,[64] preparation and use of coated micropellets in pharmacy,[65] long acting steroid preparations,[66] evaluation of long acting forms,[67] general principles of retardation,[68] and selected permeable membranes for biocompatible systems.[69]

MOLECULAR SCALE ENTRAPMENT EMPLOYING COLLOIDAL POLYMERIC DISPERSIONS

Various types of molecular dispersions were listed in Table III. The utilization of colloidal polymeric dispersions and latices as vehicles for the molecular scale entrapment of effectors[69–74] will be discussed in some detail based on certain advantages of these systems. Unlike the various methods of coating,[38] encapsulation, microencapsulation or other methods of molecular dispersion, the techniques described do not require the use of heat or fusion, and need not employ organic solvents or the involvement of a phase change. The features of the polymer latex as a unique reservior for effector entrapment are shown in Figure 3.

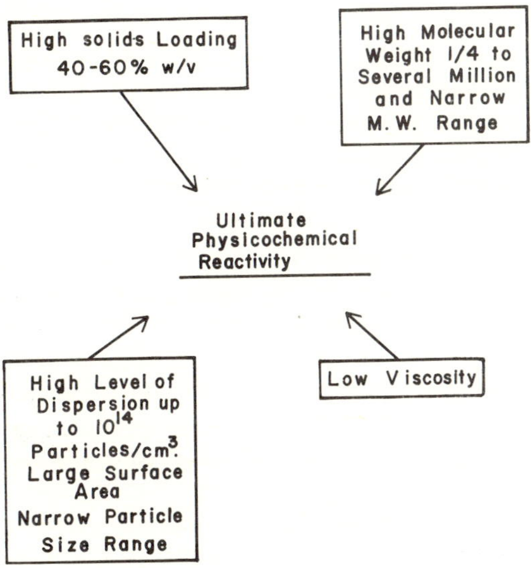

FIGURE 3 The basis of the polymer latex as a unique reservoir for drug and effector entrapment.

Polymer latices are also known as polymer emulsions. The so-called polymer emulsion in its final form is actually a colloidal solid or semisolid dispersion. The term emulsion derives from the fact that these products are made from an emulsification process wherein a liquid monomer is emulsified into an internal phase, possibly into a micelle, wherein the polymerization process occurs. When the growing polymer chain outgrows the dimension of the micelle, according to one theory, the polymerization ceases. Polymer molecular weight can, thus, be controlled within a very narrow range and the particle size of the resultant colloidal dispersion is also very uniform. The average particle size of the typical latex is in the fractional micrometer range (typically 0.1 to 0.5 um). The same polymers contained in a polymer latex, if dissolved in a suitable solvent, would produce a viscous gel at a concentration of 1 to 2% w/v. Thus, the polymer latex and similar concentrated colloidal polymeric dispersions, represent a reservoir for drug entrapment, which while largely overlooked until this time, is unmatched in physicochemical reactivity. Other unique features of the polymer latex of special interest in food and drug applications are the following: (1) Ability to remove even traces of residual monomer. Vacuum and other treatments may be employed for this purpose. (2) Freedom from heavy metal catalysts or toxic initiators. Heavy metal catalysts are not required in emulsion polymerization and the initiator can be selected from non-toxic peroxides or from volatile and removable agents. (3) The surfactants used in the manufacture of the latex can be selected from the GRAS list (generally regarded as safe) of the FDA. (4) This polymer form is potentially the least toxic and safest synthetic polymer form. Most polymers, if their molecular weight is sufficiently high to preclude absorption, are themselves non-toxic. Most of the diet of man is composed of polymeric materials (proteins, starches, cellulose). Toxicity of synthetic polymer substances is usually related to residual monomer or low molecular weight fractions or additives such as those noted which are a contaminant of the finished product.

Some of the possible mechanisms of molecular scale entrapment employing polymeric latices are shown in Figure 4. An aqueous solution of the effector when added to the colloidal dispersion, will be attracted to the surface of the colloidal particle if the ionic charge of the effector is opposite to that of the particle. Additionally, latex particles may be selected which contain ionogenic groups such as carboxyl groups which can also serve as binding sites for the effector forming direct ionic salts or hydrogen bonding sites.

Care must be taken that the number of ionogenic functional group sites on the polymer that are ionized is kept relatively low, otherwise polymer solvation will occur, and a thick gel will result. However, using the Henderson–Hasselbach equation, one can frequently calculate the number of functional groups that will be ionized at any particular pH and achieve the level of ionization of such groups that is sought. As the number of charge sites on the colloidal polymer particles are gradually neutralized, a point will occur at which coagulation of the latex occurs. Effector can be added to the point that coagulation does occur and the coagulation phenomena provides a method of concentrating and collecting the resultant "drug-polymerate." By staying below the flocculation or coagulation value, the colloidal dispersion can also be maintained. In either event, the effector will have been spread over the tremendously large surface of the colloid and associated with it,

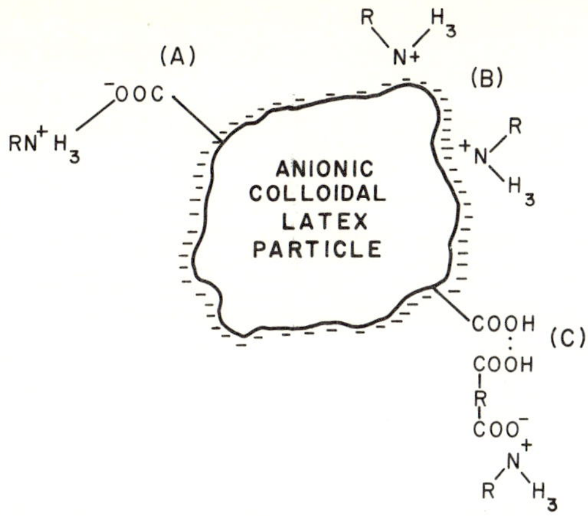

FIGURE 4 Mechanisms of molecular scale entrapment of effectors employing polymeric latices. (A) Partial ionization of functional groups on the particle may lead to salt formation, etc. (chemisorption). (B) Positively charged drug ions attracted to the negative colloid surface, reducing the surface charge and inducing coagulation (physisorption). (C) Dicarboxylic acid "bridge" hydrogen bonded to COOH of polymer, facilitating entrapment.

which prevents effector crystallization even in systems which contain as high as 30 to 40% effector in the dried product. Polymer latices may be designed or selected which have either a positive or negative charge to accommodate the charge characteristics of the effector (which if opposite to that of the colloid, will provide for maximum entrapment and binding, both extent and strength).

Non-ionic effectors may also be entrapped in these systems by a number of mechanisms. The effector may be dissolved in an organic solvent such as acetone or alcohol which is miscible with the aqueous vehicle of the typical latex. If the organic solvent solution containing drug is then slowly added to the latex, and the concentration of solvent is kept below the gelation point of the latex, it is possible to establish a situation in which drug will gradually diffuse into the colloidal polymeric particles. At this time additional solvent may be added or other mechanical means employed to gel the system after which it may be dried and redispersed or otherwise employed, with the other alternative being to retain the system in the colloidal form. Molecular dispersion of non-ionic effectors is also possible wherein no special techniques are employed to facilitate particle loading with the effector. The large number of colloidal particles and great surface of the latex disperse the effectors efficiently such that crystallization does not occur in the resultant gel.

Special advantageous features of latices for the entrapment of effectors may thus be summarized as follows: (1) Employing commercially available latices, no polymer preparation is required, i.e., preparation of special polymer solutions for coating, coacervation, etc. (2) Employing effectors which are water soluble, no organic solvents are typically required. (3) The colloidal latex form provides the ultimate in physicochemical reactivity for the entrapment of effectors. (4) No special equipment is required other than a tank and some mixing capability. (5) Continuous manufacture of large volumes of material is possible and feasible. (6) The entrapment process is a physicochemical phenomena which is readily controllable. (7) Molecular scale dispersion is readily achieved.

The actual processing steps which may be employed in the preparation of latex entrapped effectors are depicted in Figure 5. The effector solution is typically added to the polymer dispersion slowly with mixing. If the coagulation level is not achieved and a fairly dilute form of effector (low effector to polymer ratio) is sought, the resultant colloidal dispersion can be the finished product. If a dry form of the effector product is desired or a higher concentration of effector in the resultant product is the goal, inert electrolyte or coagulating solvents such as alcohol or acetone may be added to produce coagulation, and the coagulated (flocculated or gelled) system may be isolated by simple settling, centrifugation of filtration. This product may then be dried by a number of techniques, milled to an appropriate size and used in the dry form or in a subsequent suspension form or other forms as noted.

A high order of effector dispersion is listed as an ideal attribute of a controlled delivery system in Table II. The ultimate level of effector dispersion is molecular scale. In a molecular scale dispersed system there will be

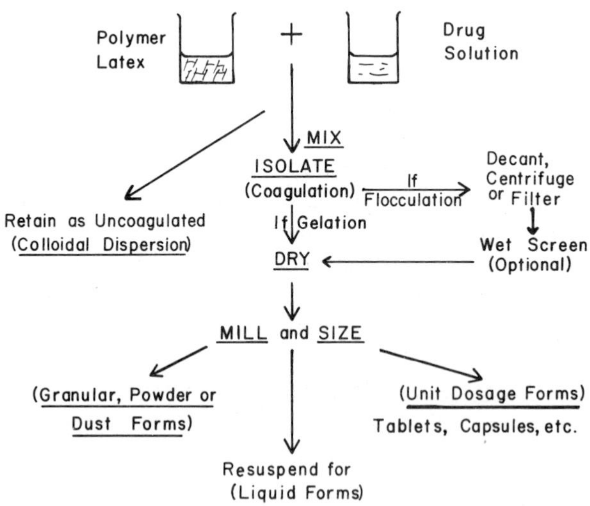

FIGURE 5 Processing steps in the molecular scale entrapment of effectors employing polymeric latices.

6×10^{23} particles per mole of effector agent (Avogadro's Number). If an effector has a molecular weight of 300 to 400, this corresponds to about 2×10^{21} particles per gram of material. This might be compared to perhaps 2×10^7 or 10^8 particles per gram at most for a finely micronized powder. An effector which is entrapped while in solution is being entrapped in its ultimate state of dispersion provided molecular dispersion is maintained in the entrapment product. Colloidal latices facilitate the maintenance of the molecular dispersion based on the fact that they contain typically about one-hundred trillion particles per cm^8 with each particle being a theoretical site for effector molecular localization. The significance of the distributive aspect of molecular scale entrapment for effectors is indicated in Table V.

TABLE V

The significance of the distributive aspect of molecular scale entrapment and dispersion for effectors

1) Unmatched uniformity of distribution of potent effectors
2) Facilitates ease of mixing, dilution, spreading and distribution
3) Permits stoichiometric reproducibility in entrapment precision and control of release profiles
4) Enables precise product design capability
5) May increase potency and bioavailability
6) Improves dosing and application accuracy and reliability
7) Better control over mechanisms of entrapment and release
8) Improves manufacturing and process control

The more potent an effector is, the lower the "dose" required to produce the intended effect. As the dose decreases the problem of uniformly dispersing fewer and fewer effector particles through an increasingly larger number of diluent particles in a uniform manner increases. The problem is one of statistical sampling. In a non-homogeneously dispersed effector system, a particular sample of a product representing one "dose" of effector to man or an insect may contain the exact required amount of effector agent, or may be inadequate to produce the intended effect, or may be an "overdose" resulting in toxic effects for the drug or represent wasteful or hazardous application in the case of an herbicide or insecticide. The bioavailability of a drug or other effector may also be expected to increase as particle dispersion is increased. Bioavailability is herein defined as the percentage of the effector presented to the host which is absorbed and which reaches a biophase. Increasing bioavailability of effectors may be expected to increase effectiveness and reliability of effector performance. More precise design capability for products is possible if the particle size of the effector can be precisely controlled and is unvarying. This is obviously most precisely controlled in the case of the solution form of the effector (molecular dispersion). A high level of dispersion and absolute control of dispersion dimensions also facilitates manufacturing process control. The accuracy and reliability of dosing of all effectors can be theoretically maximized as distribution at the molecular level is increased. Table VI illustrates typical data for the distribution of a drug in a dried molecular scale dispersed polymer latex–drug formulation for a series of ten replicated samples.

TABLE VI

Typical data for distribution of methapyrilene in the isolated dried product of a polymer latex-drug formulation[a]

Sample[b]	Amount of methapyrilene per 200 mg of isolated dried product, μg
1	20.3
2	18.6
3	19.8
4	19.3
5	20.2
6	20.0
7	19.4
8	20.3
9	19.8
10	18.9
	Average 19.7 ± 0.6

[a]This particular formulation consisted of 50 cm³ of a 0·5 mM methapyrilene hydrochloride solution and 50 cm³ of Acrysol ASE–75 (Röhm and Hass Company, Philadelphia, Pennsylvania).

[b]Samples were drawn at random from the isolated dried product.

The dilution represents a 1 to 10,000 dispersion of drug in vehicle with an average microgram concentration of 19.7 in 200 mg of sample and a standard deviation of 0.6 among the 10 samples. Utilizing the best dry-blending techniques currently available in pharmaceutical mixing, a physical mixture of the same drug also at a 1 to 10,000 dilution had at least 4 times the variability as reflected by the standard deviation values in the most completely mixed samples attainable.[75] As shown in Table VII, another unique feature of molecular dispersion systems is that as the ratio of excipient to drug is increased, i.e., the drug dilution is increased, the relative standard deviation of the system remains unaltered and probably reflects the precision of the assay.[75]

Formulation #6, which is the most dilute system, represents a 1 to 20,000 dilution in this case, which was the limiting dilution according to the chemical assay employed. When physical mixtures are evaluated in a similar manner, the relative standard deviation consistently increases as the dilution increases.

TABLE VII

Reproducibility of methapyrilene content in samples of solid dispersions obtained from a deflocculated polymer latex

Formulation[a]	Concentration of "drug solution,"[a] mM	$\bar{X}$,[b] μg	Relative standard deviation	Predicted $\bar{X}$,[c] μg	Deviation from predicted $\bar{X}$, %
1	1.00	88.1 ± 1.6[d]	1.8	94.2	6.5
2	0.50	44.9 ± 0.7	1.6	47.1	4.7
3	0.40	38.9 ± 0.7	1.8	37.7	3.2
4	0.20	19.7 ± 0.6	3.0	18.8	4.8
5	0.10	9.6 ± 0.2	2.1	9.4	2.1
6[e]	0.10	9.7 ± 0.2	2.1	9.4	3.1

[a]A formulation consisted of 50 cm^3 of methapyrilene hydrochloride "drug solution" and 50 cm^3 of Acrysol ASE–75.

[b]Represents an average amount of drug found in each of 10 randomly selected 200 mg samples drawn from the isolated dried product.

[c]Predicted from the least squares line of the linear relationship of columns 2 and 3 of this table; i.e. millimoles of bound methapyrilene = (0·685) × millimoles of methapyrilene added.

[d]Standard deviation.

[e]Micronized granulation.

The importance of this distribution aspect cannot be overemphasized, not only for achievement of accurate dosage of potent drugs but in the most effective distribution of insecticides and herbicides. The data in Table VIII illustrates the reproducibility of the molecular scale latex entrapment process.[69]

TABLE VIII

Reproducibility of the latex entrapment process

Batch number	mg of methapyrilene HCl/g of collected solid	Average of two assays
1	135.28 134.59	134.94
2	133.05 139.11	136.08
3	133.94 133.87	133.91
4	130.99 129.77	130.38
5	132.95 131.69	132.32
6	127.22 131.98	129.60

The agent entrapped in this case is the widely used antihistamine, methapyrilene. The drug was entrapped by being added to an acrylic acid–acrylate copolymer latex emulsion with the drug in solution as the hydrochloride. The 20% w/v drug solution used in this case was sufficient to produce flocculation of the colloidal latex. The flocculated part of the material was collected on a filter, dried at 50°C and powdered to 60 mesh size. As indicated in Table VIII, six replicated batches of material were prepared and two analyses were performed on each batch. Based on statistical analysis, one could expect approximately 90% of all such replicated samples to fall within ± 5.12 units of the sample mean (i.e., 132.87 ± 5.12). The coefficient of variability, therefore, is less than 5%. The data shown reflects the reproducibility of the entrapment method in general, which being a physicochemical process is similar to that of undertaking a chemical reaction, i.e., is stoichiometric. The basis of the predictability of the entrapment process is shown in Figure 6, which presents a simplified binding isotherm[75] of the drug reported in Table VII in the same polymer.

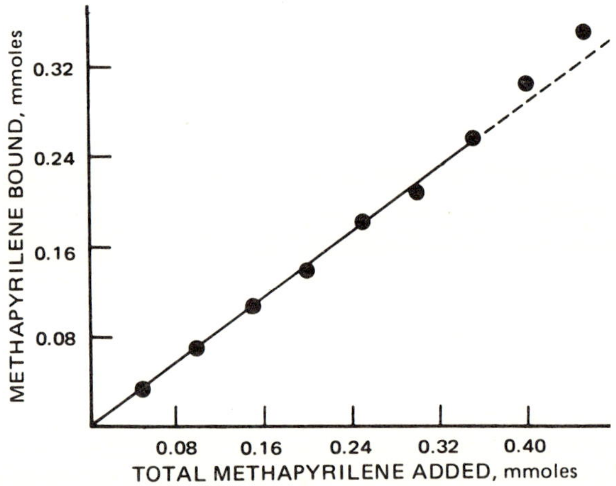

FIGURE 6 Relationship between methapyrilene bound and methapyrilene added to the latex.[75,77]

The isotherm is very linear in the subflocculation region and departs from linearity as might be expected above the flocculation region. Since the entrapment process produces linear and predictable binding isotherms, the data shown in Table VII would be expected. The deviation of drug found in the dried product from that which was predicted by the established relationship, ranged from about 2 to 6.5% with the greater deviation being at the higher concentration of drug. In the more concentrated systems, not all of the drug present is physicochemically associated with the polymer and some mechanical drug entrapment undoubtedly occurs. The mechanical entrapment is more difficult to control, but the product is still within acceptable limits.

Table IX shows the reproducibility of drug entrapment in samples of the solid dispersions obtained from flocculated systems producing higher drug concentrations.[75]

TABLE IX

Reproducibility of methapyrilene content in samples of solid dispersions obtained from a flocculated polymer latex

Formulation[a]	Concentration of "drug solution,"[a] mM	$\bar{X}$,[b] mg	Relative standard deviation
1	100	7.739 ± 0.132[c]	1.7
2	60	5.495 ± 0.181	3.3
3	15	1.778 ± 0.052	1.9
4	14	1.444 ± 0.045	3.1
5	10	0.959 ± 0.039	4.1
6	2	0.217 ± 0.009	3.9

[a] A formulation consisted of 50 cm³ of methapyrilene hydrochloride "drug solution" and 50 ml of Acrysol ASE–75.

[b] Represents the average mg of drug found in each of 10 randomly selected 200 mg samples drawn from the isolated dried product.

[c] Standard deviation.

The relative standard deviation values are again all below 5%. In summary, the reproducibility and predictibility of effector entrapment utilizing this technique, are generally within the range of ± 5%. Similar reproducibility and predictibility of effector release rates in standard dissolution release media have also been observed.

In any effector entrapment process, the efficiency of the process may be an important economic consideration. The efficiency with which the effector is entrapped, i.e., minimal waste of the agent, is usually of paramount importance since the effector is usually the high cost component in the end product. Table X shows the efficiency with which various salts of a potent antihistamine, chlorpheniramine, is entrapped when an electrolyte coagulation approach is employed.[70]

Only about 50% of the hydrochloride of the drug is swept from solution on latex coagulation while about 80–98% of the drug is entrapped when a dicarboxylic acid salt is employed. The dicarboxylic acid moiety provides a bridging mechanism to certain polymer latices as indicated in Figure 4. In the case of the hydrochloride salt, after filtering the flocculated product, the filtrate containing about half of the drug could be recycled in a subsequent entrapment procedure. However, an analytical step might be required to assay the filtrate and add the appropriate amount of new drug for a second

TABLE X

Effect of carboxylic acid anion on the efficiency of entrapment of chlorpheniramine by an acrylic copolymer emulsion

Chlorpheniramine salt	Initial concn. of drug solution % w/v	% drug bound	% drug in dry flocculant
Hydrochloride	2	54	3.3
Maleate	2	91	3.7
Oxalate	2	80	3.7
Malonate	2	90	3.9
Succinate	2	98	4.5
Succinate	5	93	10.0
Succinate	10	86	17.5
Adipate	5	96	10.4

run. In addition, the filtrate may contain surfactant or other soluble material from the original latex vehicle, which on recycling the filtrate would build up in concentration and could be undesirable in the finished product. Entrapment procedures which are efficient in a one-step operation, both in the utilization of effector and polymer are most desirable. In the coagulation of the polymer latices, by gelation or flocculation, substantially no polymer waste or loss is observed. When effector is added to a latex and no coagulation step is employed, the system approaches the ultimate in simplicity and efficiency as a process (Figure 5).

Another desirable feature in any effector microencapsulation or entrapment controlled delivery system is the ability to precisely control delivery rates to match any idealized or theoretical effector release profile. In some cases, and not uncommonly in controlled drug delivery, it may be desirable to provide an initial burst of the effector agent, followed by a constant zero order release rate. This is shown in Figure 7.[72] The goal here was to liberate about half of the 60 mg dose of phenobarbital in the first one half to one hour, and thereafter release the remaining 30 mg of drug at a constant rate, over the remainder of an 8 h period. The delivery system shown[72] had the capability of closely approaching the idealized goal with the greatest departure being some loss of linearity at the terminal time points.

Polymer selection is of course a factor in the design of any controlled delivery vehicle. By balancing the ratio of polar and non-polar groups, a virtually infinite range of materials are usually available, extending from hydrophilic materials to very lipophilic materials. Polymer selection coupled with the mechanism by which the effector is entrapped or contained in the product, along with the influence of physical factors on release rate, provide the means of formulation and design control usually at the disposal of the development scientist. Through a balance of all these factors one approaches

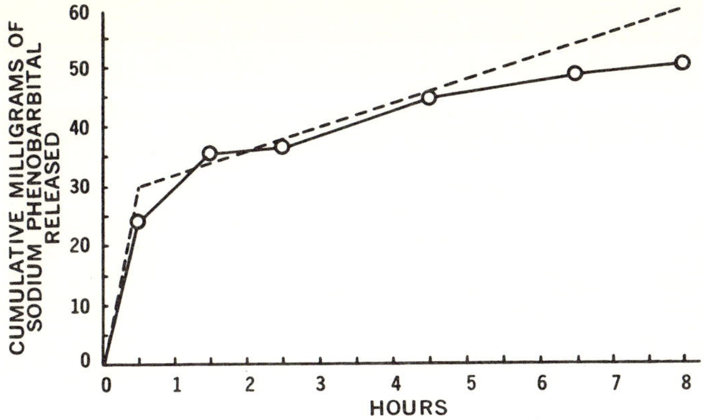

FIGURE 7 An idealized controlled delivery release rate for sodium phenobarbital (dashed line) vs. the experimental release profile of a xerogel system.[72,77]

or matches the exact release profile sought. If a mathematical relationship between the variables influencing the release rate can be established by employing such procedures as regression analysis, it is possible to mathematically design and optimize the product to a primary objective of release profile as has been reported in the pharmaceutical literature.[1]

Another consideration in the successful design of controlled release systems for effectors, which is not universally recognized by researchers in this field, is the principle of permitting a number of physicochemical factors to influence the release rate, and not to key the release characteristics to any single factor such as environmental pH, presence or concentration of moisture, heat exposure, presence and concentration of enzymes, etc. (Table II). Virtually all controlled release systems will be influenced by such physical and physicochemical factors, but a system which is keyed to any single factor will be highly dependent on that factor for performance. Table XI illustrates some of the controllable formulation and physicochemical factors which influence the drug release profile from a drug or effector system entrapped by the molecular scale method of dispersion in a polymer latex vehicle.

If a polymer is selected that is insoluble and unable to hydrate and swell appreciably at low pH (pH of 1 to 2.5), the polymer itself will be relatively inert in the stomach contents of man and certain animals, and drug release will be by desorption and/or dissolution from particle surfaces of the entrapment product. For the typical amine drug which has excellent solubility in gastric media, such initial release will thus be affected by the surface exposed and, hence, particle size. When the product gradually leaves the stomach of man and many non-ruminant animals, it enters a region of gradually increasing pH; going to a pH of 3 to 5 in the duodenum (the first small segment of

TABLE XI

Formulation and physicochemical factors which influence effector release profiles from molecular dispersion latex systems

1) Effector ionic character, solubility, salt or ester employed
2) Polymer selection
    - Functionality
    - Molecular weight
    - Monomer or comonomers used
    - Presence of crosslinking
3) Latex selection
    - Surface charge
    - Particle size
    - Concentration
    - pH
    - Colloid stability
4) Method of effector incorporation
5) Ratio of effector to polymer
6) Distribution of effector in polymerate
    - Surface concentrated
    - Uniformly distributed
7) Method of coagulation if coagulated
    - Polymerate density
    - Polymerate structure
        - Other physical properties

the small intestine), and then gradually increasing to a pH of 7 or above as it moves to the large intestine. By employing a polymer, several polymers, or more typically a copolymer, which contains the proper ratio of carboxylic acid or other ionizable functionality, it is possible to produce a product which gradually hydrates and swells to permit continued release of drug during the sojourn of the product along the gut, with the intestinal fluids becoming increasingly accessible to the drug in the particle. Ideally the polymer will completely dissolve prior to its leaving the small intestine, or passing the effector's absorption sites, to assure complete drug availability. Similar concepts may be applied to entrapment of other effectors which are gradually released based on their prolonged contact with a plant surface or in the soil as a function of such variables as exposure to sunlight or moisture. A more recent concept in controlled delivery of effectors is the subject of bioadhesion wherein the effector-carrying product has the capacity to adhere to a plant surface, insect surface or a particular region along the gastrointestinal tract or other absorption surface for a predetermined time for the purpose of better controlling durational effects or increasing absorption reliability and, hence, the effectiveness or potency of the effector. A bioadhesive system which places the effector directly on an absorbing membrane can have the potential for greatly enhancing effector utilization since the effector is not diluted as readily by environmental factors such as

GI fluids. In some instances, bioadhesion can also be employed to aid in the targeting of effectors to specific species, tissues or organs. Bioadhesion within an animal may also be employed in the future to overcome detoxification processes or bypass metabolic routes such as intestinal wall metabolism of drugs. A bioadhesive topical system to be worn as a patch behind the ear is currently under development for prolonged release of motion sickness and other drugs. Some soft contact lens systems are excellent carriers of antiglaucoma drugs, and may even be applied to other systemic drug agents.

As noted previously, potential purposes for controlling the release of effectors from polymers and microcapsules include increasing the duration of activity of effectors, increasing potency, reducing toxicity or hazard of use or improving effector reliability. Some examples of the attainment of these goals with drug effectors follow. In Figure 8 the intensity and duration of action of phenobarbital and a phenobarbital latex product are plotted as animal activity vs. time, in rats.[72]

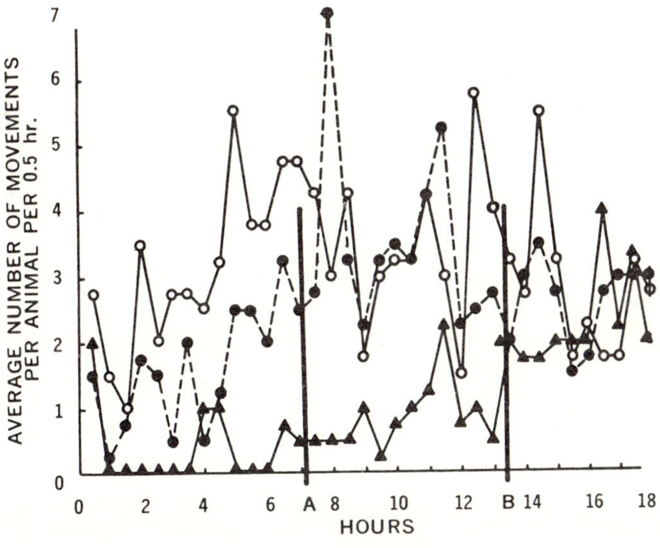

FIGURE 8 *In vivo* animal activity vs. time in male rats for a control group (open circles), conventional sodium phenobarbital drug group, 50 mg/kg (closed circles), and entrapment controlled release sodium phenobarbital group, 50 mg/kg (closed triangles).[72,77]

The rats were placed in special activity cages which were connected to a recorder to allow the measurement of animal activity. The drug was administered orally using a stomach syringe. The control animals were handled exactly like the test animals and were orally dosed with water. As can be seen, handling the animals in this manner produced a transient depression in activity for the first hour. As indicated in the figure, the test drug had a duration of activity which was significantly different from the control for about 7 h, while the drug, molecular scale dispersed in a polymer latex, had nearly twice that duration of activity (about 13.4 h). As noted in the figure, the drug dose was identical in each case. It is of further interest to note in Figure 8, that the animals receiving the polymer–drug system were more severely depressed, indicating that this form was more potent and more effective. In a separate but continuation study, large doses of drug and drug–polymer were administered to examine acute toxicity. It was found that 200 mg/kg of drug administered in solution was an approximate LD_{50} for the female rats employed, while twice the dose of drug in the drug-polymer form was an approximate LD_{40}. This indicates about a two-fold increase in safety margin for the controlled release form, which might represent a beneficial improvement for barbiturate drugs which are subject to abuse. The polymer used in Figure 8 was completely insoluble over the range of gastrointestinal tract conditions and drug release was from a xerogel product by a leach-out phenoma. As such, the system probably does not reflect the ideal product and further enhancement of safety is probably possible.

Table XII shows the results of another drug effector controlled delivery test in animals using an antihistamine. The test animals were exposed hourly to histamine vapor under standard conditions and their response noted. If the animal did not undergo a dyspepnic reaction within a standard time, the animal was considered to be protected by the drug and the duration of action was taken at the time such protection ceased. As shown, the conventional drug provided protection for an average of about 3.8 h, while the entrapped drug functioned for 8.7 h. This represented an increased period of activity at the 95% confidence interval of 4.9 $\pm$ 1.2 h. Antihistamines have depressant activity and their acute toxicity is related to this effect. An oral dose of 200 mg/kg of free methapyrilene hydrochloride was an approximate LD_{85} in rats within 30 min while the same dose of drug administered in the entrapped form was an LD_0 at 24 h. Controlled delivery of this drug clearly has the potential of improving the safety in an acute toxicity circumstance.

The ultimate test of the utility of any new controlled delivery system applied to drugs is not animal data but is *in vivo* evaluation in man. Table XIII presents such data for the commonly used sympathomimetric amine, phenylpropanolamine. The drug was entrapped in an acrylic acid/acrylate copolymer latex by electrolyte flocculation. The resultant polymerate was dried, ground to a 60 mesh powder and administered as the powder in a capsule with no further treatment. This drug is quantitatively and nearly completely eliminated unchanged in the urine. The half life of the drug alone (also given as a powder in a capsule) and of the polymerate were

TABLE XII

Measurable protection times afforded to the guinea pigs by free methapyrilene hydrochloride and a polymer–drug latex entrapment product

Free methapyrilene·HCI		Entrapped methapyrilene·HCI	
Animal number	Hours protected	Animal number	Hours protected
1	2	10	10
2	6	11	10
3	Sick	12	10
4	2	13	Sick
5	3	8	11
6	1	9	4
7	6	4	9
8	8	5	10
9	6	6	9
5	2	8	10
6	2	9	4
	$\bar{X}_1 = 3.8$ h		$\bar{X}_2 = 8.7$ h

determined from the urinary elimination data (average of five male subjects). The half life found for the drug of 4.6 h is in close agreement with the literature value.[76] The half life of the drug from the polymerate was nearly twice as long (8.0 h). It is of interest too that all the subjects receiving the polymerate (Table XIII) showed longer drug half lives than any of the values for conventional drug, and that the variability among the subjects was much less (suggesting perhaps more uniform and reliable absorption).

In summary of this section, Tables VI through XIII and Figure 8 demonstrate the real significance of the molecular scale entrapment of effectors in

TABLE XIII

Utilization of the molecular scale latex entrapment approach to increase the duration of activity of phenylpropanolamine in man

Subject weight, lb.	Biological half-life of phenylpropanolamine, h	Biological half-life of phenylpropanolamine polymer product, h
205	5.0	7.8
175	6.0	9.8
139	5.3	7.0
139	3.0	8.5
130	3.8	6.8
Average:	4.6	8.0

general (Table V) and of the molecular scale entrapment of drug effectors by polymeric latices in particular. The data of this section also shows that some of the ideal attributes of effector controlled delivery systems (Table II) and goals of controlled release effector design (Table V), including increasing durational effects while improving safety or increasing reliability or effectiveness, can in fact be achieved in practice. Areas of application of the molecular scale entrapment technology to various effectors, which have been established or are under study, are listed in Table XIV.

TABLE XIV

Areas of application of molecular scale entrapment technology to effectors and other agents

Rodenticides:	Potency enhancement
	Taste masking, bait enhancement distribution
Insecticides:	Spreading and distribution
	Bioadhesion
	Volatility and stability control
	Potency enhancement
	Durational control—increased and decreased
	Facilitate ease and safety of use
Herbicides:	Spreading and distribution
	Improve specificity
	Reduce hazard to desirable species, especially emerging crops
	Prolong effectiveness
Drugs:	Increase bioavailability
	Increase safety
	Increase effectiveness
	Increase reliability
	Improve stability
	Optimize action (combine above actions)
Cosmetics:	Long acting scents, antiperspirants, etc.
Flavors:	Flavor entrapment and control
Catalysts:	Controlled reaction processes

References

1. D. E. Fonner, J. R. Buck, and G. S. Banker, *J. Pharm. Sci.*, **59**, 1587 (1970).
2. J. R. Schwartz, J. R. Flamholz, and R. H. Press, *J. Pharm. Sci.*, **62**, 1165 (1973).
3. J. R. Buck, G. E. Peck, and G. S. Banker, Management science aids in expediting pharmaceutical product design, *Drug Development Communications*, **1**, 89–118 (1975).
4. Editors (Cover Story) *Time*, July 12, 1976, pp. 38–46.
5. D. R. Cowsar, in *Advances in Experimental Medicine*, A. C. Tanguary and R. E. Lacey, eds., pp. 1–14, Plenum Press, N.Y. (1974).
6. J. G. Wagner, *et al.*, *Int. J. Clin. Pharmacol.*, **1**, 261–285 (1968).
7. K. D. G. Edwards and M. McCredie, *Med. J. Australia*, **1**, 534 (1967).
8. L. V. Dmitryenlco and D. K. Hale, *J. Chem. Soc.*, 5570 (1965).
9. R. F. Faust, Dissertation Abstr. **27**, 2760-B (1967).

10. R. S. Shah and S. L. Bafna, *Nature*, **208**, 576 (1967).
11. S. L. Bafna and K. P. Govindan, *Ind. Eng. Chem.*, **48**, 310 (1956).
12. H. S. Bean, *et al.*, *J. Soc. Cosmetic Chemists*, **16**, 15 (1965).
13. H. S. Bean, *Pharm. J.*, **199**, 289 (1965).
14. L. H. Block and P. P. Lang, *Pharm. Acta Helv.*, **44**, 44 (1969).
15. P. P. Lamy, *Pharm. Acta Helv.*, **44**, 79 (1969).
16. S. Riefelman, *J. Pharm. Sci.*, **49**, 239 (1960).
17. J. L. Lach and T. F. Chinn, *J. Pharm. Sci.*, **53**, 924 (1964).
18. W. A. Pauli and J. L. Lach, *J. Pharm. Sci.*, **54**, 1745 (1965).
19. J. L. Lach and W. A. Pauli, *J. Pharm. Sci.*, **55**, 32 (1966).
20. J. L. Lach and J. Cohen, *J. Pharm. Sci.*, **52**, 137 (1963).
21. T. F. Chin, P. H. Chung, and J. L. Lach, *J. Pharm. Sci.*, **57**, 44 (1968).
22. D. H. Graham and L. B. Thomas, *J. Pharm. Sci.*, **51**, 988 (1962).
23. D. H. Graham, Y. M. Baker, and A. N. Njoka-Oi, *J. Pharm. Sci.*, **52** (1963).
24. D. H. Graham and Y. M. Baker, *J. Pharm. Sci.*, **52**, 893 (1963).
25. B. B. Wolfson and G. S. Banker, *J. Pharm. Sci.*, **54**, 195 (1965).
26. L. Molteni and F. Scrollini, *Eur. J. Med. Chem.-Chim. Ther.*, **9**, 618 (1974); through *Chem. Abstr.*, **83**, 48130 h (1975).
27. C. J. Cavallito and R. Jewell, *J. Pharm. Sci.*, **50**, 715 (1961).
28. B. H. Lippold and B. C. Lippolds, *Pharm. Ind.*, **36**, 432 (1974).
29. A. Heyd, D. O. Kildsig, and G. S. Banker, *J. Pharm. Sci.*, **58**, 586 (1969).
30. A. Heyd, D. O. Kildsig, and G. S. Banker, *J. Pharm. Sci.*, **59**, 947 (1970).
31. C. W. Woodruff, G. E. Peck, and G. S. Banker, *J. Pharm. Sci.*, **61**, 1916 (1972).
32. T. Higuchi and V. Stella, *Prodrugs as Novel Drug Delivery Systems*, Amer. Chem. Soc., Atlantic City, N.J. (1975).
33. M. D. Ranney, Microencapsulation Technology, Noyes Development Corp., Park Ridge, N.J. (1969).
34. C. E. Anderson *et al.*, Microencapsulation, MR Management reports, Boston, Mass. (1963).
35. M. Gutcho, Capsule Technology and Microencapsulation, Noyes Data Corp., Park Ridge, N.J. (1972).
36. J. E. Vandegaer, *Microencapsulation—Processes and Applications*, Plenum Press, N.Y. (1974).
37. R. W. Baker and H. L. Lonsdale, in *Advances in Experimental Medicine*, A. C. Tanquary and R. E. Lacey, eds., pp. 15–64, Plenum Ress, N.Y. (1974).
38. G. S. Banker, *J. Pharm. Sci.*, **55**, 81 (1966).
39. G. S. Banker, U.S. Patent 3, 896, 762, Assigned to Purdue Research Foundation (July 29, 1975).
40. A. J. Fites, G. S. Banker, and V. F. Smolen, *J. Pharm. Sci.*, **59**, 610 (1970).
41. M. S. Zora, A. J. Zimmer, and P. V. Maney, *J. Pharm. Sci.*, **53**, 487 (1964).
42. T. Higuchi, *J. Pharm. Sci.*, **52**, 1145 (1963).
43. W. Higuchi, *J. Pharm. Sci.*, **56**, 315 (1967).
44. W. A. Ritschel and R. Clotten, *Arzneim.-Forsch*, **19**, 221 (1969).
45. W. A. Ritschel and R. Clotten, *Arzneim.-Forsch*, **19**, 347 (1969).
46. J. Sjogren and L. E. Friklof, *Farm Rev.*, **59**, 171 (1960).
47. S. J. Desai *et al.*, *J. Pharm. Sci.*, **55**, 1224 (1966).
48. S. J. Desai *et al.*, *J. Pharm. Sci.*, **55**, 1230 (1966).
49. S. J. Desai *et al.*, *J. Pharm. Sci.*, **55**, 1235 (1966).
50. S. J. Desai *et al.*, *J. Pharm. Sci.*, **56**, 1542 (1967).
51. S. J. Desai *et al.*, *J. Pharm. Sci.*, **56**, 1548 (1967).
52. S. S. Kornblum, *J. Pharm. Sci.*, **58**, 125 (1969).
53. B. Farhadieh, S. Borodkin and J. D. Buddenhagen, *J. Pharm. Sci.*, **60**, 209 (1971).
54. B. Farhadieh, S. Borodkin, and J. D. Buddenhagen, *J. Pharm. Sci.*, **60**, 212 (1971).
55. F. Theeuwes, *J. Pharm. Sci.*, **64**, 1987 (1975).
56. W. L. Chiou and S. Riegelman, *J. Pharm. Sci.*, **60**, 1281 (1971).
57. A. P. Simonelli, S. C. Mehta and W. I. Higuchi, *J. Pharm. Sci.*, **58**, 538 (1969).

58. F. Sjuib, A. P. Simonelli, and W. I. Higuchi, *J. Pharm. Sci.*, **61**, 1381 (1972).
59. W. L. Chiou and S. Riegelman, *J. Pharm. Sci.*, **59**, 937 (1970).
60. W. L. Chiou and S. Riegelman, *J. Pharm. Sci.*, **62**, 498 (1973).
61. L. B. Wingard Jr., Insolubilized Enzymes, pp. 179 (1974); through *Chem. Abstr.*, **82**, 90018g (1975).
62. J. G. Wagner, *Biopharmaceutics Relevant Pharmacokinetics*, pp. 148, Drug Interscience Pub., Hamilton, Ill. (1971).
63. R. DeNeve, *Farm. Tijdschr, Belg.*, **51**, 282 (1974).
64. M. S. Socias and A. Cerezo, *Cienc. Ind. Farm.*, **5**, 114 (1973).
65. G. Rothgang, *Informationsdienst. Arbietsgem. Pharm. Verfahrenstech*, **20**, 39 (1974).
66. A. Vermeulen, *Acta Clin. Belg.*, **30**, 48 (1975).
67. A. Grzesiczak, *Farm. Pol.*, **30**, 539 (1974).
68. R. Huettenrauch, *Pharmazie*, **30**, 3 (1975).
69. H. Goodman and G. S. Banker, *J. Pharm. Sci.*, **59**, 1131 (1970).
70. C. T. Rhodes, K. Wai, and G. S. Banker, *J. Pharm. Sci.*, **59**, 1578 (1970).
71. C. T. Rhodes, K. Wai, and G. S. Banker, *J. Pharm. Sci.*, **59**, 1581 (1970).
72. J. C. Boylan and G. S. Banker, *J. Pharm. Sci.*, **62**, 1177 (1973).
73. G. S. Banker, U.S. Pat. 3,608,063, Assigned to Purdue Research Foundation (September 21, 1971).
74. G. S. Banker and H. Goodman, U.S. Patent 3,629,392, Assigned to Purdue Research Foundation (December 21, 1971).
75. A. B. Larson and G. S. Banker, *J. Pharm. Sci.*, **65**, 838 (1976).
76. K. R. Heimlich and D. R. MacDonnell, *J. Pharm. Sci.*, **50**, 232 (1961).
77. Reprinted with the permission of the American Pharmaceutical Society.

DISCUSSION

Dr. R. Nevin (*Eli Lilley and Co., Indianapolis, IN*): Have the emulsions which you studied all been commercial emulsions?

Dr. G. Banker: No, we're also working with experimental emulsions. I can't discuss these and tell you how we are making them, because the patents still apply. These include polymers which have complete FDA acceptance.

The polymers which you saw here were, for the most part, acrylic/acrylate copolymers that contained 0 to 50% acrylic acid. The acrylic polymers are available and marketed as the Eudragit Products by Röhm Pharma in Germany. They are being used commercially in Europe, and they have a master file with the FDA in the United States.

One of the problems which has been described with acrylic polymers, but has not been implicated in the latex, is that if there is monomer present lower bowel lesions may occur (test animal results). This has not been demonstrated for the latex probably because of the very low amount of residual monomer.

Dr. K. Lehmann (*Röhm Pharma GMBH, Darmstadt, Germany*): Are your particles embedded or encapsulated by your flocculation process?

Dr. G. Banker: We can either embed or encapsulate the drug. By using organic solvents we can load drugs into the beads in solution. Most of the drug is on the surface of the particle when we use the electrolytic flocculation process in which the drug is added in solution as the electrolyte. An electron micrograph shows no evidence of crystallinity in these samples, and there is no x-ray pattern for the drug. This is good evidence that the drug is molecularly dispersed.

Dr. H. Hopfenberg (*North Carolina State Univ. Raleigh, NC*): What was the basis for assuming that you had to have a linear isotherm for distribution of the drug between the continuous phase and the particular phase?

Dr. G. Banker: We don't, and I didn't mean to imply that. What I showed you was the region that looks like this: It's linear, then it bends a little this way, and then it goes off that way. At some point there is so much drug present that the polymer can no longer accommodate it, then you will get particulate crystalline drug.

When you're in the linear phase, the rate of addition, the temperature of addition and all these variables have little consequence on the ability to predict the drug content of the finished product.

Another question might be how much drug can be entrapped on a molecular scale. This depends upon how the drug is entrapped, i.e., is it solely on the surface, is it in the interior of the bead or is it bulk. We found that the resultant product could contain 30 to 40% drug for the basic amine drugs and still have molecular dispersion.

V. Stevens (*Dow Chemical Co., Midland, MI*): Do you have any problems with drug entombment?

Dr. G. Banker: There can be problems with drug entombment. We've studied this using artificial membranes like the Visking cellulose membrane and hydrophobic membranes. The appropriate membrane must be selected depending upon whether your drug is hydrophilic or hydrophobic.

One can quickly study whether the drugs come out of the particles under the pH and temperature conditions chosen.

If you use a polymer which will not hydrate or dissolve in gastric fluid and load the drug inside the bead, you won't get it back until the particle leaves the stomach. One normally would like to get 80 to 90% of the drug back predictably and reliably.

V. Stevens: Can we assume that your system is a thermosettable, crosslinked system?

Dr. G. Banker: Yes. I should state that the gastric retention systems are lightly crosslinked. If these are heavily crosslinked, they won't swell. Swelling increasing up to a given degree of crosslinking and then decreases upon further crosslinking. We work at the region where there is maximum swelling.

Dr. M. Goodman (*Univ. of California/SanDiego, LaJolla, CA*): Have you considered any of the problems in cattle feeding where the rumen seems to take a good part of the nutrient?

Dr. G. Banker: We've thought about ways of getting nutrients past the rumen into the stomach, but we haven't worked in that area yet.

Dr. M. Goodman: It seems like your work could have direct applications in this field.

Dr. G. Banker: One of the problems in working with ruminant animals is that one has to use just the opposite kind of polymers that is used for man. In the ruminant animal, one needs a polymer which will resist base and release in acid instead of resisting acid and releasing in base. The polymers which would be suitable for most ruminants are toxic. However, some of the polymers which we are working with, especially the crosslinked ones, might have application in this area.

Dr. K. Lehmann: You have flocculated a polymer containing carboxylic groups with a drug containing amino groups. Have you seen any differences in the release patterns for these preparations and those of a drug fixed on an ion exchange resin?

Dr. G. Banker: There can be very large differences if the drug is also in a solution phase in the polymer. If the drug is purely electrolytically entrapped, then the system has the same sensitivities as the ion exchange resin and the same shortcomings.

Dr. L. Bangs (*Dow Chemical Co., Midland, MI*): Have you worked with any styrene-based materials?

Dr. G. Banker: No. However, the functional groups of styrene lattices would facilitate drug binding.

Preparation and Evaluation of a Bioerodible Naltrexone Delivery System

RICHARD C. CAPOZZA, LEE SENDELBECK and WAYNE J. BALKENHOL

ALZA Corporation, 950 Page Mill Road, Palo Alto, CA94304, U.S.A.

The release profile *in vivo* of polymer/^{3}H-naltrexone systems was determined by measuring urinary excretion of ^{3}H in rats following implantation of systems (two systems/rat) in the subcutaneous tissue. After a 24-h transient, urinary excretion was approximately constant between 1 and 28 days, at a rate equivalent to 3–5 μg naltrexone/h.

Groups of 10 rats bearing polymer/drug-containing systems or placebo polymer systems (one system/rat) were subjected to the tail-flick test for narcotic antagonist activity at 4, 14, 24 and 34 days post-implantation. Immediately thereafter the rats were sacrificed, the implant sites examined grossly, and the systems recovered for analysis of drug content. Additional groups of 10 rats were sacrificed at 43 days post-implantation. Ninety-eight percent of the drug had been released by 24 days post-implantation. During the first 24 days, average release rates ranged from 2.5 to 5.5 μg/h, which is equivalent to a continuous rate of 6 to 15 μg/h/kg body weight. Narcotic antagonist activity ranged from 88% (at 14 days) to 62% (at 24 days) against an ED_{80} dose of morphine. The results agree well with an experiment in which intravenous infusion of 4 and 16 μg/kg/h for 6 h resulted in 54 and 89% antagonism, respectively, against an ED_{80} dose of morphine. No significant pharmacological activity was observed at 34 days post-implantation. At 34 days, the material recovered at the implant site, including the thin fibrous membrane, amounted to $<2\%$ of the original weight of the system. No recoverable material was present at 43 days post-implantation.

Local tissue responses were generally mild in both treatment groups, although at 14 and 24 days the response to drug-containing systems was consistently higher than the response to placebo systems. Responses in both treatment groups were minimal at 34 and 43 days (10 to 20 days after erosion of the system was nearly complete).

INTRODUCTION

A rational approach to the treatment of narcotic addiction requires the development of a therapeutic program which (1) facilitates compliance of the patient with the anti-addictive drug administration regimen over an extended time-period, (2) minimizes the frequency of contact with the physician (or paramedical personnel) during the course of therapy, (3) assures continuity of therapy throughout the treatment program, and (4) eliminates the risk of misuse or abuse of the anti-addictive medication by the patient or others.

Experience with methadone and other "substitutional addictive" substances has, over the recent past, made it clear that this approach to narcotic addiction control is substantially less than universally satisfactory because of the inability of this treatment to meet the criteria of compliance, continuity, and safety enumerated above. On the other hand, the emergence of a number of true narcotic antagonists (exemplified by naloxone, cyclazocine, naltrexone and diprenorphrine) has provided the promise for development of an alternate strategy of addiction management, which obviates methadone's addictive problems. The dosage requirements of these compounds to antagonize the effect of morphine-type drugs are very small (of the order of 1 mg/day or less). Since they are poorly absorbed via the G.I. tract, they must be given parenterally. However, these drugs have relatively short pharmacologic half-lives. Therefore, if they are given via conventional parenteral injection they must be administered very frequently to be effective.

Considering only the classical methods of drug administration, one would be forced to search for antagonist compounds which have significantly longer half-lives and have very large therapeutic indices, i.e., display little toxicity when administered in doses far in excess of the therapeutically effective amounts. Since finding agents which possess a quantum jump in the magnitude of either of these properties is likely to be difficult and time consuming, it would seem appropriate to seek novel means of delivering the existing drugs.

The unique problems associated with the sustained treatment of narcotic addiction on an outpatient basis, coupled with the special properties of now-available narcotic antagonists, provide an unusually attractive opportunity for the application of the principles of *controlled* long-term drug administration via a bioerodible subcutaneous implant.

Our objective was to prepare a biodegradable, biocompatible naltrexone delivery system which could be implanted subcutaneously and release the narcotic antagonist for approximately one month at relatively constant rates sufficient to block the euphoric action of morphine-based drugs.

MATERIALS AND METHODS

Materials

Naltrexone and naltrexone-15,16-^{3}H were obtained from the Research Triangle Institute. ^{3}H-Naltrexone was diluted with naltrexone by dissolving both independently in boiling THF or methanol and then mixing the two solutions together. The mixture was evaporated to dryness on a rotary evaporator at 70°C, 12 mm Hg for 2 h. The dried material was ground with a mortar and pestle, sieved through a 125 μm sieve, and redried at 70°C for 2 days *in vacuo*.

Morphine sulfate $5H_2O$: Mallinkrodt, USP, 90 mg dissolved in 10 cm^3 sterile saline.

One member (C101ct) of a family of biocompatible, bioerodible polymeric materials was used. These materials can be formulated with drugs under a variety of physical conditions into a wide variety of medically useful drug delivery systems with predictable rates of bioerosion. Because these materials are novel, their chemistry cannot be divulged until the filed patent applications issue. In general, the materials are hydrophobic and undergo a surface chemical reaction with the surrounding aqueous media which results in erosion and concomitant release of drug such that no residue remains after total drug depletion.

The polymer used in these studies has had satisfactory toxicological evaluation in connection with other applications.

Animals

Male Spraque Dawley albino rats were obtained from Simonsen Laboratories. They were housed, two per cage, at ALZA one week prior to implantation and fed Purina Rat Chow and water *ad libitum*.

Methods

Preparation of systems The polymer was heated to its molten state on a stainless steel platen in a dry nitrogen atmosphere. To the molten polymer was added and mixed ^{3}H-naltrexone, 20% by weight, specific activity 272 μCi/g, using a stainless steel spatula. Nontritiated naltrexone was mixed into the polymer under similar conditions. The composites were pressed into films with a Carver Press using a spacer between Teflon® sheets. Rod-shaped systems ($1 \times 1 \times 14$ mm, 0.58 cm^2, and weighing approximately 15 mg) were punched from this sheet using a heated automatic punch. The rods were placed between Teflon spacers in glass vials and capped under a helium atmosphere. The packaged systems were placed in a polyfoil package and irradiated with ^{60}Co, 2.5 meg rad.

Determination of naltrexone content of ^{3}H-containing systems Polymer/^{3}H-naltrexone systems were dissolved in acidified methanol and diluted with water. An aliquot was added to Aquasol® fluor (New England Nuclear) and counted in a liquid scintillation counter. A previously prepared quench curve was used to convert cpm to dpm. This counting procedure to determine the amount of ^{3}H-naltrexone was followed, where applicable, in the experiments that follow.

Determination of release rate of ^{3}H-naltrexone from the systems in vitro Systems containing tritium-labelled drug were allowed to erode in phosphate buffer (0.1 M, pH 7.0) maintained at 37°C. Aliquots of the erosion media were drawn at intervals of 1 to 3 days. The systems were placed in fresh erosion media after each sampling. Each aliquot of erosion medium was analyzed for the amount of tritiated drug present as before.

Procedure for implantation of systems in rats Using aseptic surgical techniques, the systems were placed in the subcutaneous tissue of the abdominal region by means of a trocar fitted with a 15 G needle. The cutaneous needle perforations were closed with wound clips.

Analysis of urinary excretion of ^{3}H-naltrexone and metabolites Following implantation with systems, the rats were housed individually in stainless steel metabolism cages equipped to separate urine and feces. Total urine output was collected at 1 to 3 day intervals. The volume of each sample was measured and an aliquot was counted as before.

Determination of naltrexone (nonradioactive) content of systems and the average release rate of naltrexone from systems (residual drug content method) Each system was dissolved in THF, then hydrolyzed with HCl/methanol. The samples were diluted with 0.1 M phosphate buffer, pH 8.0, and an aliquot was analyzed by reverse phase liquid chromatography using a μ Bondapak C column (Waters Associates). The mobile phase was 60% methanol in 0.5 M phosphate buffer at pH 8.0. The flow rate was 2.5 cm^3/min, and the detector was set at 254 nm.

The average naltrexone release rate was calculated by comparing the naltrexone content of noneroded systems with the residual drug content of systems recovered at various times after implantation in rats. An example of the calculation of the release rate for a given interval is shown below:

$$\text{Release rate for 4–14 days } (\mu\text{g/h}) = \frac{\begin{pmatrix}\mu\text{g naltrexone}\\ \text{released 0–14 days}\end{pmatrix} - \begin{pmatrix}\mu\text{g naltrexone}\\ \text{released 0–4 days}\end{pmatrix}}{240 \text{ h}} \quad (1)$$

Procedure for cannulation of right external jugular vein of the rat A heparin-filled polyethylene cannula (Intramedic PE50) was inserted into the right external jugular vein under pentobarbital anesthesia. The cannula was threaded through a subcutaneous tunnel and out through an intrascapular incision in the skin of the back. During the infusion, the cannula was protected from the rat's claws and teeth by a metal coil spring attached to wound clips at the skin incision. A Liquid Infusion System designed and built at ALZA was used to deliver approximately 0.5 cm^3/h of naltrexone in saline.

Tail-flick procedure for assay of narcotic antagonist activity This procedure measures the effectiveness of naltrexone in blocking morphine induced attenuation of heat-inflicted pain.[1,2]

Results

Two major experiments were conducted to evaluate the polymer/naltrexone systems. The first experiment compared the *in vitro* and *in vivo* functionality of polymer/naltrexone systems. The *in vivo* functionality study was carried out by measuring the ^{3}H excreted in the urine of rats implanted with polymer/naltrexone systems. The second experiment evaluated the pharmacological activity of the polymer/naltrexone systems implanted in rats and the tissue response to these systems.

IN VITRO FUNCTIONALITY AND URINARY EXCRETION

^{3}H-Naltrexone content of polymer/^{3}H-naltrexone systems, rods 1 × 1 × 14 mm

The average ^{3}H-naltrexone content of 5 irradiated systems used in the urinary excretion study was 3.34 ± 0.17 mg, or 20.8% by weight. Each system contained approximately 0.9 μCi ^{3}H.

^{3}H-naltrexone released from the systems *in vitro*

Five irradiated systems were eroded *in vitro* as described above. As shown in Figures 1A and 2A a brief transient occurred during the first 24 h, in which approximately 7% of the drug contained in the system was released. The transient was followed by a constant rate of ^{3}H-naltrexone release of about 3 μg/h during the next 9 days. After the tenth day, drug was released at a higher and variable rate until erosion was nearly complete (21–24 days). Thereafter, the systems maintained a release rate of less than 1 μg/h until they were recovered from the erosion bath at 33 days. Six percent of the initial naltrexone content remained in the system after 33 days erosion *in vitro*.

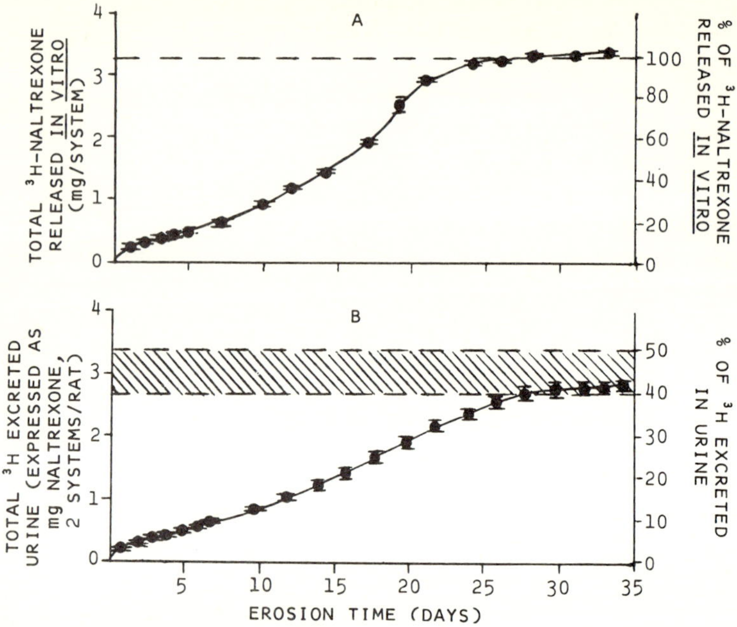

FIGURE 1 Cumulative release of ^3H-naltrexone *in vitro* (A) and urinary excretion of ^3H in rats. (B) polymer/^3H-naltrexone (20% w/w) systems, rods, $1 \times 1 \times 14$ mm. (A) Each value is the mean ± standard deviation for 5 systems eroded *in vitro*. The dashed line represents the total naltrexone content; (B) each value is the mean ± standard deviation for 5 rats, 2 systems in the subcutaneous tissue of each rat. The dashed line represents the range of expected total ^3H excretion in urine.

Urinary excretion of ^3H in rats implanted with systems

Two irradiated systems were implanted in each of 5 rats weighing 130–150 g, and urinary excretion of ^3H was followed as described above. The pattern of urinary excretion of drug and/or metabolites (Figures 1B and 2B) follows closely that of naltrexone release from systems *in vitro* during the first 10 days of erosion. In contrast to the *in vitro* drug release pattern, however, the rate of urinary excretion of ^3H increased only slightly after 10 days.

The excretion rate between 1 and 28 days post-implantation (two systems/rat) was equivalent to 3–5 μg naltrexone/h; it dropped rapidly after 28 days, to a fraction of 1 μg/h. Two percent of the initial ^3H-naltrexone content was found in the systems recovered from the rats at termination of the experiment 34 days post-implantation. Recovery of ^3H in the urine amounted to 43% of the ^3H released from the systems during the 34-day period. It is presumed that the remainder was excreted in the feces.

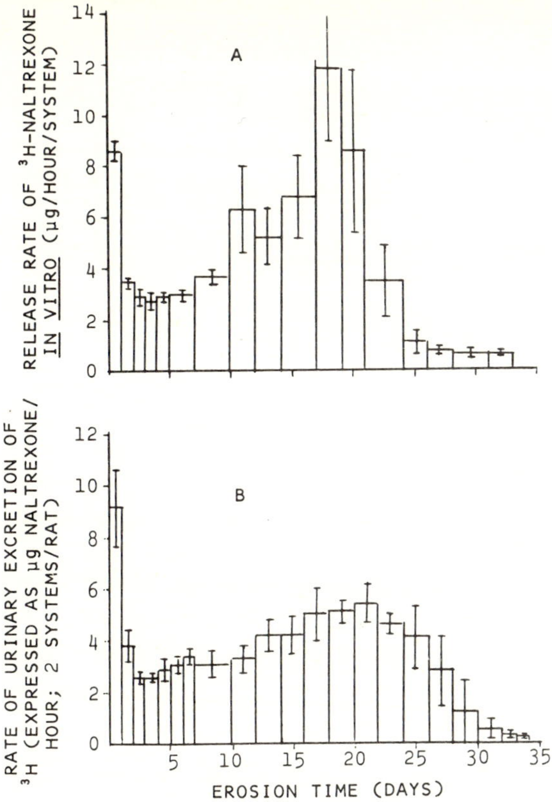

FIGURE 2 Rate of release of ^{3}H-naltrexone *in vitro* (A) and urinary excretion of ^{3}H in rats (B). Polymer/^{3}H-naltrexone (20% w/w) systems, rods, $1 \times 1 \times 14$ mm. (A) Each value is the mean ± standard deviation of 5 systems eroded *in vitro*; (B) each value is the mean ± standard deviation for 5 rats, 2 systems in the subcutaneous tissue of each rat.

PHARMACOLOGICAL ACTIVITY OF NALTREXONE AND TISSUE RESPONSE TO SYSTEMS

Six hour intravenous infusion of naltrexone in rats

Naltrexone was infused intravenously into rats at rates of 4 and 16 µg/kg/h for 6 h, at which time blood levels of the drug were assumed to have reached steady state. At 6 h, both dosages were effective against the analgesic effect of morphine, as measured by the tail-flick assay, but the higher dose was required to achieve 88% antagonism.

It should be noted that morphine, at a dose of 6.5 mg/kg, did not result in the expected 80% agonist effect in control animals, but instead, had a

63.5% effect. Thus, a slightly lower level of antagonism by naltrexone could be expected against an 80% effective dose of morphine. (See Table I.)

TABLE I

Analgesic effect of morphine at the end of a six hour intravenous infusion of the narcotic antagonist naltrexone in rats[a]

Dose[b] μg/kg/h	% Agonism[b]	% Antagonism[b]
0	63.5 ± 11.3	0
3.8 ± 0.1	29.2 ± 4.5	54.0 ± 10.7
16.6 ± 0.6	7.2 ± 2.1	88.7 ± 4.1

[a] 5 rats/group.
[b] Mean ± standard error

Naltrexone content of polymer naltrexone systems, rods 1 × 1 × 14 mm

The average naltrexone content, as determined by LC analysis, of systems used in the study of pharmacological activity was 2.42 ± 0.10 mg, or 17.8% by weight.

Release rate and pharmacological activity of naltrexone from the systems

Rats (100) were divided into 5 groups of 20 rats each, and implanted with one placebo or one drug-containing system in each rat, as shown in Table II.

Groups of 20 rats were challenged with an ED_{80} dose of morphine (9 mg/kg) and the tail-flick test was performed, according to the schedule shown in Table II.

TABLE II

Grouping of rats for pharmacological activity studies

Group	Sacrifice: days post-implant	Age in days at		Number of rats receiving	
		Implant	Sacrifice	Polymer/ naltrexone	Polymer
A	4	73	77	10	10
B	14	63	77	10	10
C	24	53	77	10	10
D	34	43	77	10	10
E	43	34	77	10	10

The percent antagonism of naltrexone released from the systems was calculated by comparing the tail-flick response times of control rats (bearing placebo systems) and treated rats (bearing drug-containing systems).

Immediately following the tail-flick test, each rat was sacrificed by overdose of ether.

At autopsy, the location of the system in the subcutaneous tissue was noted and the area was examined for any gross manifestation of tissue reaction.

As shown in Table III almost all the drug had been released from the systems by 24 days post-implantation. At 34 days, the material recovered at the implant site, including the thin fibrous membrane, amounted to only 2% of the original weight of the system. This quantity was insufficient for drug analysis and it was assumed that all the drug had been released. At 43 days only a small brown speck marked the implant site. Placebo systems had completely eroded by 34 days, and little or no evidence of the implant sites could be found at 34 or 43 days.

During the first 24 days, the average rate of release of naltrexone from the drug-containing systems varied from 2.5 to 5.5 μg/h, or approximately 6 to 15 μg/h/kg body weight (Table III and Figure 3). The pattern of release

TABLE III

Release rate and pharmacological activity of naltrexone from polymer/naltrexone (20% by wt) systems, rods 1 × 1 × 14 mm, implanted subcutaneously in rats

System	Days post-implant	Number of rats	% naltrexone released	μg naltrexone released	Average release naltrexone rate (μg/h)	% agonism[a]	% antagonism[a]
Placebo	4	10	—	—	—	83.1 ± 8.9	—
Naltrexone	4	10	17.6	420 ± 98	0–4 Days 4.5 ± 1.3	22.3 ± 4.3	73.2 ± 5.9
Placebo	14	10	—	—	—	95.3 ± 2.4	—
Naltrexone	14	10	42.6	1050 ± 82	4–14 Days 2.5 ± 0.5	11.7 ± 3.0	87.7 ± 3.2
Placebo	24	10	—	—	—	80.0 ± 9.1	—
Naltrexone	24	10	97.6	2377 ± 26	14–24 Days 5.5 ± 0.3	30.0 ± 8.7	62.5 ± 11.7
Placebo	34	10	—	—	—	70.2 ± 9.0	—
Naltrexone	34	10	100.0	2402 ± 28	24–34 Days 0.1 ± 0.05	59.1 ± 11.1	15.8 ± 19.1

Each value is the mean ± standard error, 10 rats.

[a] Tail-flick response to 9.0 mg/kg morphine sulfate, previously determined to be ED_{80}.

was similar to that observed when urinary excretion of drug and/or metabolites was determined following implantation of similar systems (Figure 2), although in the latter experiment the proportion of drug excreted in the interval 14–24 days was lower and the functional lifetime of the system was a few days longer (28 days *vs.* 24 days).

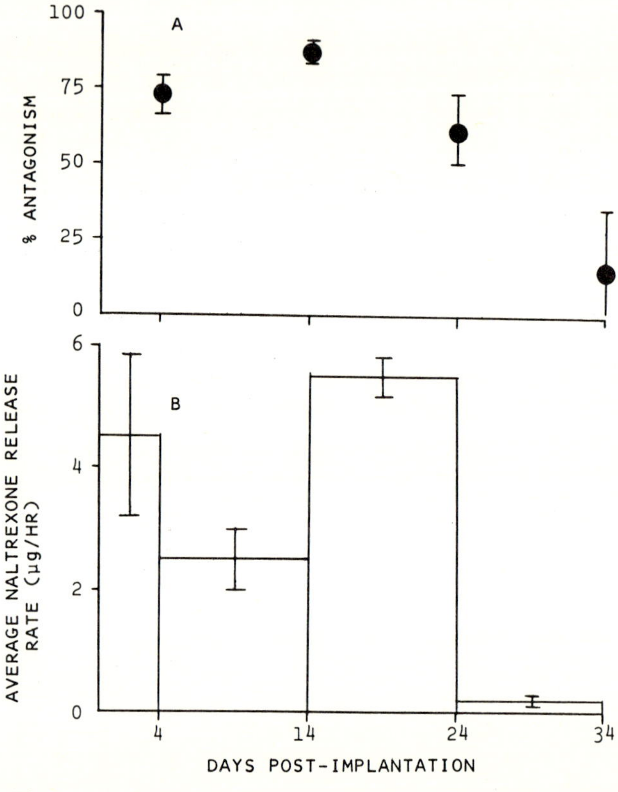

FIGURE 3 Pharmacological activity (A) and average release rate (B) of naltrexone from polymer/naltrexone (20% w/w) systems (rods, 1 × 1 × 14 mm) implanted subcutaneously in rats. Each value is the mean ± standard error for 10 rats, one system per rat.

A moderately high level of antagonist activity (62–88%) to an ED_{80} dose of morphine was exhibited through 24 days post-implantation (Table III and Figure 3). The data agree well with the continuous IV infusion experiment in which infusion of 4 and 16 μg/kg/h resulted in 54 and 89% antagonism, respectively, against an ED_{80} dose of morphine.

No significant antagonist activity was observed at 34 days, at which time the systems were almost completely eroded.

Local tissue response of rats implanted with placebo and drug-containing systems

Body weight changes during the 43 days of study are shown in Table IV. At 4, 14, 24, 34, and 43 days post-implantation all groups gained weight without notable differences between drug and placebo groups.

TABLE IV

Toxicological evaluation of rats implanted subcutaneously with polymer (87.5/12.5) placebo and polymer/naltrexone (20% by wt) systems, rods 1 × 1 × 14 mm

System	Days post-implant	Number of rats[a]	Body weight[b] (g)		Number of implant sites with response rated[c]			Average tissue response score
			Initial	Final	Mild	Moderate	Severe	
Placebo	4	10	393 ± 9	403 ± 8	10	0	0	0–1
Naltrexone	4	10	380 ± 8	384 ± 8	10	0	0	0–1
Placebo	14	10	358 ± 5	405 ± 7	10	0	0	1
Naltrexone	14	10	367 ± 8	406 ± 8	9	1	0	2
Placebo	24	10	264 ± 4	372 ± 9	10	0	0	0–1
Naltrexone	24	10	264 ± 5	355 ± 6	8	2	0	1–2
Placebo	34	10	218 ± 4	413 ± 8	10	0	0	0–1
Naltrexone	34	10	210 ± 3	391 ± 8	10	0	0	0–1
Placebo	43	10	125 ± 2	378 ± 8	10	0	0	0–1
Naltrexone	43	10	124 ± 2	398 ± 7	10	0	0	0–1

[a] One system per rat.
[b] Each value is the mean ± standard error.
[c] See Appendix for detailed scoring system.
mild response: score of 0 to 2–3.
moderate response: score of 3 to 3–4.
severe response: score of 4 to 5.

The tissue reactions at the sites of implantation of placebo and drug-containing systems are summarized in Table IV. The sites were evaluated according to the scoring system presented in the appendix. In general, a mild response consisted of a slight increase in vascularity and/or formation of a thin fibrous membrane around the system. In the moderate response, the same elements were present but to a greater degree. Fifty (50) of the 50 responses to placebo systems were rated mild. Forty-seven (47) of the 50 responses to naltrexone-containing systems were mild, and 3 moderate. At 14 and 24 days, the tissue response to the drug-containing systems was consistently greater than the response to placebo systems, as evidenced by the higher average tissue response scores in the naltrexone-treated group. By 34 and 43 days the tissue response was minimal in both treatment groups. It can be concluded that there is no limiting acute local irritation when naltrexone is delivered at average rates of up to 5.5 μg/h.

Acknowledgements

The work upon which this publication is based was performed pursuant to Contract No. ADM-45-157 with the National Institute on Drug Abuse, Department of Health, Education, and Welfare under the direction of Robert E. Willette, Ph.D.

We gratefully acknowledge the valuable technical assistance of Arthur Hall and Jim Day.

References

1. W. L. Dewey, J. W. Synder, L. S. Harris, and J. F. Howes, *J. Pharm. Pharmac.* **21**, 548–550 (1969).
2. W. L. Dewey and L. W. Harris, *J. Pharmacol. Exp. Ther.* **179**, 652–659 (1971).

Appendix

SYSTEM FOR SCORING OVERALL GROSS TISSUE RESPONSE

0: None.
1: Minimal proliferation of connective tissue and/or increase in vascularity.
2: Mild proliferation of connective tissue and/or increase in vascularity.
3: Moderate proliferation of connective tissue and/or increase in vascularity and/or minimal amount of fluid in capsule and/or involvement of tissue adjacent to site.
4: Moderate to severe proliferation of connective tissue and/or increase in vascularity and/or moderate amount of fluid or blood in capsule and/or involvement of tissue adjacent to site.
5: Severe proliferation of connective tissue and/or increase in vascularity and/or involvement of adjacent tissue; significant amount of fluid, pus or blood in capsule.

DISCUSSION

Dr. W. Walles (*Dow Chemical Co., Midland, MI*): Can you say anything more about the composition of the erodible polymer?

Dr. R. Capozza: The polymer is a proprietary material. Patent applications have been filed, but unfortunately, until the patents are allowed I cannot comment on the composition of the materials.

Dr. H. Ringsdorf (*Univ. of Mainz, Mainz, Germany*): Can you make any remarks regarding the fate of the polymer and its excretion?

Dr. R. Capossa: I can't tell you in detail because we have carried out limited excretion studies.

Dr. R. Phares (*Alza Corporation, Palo Alto, CA*): The polymer really is a member of a family of polymers. We've done some work on another member of this family in a preliminary experiment using a ^{14}C labelled polymer. More than 90% of that polymer was very rapidly lost in the body. The rest of it appeared to be fairly uniformly distributed throughout all body tissue and was not concentrated in any particular location. I stress that this was from only one very preliminary experiment with only a single acute administration of the polymer. We've just received the ^{14}C labelled polymer which was used in the experiments reported today, and this polymer will be the object of a very intensive metabolism distribution elimination study in animals. I believe we'll see the same thing as with the other member of this family.

Dr. R. Capozza: As I recall, the 90% that was eliminated was eliminated in the expired air.

Dr. R. Phares: Yes, it was conjectured that the 90% which was eliminated very rapidly went out in the form of CO_2 in the expired air. However, at that time we had not expected that route of elimination, and we were not collecting expired air. But 90% of the ^{14}C activity could not be found in the urine, feces, or body tissues.

Dr. H. Ringsdorf: This points to the structure of the polymer, doesn't it?

Dr. R. Phares: In what way?

Dr. H. Ringsdorf: There are not many polymers which rapidly degrade to CO_2 under body conditions.

Dr. R. Capozza: Poly(glycolic acid) is one such polymer.

Dr. H. Ringsdorf: Sure, and there are others.

Dr. R. Phares: I think we can say a little about the mechanism, and that is the fact that the backbone breaks, and the polymer is degraded to low molecular weight material, as opposed to the breaking of crosslink bonds.

Dr. K. Doebel (*Ciba–Geigy Corporation, Summit, NJ*): Is naltrexone identical to naloxone?

Dr. R. Capozza: Naltrexone (1) has the following structure:

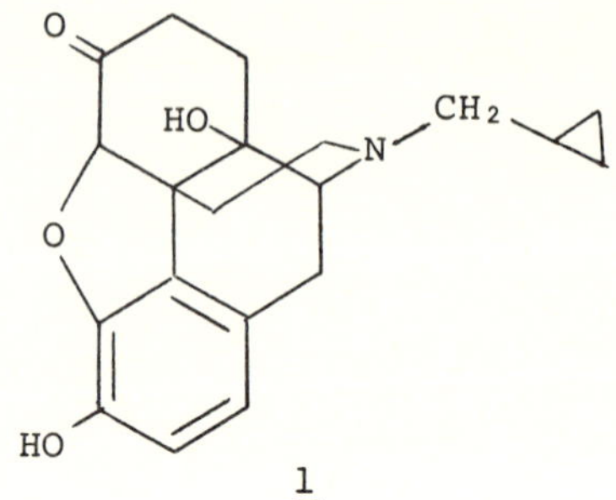

1

When the *N*-cyclopropylmethyl group on naltrexone is replaced with an *N*-allyl group the molecule becomes naloxone (2).

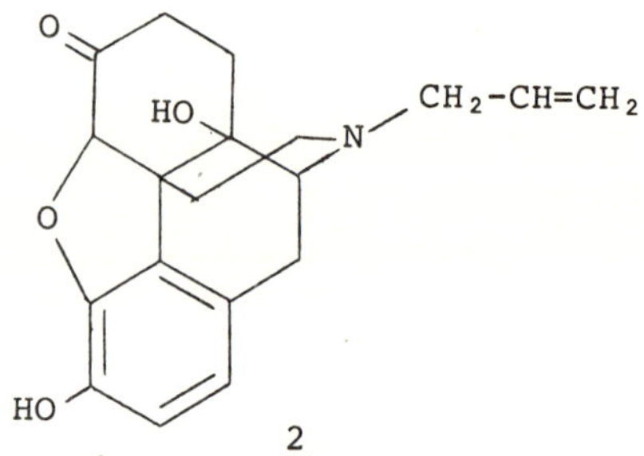

2

Dr. H. Hopfenberg (*North Carolina State Univ., Raleigh, NC*)**:** Would you care to speculate regarding the mechanism responsible for the acceleration of release with time?

Dr. R. Capozza: We believe the polymer hydrolizes via an autocatalytic mechanism. Thus, one would expect autocatalytic drug release kinetics if the system were eroding via a surface type mechanism. As you pointed out in your talk on Monday, erodible systems in the shape of a rod should deliver drugs at an ever decreasing rate since the surface area is continually decreasing. We are able to obtain relatively constant drug release dynamics because the autocatalytic polymer hydrolysis fortuitously compensates for the shape limitations.

Dr. H. Hopfenberg: I think that is a very important point because the geometric limitations of the implant typically dictate that you go to a sphere or cylinder or combinations thereof. Thus, even a perfectly zero order eroding device gives you overall non-zero order kinetics unless you get compensating non-zero order erosion. The important point is that you don't want zero order erosion but non-zero erosion of the type that you have seen.

Sustained Delivery of a Narcotic Antagonist from Lactic/Glycolic Acid Copolymer Implants†

DONALD L. WISE and ARTHUR D. SCHWOPE

Dynatech R/D Company, Cambridge, MA02139, U.S.A.

STEPHEN E. HARRIGAN and DUNCAN A. McCARTY‡

Parke-Davis Company, Ann Arbor, MI, U.S.A.

and

JOHN F. HOWES

SISA, Inc., Cambridge, Massachusetts, U.S.A.

Narcotic antagonism was demonstrated for designed periods of 15 and 30 days using two biodegradable polymeric implant systems which released naltrexone (base). Evaluation was carried out using monkeys trained to self-administer morphine and by morphine challenge of mice using the tail-flick test. The results of the challenge were supported by monitoring the tritium from the labelled drug found in the feces and urine of the animals, as well as by *in vitro* testing. Delivery systems were removed from the animals at selected intervals after implantation for evaluation. The delivery vehicles were synthesized from L(+)-lactic (L(+)) and glycolic (G) acids. Specifically, copolymers of weight fractions 90L(+)/10G and 75L(+)/25G, with molecular weights less than 50,000, were blended with ^{3}H-naltrexone (base) at weight loadings of 70 and 50%, respectively. The drug/polymer matrix was molded into 1.5 mm diameter, spherical beads. The beads prepared using 75L(+)/25G copolymer were coated with a thin (3 mil) coating of this pure polymer. Beads were implanted subcutaneously using a trochar. Release rate was found to be approximate first order until all drug was exhausted—this being 30 days for the 90L(+)/10G 70% drug system and 15 days for the coated 75L(+)/25G 50% drug system. Gel permeation chromatography was used to characterize the biodegradability pattern of implants removed from the animals. Histopathological examination of the immediate implant site found little or no adverse tissue reaction.

† This work was carried out under contract with the National Institute on Drug Abuse under the direction of Robert E. Willette, Ph.D.

‡ Deceased.

INTRODUCTION

Narcotic antagonists have been clinically evaluated for use as both the sole therapeutic agents and adjuncts in the post-detoxification stage of rehabilitation programs for opioid dependent persons.[1–3] Presently available antagonists are relatively short acting requiring daily or more frequent administration. Since chemical intervention to prevent relapse to drug seeking behavior may be required for months, a formulation that extends the duration of the efficacy of an antagonist in the patient is desirable. In related work, it has been demonstrated *in vivo* and clinically that implanted, steroid-containing poly(dimethylsiloxane) capsules are effective means of attaining the goals of extended, controlled drug delivery.[4,5] However, following exhaustion of the active agent, the surgical removal of the residual poly(dimethylsiloxane) shell may be required.

In order to obviate the removal problem while maintaining the advantages of long-term drug delivery via implants, the development of formulations based on biodegradable polymers is in progress. Lactic/glycolic acid polymer is one type of such polymers.[6–8] Recent work has shown that implants composed of poly(L(+)-lactic-co-glycolic acid) copolymers and the narcotic antagonist, naltrexone, do provide long-term, controlled drug delivery to test animals with no apparent objectionable reaction of the implantation site.[9]

The objective of this paper is to present the results of recent evaluations of two naltrexone-poly(L(+)-lactic-co-glycolic acid) formulations in monkeys and mice. In addition to release rate and efficacy data, analyses of the polymer molecular weight and tissue response to the implants as functions of implantation time are presented.

EXPERIMENTAL

Implant preparation

Glycolide synthesized from glycolic acid (Fisher Scientific) and L(+)-lactide (Henley and Co., Incorporated) were bulk polymerized into either 75/25 or 90/10 w/w poly(L(+)-lactic-co-glycolic acid), PLGA, copolymers using the non-metallic, acidic catalyst *p*-toluene sulfonic acid. Molecular weights of the polymers ranged from 30,000 to 60,000 as determined by light scattering, membrane osmometry, and gel permeation chromatography (GPC).

The narcotic antagonist used throughout this study was naltrexone base (Research Triangle Institute and Endo Labs). The naltrexone was tritium labelled (1 mCi/g) at the 15 and 16 positions.

The spherical, 1.5 mm diameter preparations (beads) were prepared by transfer molding vacuum oven-dried, solvent-blended mixtures of chemical and polymer. The weight of the implants ranged from 3.2–3.5 mg. Implants 70% w/w ^{3}H-naltrexone were prepared with 90/10 PLGA and implants 50% w/w ^{3}H-naltrexone base were prepared with 75/25 PLGA. In the case of the 75/25 PLGA formulation, a thin polymer (3 mil) coating of the same polymer as the matrix was added by dipping the spheres into a 10% w/w polymer solution. The naltrexone base was observed to be somewhat soluble in the copolymers.

In vitro

A known weight of beads was placed inside a Whatman cellulosic extraction thimble and suspended in a test tube containing 50 cm^3 of phosphate-buffered pH 7 solution. The test tube was maintained in a shaker bath (American Optical #2156) at 37°C. The solution was sampled daily and was changed twice weekly, or more often, if the naltrexone base concentration approached 20% of saturation. The release rate was determined from the measurement of the naltrexone concentration of the leaching solution by means of a Beckman LS 100 liquid scintillator using Riaflour® cocktail (New England Nuclear).

In vivo

Primates Primate testing of the implants was carried out at Parke-Davis and Co. In the evaluation of the efficacy and release rate characteristics of the two formulations, four rhesus monkeys (*Macaca mulatta*) were housed individually and restrained with a steel arm and harness system.[10] Each monkey was surgically prepared with a silicone rubber catheter in the internal jugular vein. The distal end of the catheter exited through the skin of the back and connected to a tube which passed through the hollow restraining arm to pumps mounted outside the cage.

A transilluminated button served as the manipulandum. One button press activated a reciprocating pump to deliver 0.5 cm^3 of solution through the catheter over 5 sec followed by a 5-s refill. Button presses occurring during the infusion cycle had no consequence. The availability of drug solutions was limited to a 15-min period every 4 h. Room lights were controlled automatically to provide 12-h diurnal cycles. Drug solutions were changed in a rotating sequence of 3 days on morphine (8 μg/kg/injection), 3 days on methamphetamine (4 μg/kg/injection) and 3 days on saline. Methamphetamine hydrochloride and morphine sulfate were prepared in sterile 0.9%

saline. All drug dosages are expressed as the base. At the time of this study, all monkeys had been exposed to this schedule for several months.

Implantation was carried out on the first day of saline availability. A specific dose (10 mg/kg) of drug-loaded material was inserted subcutaneously to a single site through a trochar introduced through an incision in the lower abdominal area of each monkey. Self-administration behavior was then monitored on the rotating schedule. The number of injections taken in each of the three access periods during the daytime were used for making comparisons between treatments, since drug self-administration is generally quite variable at night. Elapsed time in this study was counted as days between the injection of the beads and the middle day of each subsequent 3 day drug disposure. Statistical comparisons were made with an analysis of variance for crossover studies and the Newman–Keuls' multiple range test. Urine and fecal outputs were monitored for tritium using liquid scintillation and combustion analysis.

At the conclusion of measurable narcotic activity, residual implant material was removed and evaluated for residual tritium and polymer degradation by means of liquid scintillation and gel permeation chromatography, respectively. In addition, tissue samples were taken from the injection site for microscopic examination.

In addition, 40 beads of the 70% naltrexone base—90/10 PLGA formulations were implanted (20/site) in each of four unrestrained monkeys. On a weekly schedule the beads and immediate sites of one implantation area of two animals were extracted from the animals so that after 4 weeks all beads were extracted. The tissue around the implants was removed and subjected to histological examination. All the beads were dried, dissolved in THF, and the molecular weight of the polymer determined using GPC.

Mice The testing in mice was performed at SISA, Incorporated. Male, albino, Charles River mice (18–22 g) were used to evaluate duration of efficacy (morphine challenge) and the rate of naltrexone release (urine analysis). Three beads were implanted in each anesthetized (Penthrane®, Abbot) mouse through a small slit in the scapular region. The wound was closed with a wound clip.

Ten implanted mice were maintained in individual metabolism cages and the urine collected daily. The cages were rinsed and the washings used to dilute the urine to a specific volume from which a 1 cm^3 aliquot was taken and added to 10 cm^3 of scintillation cocktail (Aquasol®, New England Nuclear). The tritium concentration of the final solution was determined by means of a Beckman LS-230 liquid scintillator and a quench curve.

The mouse tail flick procedure[11,12] was used to measure the efficacy of the implant. Ten-animal groups were challenged weekly with an ED_{80} dose

(14.0 mg/kg ip) of morphine sulfate. No group was used more than once. The analgesic response was calculated as the percentage of the maximum possible response time as compared with the control mice.[13]

At the end of each experiment the animals were sacrificed, the beads recovered, and the implant site inspected for signs of encapsulation, infection or irritation.

Results

70% naltrexone base, 30% 90/10 PLGA implant The release profile of the 70% w/w naltrexone base, 30% w/w 90/10 PLGA system was determined by measurements of the tritium levels of (1) leaching bath solution, (2) the urine of implanted mice, and (3) the urine, feces and residual implant of implanted monkeys. The over-all summary of the tritium recovered from four monkeys after 52 days of implantation is presented in Table I.

TABLE I

Summary of tritium recovered after 52 days from monkeys implanted with the 70% w/w ^{3}H-naltrexone base—30% w/w 90/10 PLGA system.

Monkey number	Implanted			Recovered						
				Urine		Feces		Residue		Total
	# Beads (recovered)	mg	DPM/10^7	DPM/10^7	%	DPM/10^7	%	DPM/10^7	%	%
6758	14	48.0	7.4	5.1	69	1.2	16.6	0.2	2.5	88
6749	16	54.5	8.5	5.2	62	1.7	20.0	0.2	2.0	84
6753	17	57.0	8.8	4.9	55	1.65	19.0	0.1	1.0	75
443	19	65.0	10.1	5.6	55	1.7	17.0	0.1	1.0	73

An average of 80% implanted ^{3}H—60% in the urine, 18% in the feces, and 2% in the residual implants—was accounted for by the analyses described above. The balance was considered lost due to the fact that the animals were not maintained in metabolism cages. In an independent test, monkeys were placed in a restraining chair to measure total clearance of radioactive label. Monkeys injected subcutaneously with 5 mg/kg of naltrexone base while in the restraining chair were found to release (largely urinary) approximately 95% of the input label within 72 h.

The cumulative release of naltrexone as measured by the tritium content in the urines of four monkeys and the feces of one monkey as a function of time are presented in Figure 1. As a means for comparison, the results are presented as percentages of tritium recovered rather than as percentages of tritium implanted. Essentially all the tritium was excreted within 35 days of implantation. The total fecal and urinary recoveries are presented in Figure 2 along with the results of mouse and *in vitro* testing of the system. Except for a slightly faster initial release, the *in vitro* results are identical to the *in vivo* recoveries. It is to be noted that, typically, the proportion of naltrexone and naltrexone metabolites found in the urine and the feces of mice was 1:1; therefore, double the measured urine values for mice are presented in Figure 2.

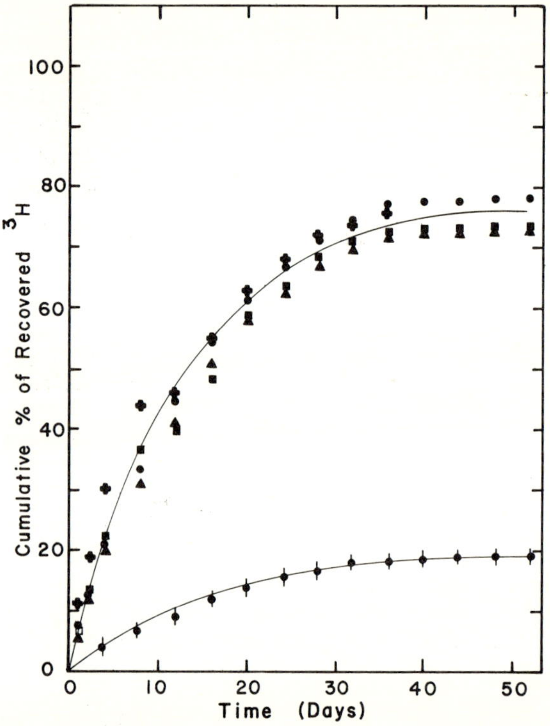

FIGURE 1 Recovery of tritium from monkeys implanted with 70% w/w ^3H-naltrexone base, 90/10 PLGA beads. (●) Monkey # 6758, urine; (♦) Monkey # 6758, feces; (■) Monkey # 6749, urine; (▲) Monkey # 6753, urine; (+) Monkey # 443, urine.

In addition to the naltrexone release profiles, the results of the morphine challenge testing of the system are presented in Figure 2. The degree and duration of narcotic antagonism provided by the implants was determined weekly by comparing the tail-flick reaction times of mice before and after an injection of morphine. The degree and duration of chemical intervention to the desire for morphine in primates was determined by comparing the number of morphine self-administrations before and after implantation. In

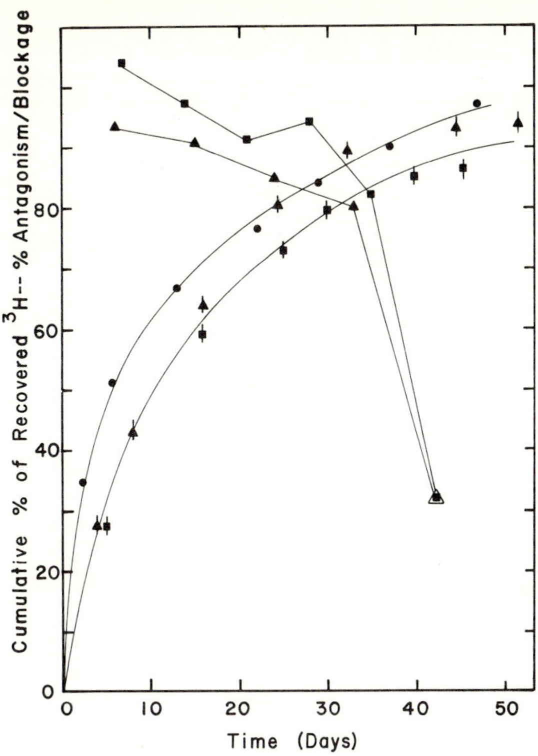

FIGURE 2 Performance of 70% w/w ^{3}H-naltrexone base, 90/10 PLGA beads in mice, monkeys and *in vitro*. (■) Antagonism to morphine challenge in mice; (■) *in vivo*, mice, urine values doubled; (▲) blockage to morphine self-administration in monkeys; (▲) *in vivo*, monkeys, urine and feces; (●) *in vitro*, pH 7 buffer at 37°C.

both *in vivo* tests, the implants provided antagonistic levels of naltrexone to the animals for 30 days.

Polymers of lactic and glycolic acids hydrolyze in aqueous environments. Gel permeation chromatography was used to analyze the polymer of the excised implants. An approximately 50% reduction of the molecular weight was realized after 52 days of implantation. (See Table II.)

Dip-coated, 50% naltrexone base, 50% 75/25 PLGA implant Similar to the analyses of the 70% naltrexone system, a 1.5 mm diameter, spherical, dip-coated implant of 50% w/w naltrexone base and 50% w/w 75/25 PLGA was evaluated *in vitro* and using mice and monkeys. The over-all summary of the release analysis of the monkey experiment is presented in Table III. An average of 82% of the implanted tritium was recovered after 40 days

TABLE II

Molecular weight of the 90/10 Poly (L(+) lactic-co-glycolic acid) used in the 70% w/w naltrexone base beads removed from monkeys.

Removal schedule from monkeys	Molecular weight, $\bar{M}_w$ (material from all recovered beads)
prior to implantation	37700
7 days (in 1 monkey)	27100
14 days (in 1 monkey)	22100
21 days (in 1 monkey)	19900
28 days (in 1 monkey)	18800
54 days (in 2 monkeys)	18200, 14900

(64% in the urine, 18% in the feces, and less than 1% in the residue). As above, the balance of the tritium was considered lost because of incomplete collection.

TABLE III

Summary of tritium recovered after 40 days from monkeys implanted with dip-coated 50% w/w ^3H-naltrexone base—50% w/w 75/25 PLGA beads.

Monkey number	Implanted			Recovered						
				Urine		Feces		Residue		Total
	# Beads recovered	mg	DPM/10^7	DPM/10^7	%	DPM/10^7	%	DPM/10^7	%	%
444	28	100	1.8	1.1	61	0.32	18	0.01	0.5	79.5
6731	22	78	1.4	0.96	68	0.28	20	0.01a	1	89
6766	23	83.8	1.5	1.1	72	0.25	17	—	1	89
6734	24	83	1.55	0.85	55	0.27	17	0.01	0.4	72

[a] ^3H of residue not determined.

The cumulative percentage recoveries of tritium in the urines of four animals and the feces of two animals are presented in Figure 3. As above, the results are presented as percentages of recovered tritium rather than percentages of implanted tritium. In Figure 4, the total of the averaged monkey urinary and fecal recoveries are shown along with the results from mice and *in vitro* tests. It is seen that the values of the tritium recovery rates from monkeys and leaching baths are essentially identical and that the pattern of those two rates is similar to that of the cumulative recovery frou mice, as calculated by doubling the urine tritium levels.

Also in Figure 4 are the results of the morphine challenge tests of monkeys and mice implanted with the 50% w/w naltrexone base system. It is seen

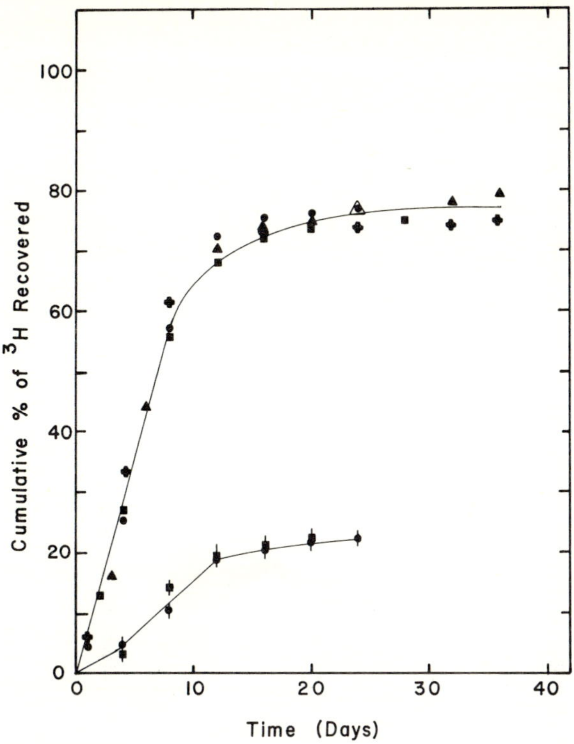

FIGURE 3 Recovery of tritium from monkeys implanted with dip-coated 50% w/w ^{3}H-naltrexone base, 75/25 PLGA beads. (●) Monkey # 444, urine; (●) Monkey # 444, feces; (■) Monkey # 6731, urine; (■) Monkey # 6731, feces; (▲) Monkey # 6766, urine; (✚) Monkey # 6734, urine.

that the implants provided antagonistic levels of naltrexone for approximately 15 days.

In Table IV, the results of the GPC analyses of polymer from implants excised from animals over a 42-day period are presented. Over that period a 60% reduction in the molecular weight of the 75/25 PLGA occurred.

Implant site tissue reaction evaluation (primates) Gross inspection of the injection sites revealed negligible tissue reaction. The beads were generally clustered together in a small sac, easily isolated for biopsy. For the first two weeks, the beads recovered were firm and did not adhere to the subcutaneous tissue. In the third and fourth weeks, beads appeared somewhat porous, were quite fragile, and were difficult to separate from surrounding tissue.

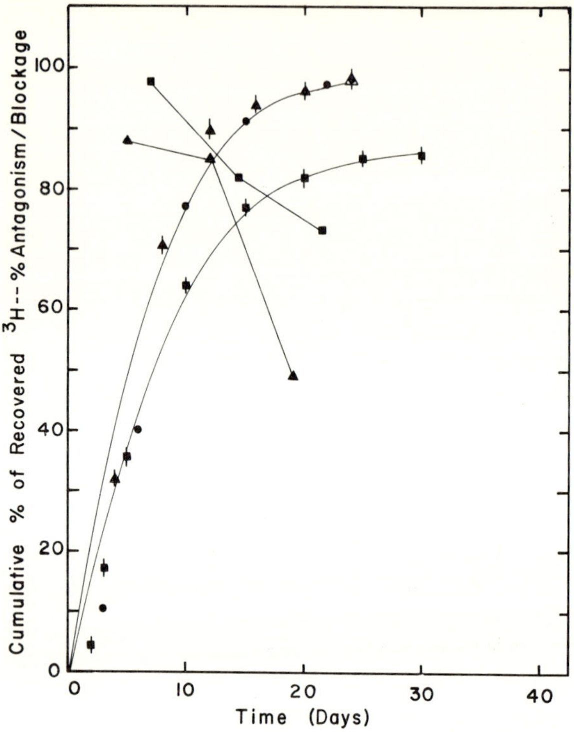

FIGURE 4 Performances of dip-coated, 50% w/w ^3H-naltrexone base, 90/10 PLGA beads in mice, monkeys and *in vitro*. (■) Antagonism to morphine challenge in mice; (■) *in vivo*, mice, urine values doubled; (▲) *in vivo*, monkeys, urine and feces; (▲) blockage to morphine self-administration in monkeys; (●) *in vitro*, pH 7 buffer at 37°C.

TABLE IV

Molecular weight of the 75/25 PLGA, coated, 50% w/w naltrexone base implants tested in mice and monkeys.

Removal schedule from animals	Molecular weight, $\bar{M}_w$ (material from all recovered beads)
Prior to implantation	29,100
7 days post implantation in mice[a]	14,800
14 days post implantation in mice[a]	10,700
28 days post implantation in mice[b]	12,300
42 days post implantation in monkeys[b]	11,200, 10,600

[a] $\bar{M}_w$ from 18 beads—3 beads taken from 6 mice.
[b] $\bar{M}_w$ from 20 beads from each of 2 monkeys.

Weekly histopathological examination of the depot sites of animals implanted with both the 90/10 PLGA and 75/25 PLGA systems had similar results: the initial basic reaction was granulomatous in nature and consisted of discrete foci of dense accumulations of mononuclear leukocytes, foreign body giant cells and epithelioid cells. These foci were at times associated with small refractile or eosinophilic foreign bodies in their midst. No necrosis was evident. The most florid reactions were observed in the 1- and 2-week implants. In the 3-week implants, there was evidence of diminution of the cellular response and increased collagenous connective tissue. In the 4-week implants, the impression of further diminution of cellular reaction, but with little, if any, collagenous reaction, was sustained. When foreign refractile material, presumably fragments of the implanted material, was still present in the longest implants, there were many instances in which only a small amount of collagenous connective tissue was present about the foreign material without an inflammatory cellular response.

Discussion

Two lactic/glycolic and polymer-based systems which provide the sustained, matrix-controlled release of naltrexone base to neutral aqueous environments have been developed. Evaluations of both systems as subcutaneous implants have been completed using mice, monkeys, and leaching baths. Release rate pattern and duration were determined by means of tritium labelled naltrexone. The degree and duration of antagonism resultant from the release of naltrexone was determined by direct morphine challenge of the animals.

The composition of copolymer constituent of the implant was found to be one of the major factors governing the release rate of naltrexone base. It was found that the duration of release from the 90/10 PLGA spheres was almost twice that from the 75/25 PLGA implants. Such was the case in spite of the fact that two distinctive features of the 75/25 PLGA system—lower loading and the pure polymer coating—typically extend rather than shorten the duration of drug delivery.

Agreement between independent evaluations of each system was found. Three general observations were apparent. One, since pure, injected naltrexone base was passed from an animal in 24–48 h, the release rates were matrix controlled. Two, the release rate of naltrexone base from PLGA was not species dependent. The release rates to all three media—monkeys, mice, and leaching baths—were essentially the same. Three, when maintained at/above a specific level, naltrexone was an effective antagonist to administered morphine (e.g., the mice tests) and to available morphine (e.g., the monkey tests). For each challenge test, when the implant was depleted of naltrexone—as determined by urinary and fecal analyses—the blockade quickly dissipated. Further, no significant adverse tissue reaction to the depot was observed grossly or microscopically during the implantation period. Such a result is especially meaningful since, at the same time drug was being released, the copolymers were decomposing to the extent that the molecular weights had decreased by more than 50% in 40–50 days of the tests.

References

1. W. R. Martin, C. W. Gorodetzky, and T. K. McClane, *Clin. Pharmacol. Ther.*, **7**, 455 (1966).
2. A. Schecter, As reported to the Committee on Problems of Drug Dependence (1975), Printing and Publishing Office, National Academy of Sciences, Washington, D.C.
3. A. Goldstein, *ibid.*
4. H. Croxatto, S. Diaz, R. Vera, M. Etchart, and P. Atria, *Am. J. Obst. and Gynec.*, **105**, 1135 (1965).
5. J. Folkman and D. M. Long, *J. Surgical Res.*, **4**, 139 (1964).
6. T. M. Jackanicz, H. A. Nash, D. L. Wise, and J. B. Gregory, *Contraception*, **13** (1973).
7. L. C. Anderson, D. L. Wise, and J. F. Howes, *Contraception*, **13**, 375 (1976).
8. N. Mason, C. Thies, and T. J. Cicero, *J. Pharm. Sci.*, **65**, 847 (1976).
9. A. D. Schwope, D. L. Wise, and J. F. Howes, *Life Sciences*, **17**, 1877 (1975).
10. G. Deneau, T. Tanagita, and M. H. Seevers, *Psychopharmacologia*, **16**, 30 (1969).
11. W. L. Dewey, J. W. Snyder, L. S. Harris, and J. F. Howes, *J. Pharm. Pharmac.*, **21**, 548 (1969).
12. W. L. Dewey and L. W. Harris, *J. Pharmacol. Exp. Ther.*, **179**, 652 (1971).
13. L. S. Harris and A. K. Pierson, *J. Pharmacol. Exp. Ther.*, **143**, 151 (1964).

DISCUSSION

D. K. Lehmann (*Rohm Pharma GMBH, Darmstadt, Germany*): Would you please describe the method and apparatus that you used to mold the spherical particles?

Dr. D. Wise: We used a specially designed mold that has something like an "X" shape with 88 small holes machined in it. Upon heating, the polymer/drug material comes in through the center and is forced out through connecting links. It's much like molding a child's toy.

Dr. K. Lehmann: Can this procedure be used on a production scale?

Dr. D. Wise: We have made several thousand beads in one week in the laboratory. This procedure will be sufficient for the next several years, even for the clinical and toxicological tests. So we've had no basis to automate this process.

Dr. J. Robinson (*Ciba–Geigy Corp., Summit, NJ*): In the *in vitro* release experiments, do you see the same kind of cratering as in the *in vivo* experiments, and do you observe the same kind of molecular weight reduction of the residual polymeric material?

Dr. D. Wise: Yes, it seems to be identical. This leads us to believe that the mechanism is hydrolysis.

Dr. J. Robinson: This might then account for the species independence.

Dr. R. Capozza (*Alza Corp., Palo Alto, CA*): From the urinary excretion data in mice, how much of the tritium is excreted in the urine? It appears to be quite high.

Dr. D. Wise: On repeated testing we found that the urinary:fecal ratio was always 50:50.

Dr. R. Capozza: It appears that the drug is dissolving without matrix erosion. Is water getting into the beads and leaching out the drug but not the polymer matrix? This process could explain both the resulting porous structure and the observed molecular weight data.

Dr. D. Wise: We have some questions as to what is the actual release mechanism. Clearly we have a system in which there is diffusion; we also have polymer hydrolysis which must be addressed.

We have done some calculations with some other systems which give us an insight into this, and we think that there are three regimes. First, there appears to be an initial release from the surface, due perhaps to a blooming or a drug crystal wash-off. Secondly, there is diffusion of drug which may not be completely dissolved in the polymer. The third regime is diffusion from the drug actually dissolved in the polymer; however, the release appears to be also related to polymer hydrolysis.

Dr. H. Hopfenberg (*North Carolina State University, Raleigh, NC*): Have you measured the water absorption kinetics in these spheres independently and attempted to correlate water absorption with drug release?

Dr. D. Wise: Yes, we have. However, we never have been very confident about the data because we are measuring the very small weight increase of the beads as such. We haven't really pursued this from a theoretical point of view.

Dr. H. Hopfenberg: In my opinion, your data indicate a zero-order release process from spheres. Do you agree?

Dr. D. Wise: It doesn't appear to be.

Dr. H. Hopfenberg: I'm not sure of that. Remember the equation I wrote on the board with the "magic 3." If it was a perfectly eroding system where the rate determining step was a chemical relaxation at a well-defined boundary, curves identical to yours would result. This is because the area or shell volume is ever decreasing as you go into the beads. Can you prepare slab-like specimens?

Dr. D. Wise: Yes, we can.

Dr. H. Hopfenberg: Do you observe zero-order release from these specimens? The important point, as I suggested in my paper, is that zero-order release from spherical systems is not expected from polymers which erode and provide zero-order release from slab geometries.

Dr. D. Wise: We have not prepared nor evaluated slab-like specimens from this particular formulation.

Dr. R. Nevin (*Eli Lilly and Co., Indianapolis, IN*): Measure of elimination through feces or urine does not necessarily correlate with the actual release of drug from the matrix material itself. Once material starts releasing from the matrix, if it is not removed from the site quickly, then the concentration factors which you give no longer apply. For example, the released material could go into fat tissue and be subsequently distributed from there. Therefore, the fat tissue might be the distribution center and not the beads themselves.

Dr. D. Wise: Perhaps we don't have as much data on that as we should. However, in other systems we have done total necropsy, because there has been concern about a type of "hide-out" concept. No drug residue was observed in these experiments.

Dr. H. Ringsdorf (*Univ. of Mainz, Mainz, Germany*): Dr. Nevin's comment is an important point. One should, for example, compare release rates between *in vivo* and *in vitro* experiments to establish whether uptake into the lipid fluid is important in controlling release rates.

Dr. W. Good (*Ciba–Geigy Corp., Ardsley, NY*): I wish to comment on your use of elimination data to represent the release of drug. I believe that this procedure may be very dangerous. All pharmo-kinetic models are based on first order mechanisms within the body. Therefore, it is unlikely to obtain zero-order elimination with such mechanism, even if the release is zero-order.

I would suggest that you recover the beads at various time intervals of *in vivo* testing and examine the beads themselves to determine the order of the release process.

Dr. D. Wise: This is true.

Dr. M. Goodman (*University of California/San Diego, LaJolla, CA*): I interpret the Dynatech results as a positive initial step in relating *in vitro* results to preliminary *in vivo* testing. This work is by no means the complete story, and of course additional work is necessary to establish the details that would explain all of the steps from release from the matrix to the actual effect and then to the total elimination. This follow-up work is a whole new area of study which really should not be expected out of the work presented this morning by Don Wise.

Dr. D. Lewis (*Southern Research Institute, Birmingham, AL*): You mentioned the addition of a second polymer to the matrix. What was this polymer and how were the samples prepared?

Dr. D. Wise: The one system you refer to was prepared using polymer synthesized from 75L(+) lactide/25 glycolide, and then blended with naltrexone to form a bead of 50% drug. After we had molded the bead, we just dipped it into polymer/solvent (the same 75L(+)/25G type polymer). The resulting film around the bead was just about a mil or so of pure polymer.

Dr. N. Kaplan (*University of California/San Diego, LaJolla, CA*): You mentioned that you used the L-lactide. Is it necessary to use the L-lactide? Could you have used the D-isomer instead?

Dr. D. Wise: We find that the L(+) lactide is more crystalline and does not hydrolyze as fast as the racemic mixture. The D-isomer is much more expensive, and we have never worked with it.

Release of Naltrexone Pamoate from Injectable Microcapsules

STEPHEN E. HARRIGAN and DUNCAN A. McCARTHY†

Park-Davis Company, Ann Arbor, Michigan, U.S.A.

RICHARD REUNING

The College of Pharmacy, Ohio State University, Columbus, OH43210, U.S.A.

and

CURT THIES

Biological Transport Laboratory, Washington University, St. Louis, MO63130, U.S.A.

The concept of long-acting injectable pharmaceutical formulations has intrigued researchers for some time. One way to prepare such formulations is to synthesize drug-filled microcapsules from polymers that are bioabsorbable and biocampatible. This paper will summarize the current status of an effort to prepare such capsules by using *dl*-poly(lactic acid). Physical properties and *in vitro* behavior of the capsules will be discussed. Results of *in vivo* experiments with trained monkeys will also be presented. The *in vivo* experiments involved $\leqslant$ 106 μ capsules containing initially 50 wt. % naltrexone pamoate, a narcotic antagonist. These capsules provide therapeutic action *in vivo* for 42 to 51 days when injected through a 20 gauge needle as a 2% aluminum monostearate/peanut oil suspension. Additional data obtained with capsules injected into monkeys as an aqueous suspension will be presented. Preliminary results of monkey blood analyses will also be discussed. Throughout the presentation, problems associated with the development of a clinically useful injectable depot drug preparation will be stressed.

1 INTRODUCTION

The concept of sustained-release microencapsulated pharmaceuticals is well established.[1] However, relatively little effort has been made to fabricate injectable microcapsules, perhaps because the number of bioabsorbable and biocompatible coating materials is limited. One promising coating material, *dl*-poly(lactic acid), is being used to fabricate injectable microcapsules. The first capsules prepared contained cyclazocine, a narcotic antagonist. They

† Deceased.

were relatively large and released their payload *in vitro* and *in vivo* much more rapidly than desired.[2] Recent improvements in encapsulation technology have yielded smaller and longer-lasting capsules.[3] The primary purpose of this paper is to report the *in vitro* and *in vivo* performance of such a capsule preparation containing the pamoate salt of naltrexone, a narcotic antagonist.

2 EXPERIMENTAL

2.1 Microcapsules

Microcapsules containing naltrexone pamoate were prepared from *dl*-poly(lactic acid) supplied by Dr. Donald Wise of the Dynatech R/D Company (Cambridge, Mass.). The capsules were $< 106\ \mu$ in diameter. They contained 50 wt. % naltrexone pamoate which corresponds to a naltrexone free base content of 32 wt. %.

2.2 In vitro testing

In vitro release properties of the capsules were evaluated at 37°C in pH 7.4 phosphate buffer on a rotating-bottle apparatus (40 rpm) (Ernest Menold, Lester, Pa.). The apparent naltrexone pamoate concentration in the extracting solution was established spectrophotometrically by using the 288 nm band characteristic of naltrexone pamoate. Control experiments showed that the 288 nm absorbance of unencapsulated naltrexone pamoate kept $\leqslant 700$ h on the rotating bottle extractor changed < 10–15%. Thus, the spectrum of naltrexone pamoate does not change substantially under the conditions used for *in vitro* capsule evaluation. Results are plotted as apparent percent naltrexone pamoate released by the capsules *vs.* time. Each data point reported was obtained with a separate test sample. All test samples consisted of 15 mg of capsules suspended in 75 cm^3 of buffer. Complete extraction of naltrexone pamoate from the capsules would yield a drug concentration of 10.0 mg/100 cm^3 buffer, less than 15% of the 77 mg/100 cm^3 saturation solubility of naltrexone pamoate in pH 7.4 phosphate buffer at 37°C.

2.3 In vivo testing

Pharmacological *in vivo* evaluations of the microcapsules were carried out in male Cox Swiss Webster mice (Lab Supply Inc., Elkhart, Indiana) and Rhesus monkeys (*Macaca mulatta*).

Mice used in this study were randomly chosen from the same delivery batch and placed in one of three groups (5 animals per group). Group A received encapsulated drug and agonist (morphine sulfate); Group B received vehicle and agonist; Group C received vehicle only. A naltrexone pamoate capsule dose corresponding to 40 mg naltrexone free base/kg mouse was administered subcutaneously through a 20 gauge, 1-in. needle at various times prior to testing. Peanut oil containing 2 wt. % aluminum monostearate was the suspension medium. All injections in mice were caudal to the left front leg in order to prevent leakage from the injection site.

Mice of all three test groups were housed together and tested blindly against each other. Mice in Groups A and B were pretreated 45 min before testing with 7.1 mg/kg morphine sulfate given intraperitoneally.

The mouse tail-flick test procedure of Dewey, *et al.*[4], modified to permit application of a constant temperature to the mouse tail,[3] was used to assess *in vivo* effectiveness of the capsules containing the narcotic antagonist. The method measures the degree of antagonism of a standard morphine response and results are plotted as percent antagonism *vs.* days after capsule injection. Percent antagonism is calculated as follows:

$$\%\text{ antagonism} = (100)\left\{\frac{\begin{matrix}\text{Mean response time,} \\ \text{Group B}\end{matrix} - \begin{matrix}\text{Mean response time,} \\ \text{Group A}\end{matrix}}{\begin{matrix}\text{Mean response time,} \\ \text{Group B}\end{matrix} - \begin{matrix}\text{Mean response time,} \\ \text{Group C}\end{matrix}}\right\}. \quad (1)$$

The capsules were also evaluated in four Rhesus monkeys housed individually and restrained with a steel arm and harness system.[5] Each monkey was surgically prepared with a silicone rubber catheter in the internal jugular vein. The distal end of the catheter exited through the skin of the back and connected to a tube which passed through the hollow restraining arm to pumps mounted outside the cage.

The monkeys were trained to self-administer drug solutions. A transilluminated button served as the manipulandum. One button press activated a reciprocating pump to deliver 0.5 cm^3 of solution through the catheter over 5 s followed by a 5-s refill. Button presses during the infusion cycle had no consequence. The availability of drug solutions was limited to a 15-min period every 4 h. Room lights were controlled automatically to provide 12-h diurnal cycles. Drug solutions were changed in a rotating sequence of 3 days on morphine (4 μg/kg/injection), 3 days on methamphetamine (4 μg/kg/injection) and 3 days on saline. These drug dosages are expressed as the base. At the time of this study, all monkeys had been exposed to the above schedule for several months.

On the morning of the first day of a saline presentation, the monkeys were given the microcapsules. A naltrexone pamoate capsule dose corresponding to 10.5 mg naltrexone free base/kg was injected intramuscularly into each monkey through a 20 gauge needle with 2 wt. % aluminum monostearate in peanut oil as the suspension vehicle. Approximately 1 cm^3 of the capsule suspension was given into the rectus femoris muscle of the thigh. Self-administration behavior was then monitored on the rotating schedule until the monkeys showed rates of morphine intake equivalent to their control values.

Because drug self-administration is generally quite variable at night, only the number of injections taken in each of the three access periods during the daytime were used for making comparisons between treatments. Results are plotted as number of self-administered drug injections per access period *vs.* days after injection. Elapsed time was counted as days between the injection of the microcapsules and the middle day of each subsequent three-day drug exposure. Statistical comparisons were made with an analysis of variance for crossover studies and the Newman-Keuls' multiple range test.[6]

Periodic blood samples were taken from each monkey throughout this study and stored frozen until immediately before naltrexone analyses were performed. The gas chromatographic–electron capture assay procedure for determining naltrexone in plasma is described elsewhere.[7,8]

3 RESULTS AND DISCUSSIONS

Figure 1 is a time plot of the 288 nm absorbance of *in vitro* capsule extracts.

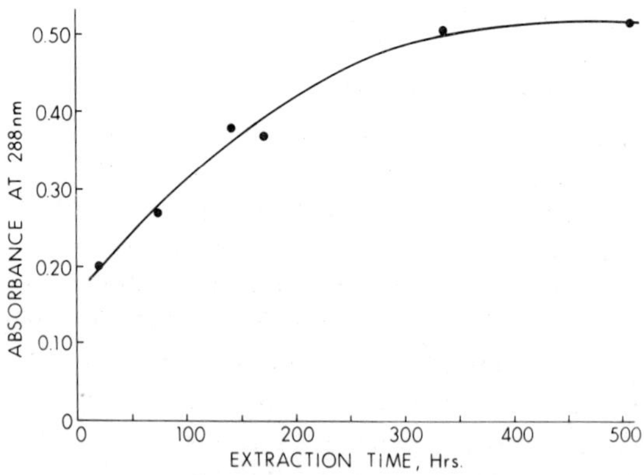

FIGURE 1 Time plot of the 288 nm absorbance of extracts from < 106 μm *dl*-poly(lactic acid) microcapsules that have a theoretical drug payload of 50 wt. %. Extractions run at 37°C and pH 7.4.

If the capsules released their entire naltrexone pamoate payload, the absorbance would be approximately double the maximum value recorded in Figure 1. Thus, these data suggest that the capsules released but a fraction of their payload over a 3 week extraction period. The shape of the release curve suggests that infinite extraction time would be needed to achieve complete naltrexone pamoate extraction. However, the capsule wall should ultimately undergo hydrolytic breakdown thereby allowing total drug release at an accelerated rate.

The data in Figure 1 are not reported in terms of amount of naltrexone pamoate released per unit capsule weight because it has not been resolved how accurate a representation of naltrexone pamoate release rates the spectroscopic analyses are. The naltrexone and pamoic acid fragments of naltrexone pamoate could diffuse from the capsules in a nonstoichiometric ratio, thereby greatly complicating quantitative evaluations. It has not been firmly established yet to what extent this occurs. Nevertheless, the data shown indicate that the capsules release their drug payload slowly and provided justification for carrying out *in vivo* evaluations.

Table I summarizes the mouse tail-flick data for the capsules used in this study. The percent analgesic antagonism achieved varies considerably, but measurable antagonism was maintained up to 5 weeks after injection.

TABLE I

Percent analgesic antagonism-time profile for naltrexone pamoate capsules injected subcutaneously into mice

Time after injection, weeks	Analgesic antagonism, %[a]
2	68(16)
3	28(6)
4	51(8)
5	37(9)
6	11(10)

[a] Mean of five mice; standard deviations in parenthesis.

Figure 2 shows that morphine self-administration by monkeys containing the naltrexone pamoate capsules is suppressed for approximately 51 days after injection. Onset of peak antagonist activity was delayed with the mean morphine intake on day 15 being lower than that on day 6. Morphine intake of days 15, 24, and 33 was about 20% of control. On day 42, morhpine self-injections declined to a mean of 3.5 ± 0.9 (SEM) injections per access period. Increasing morphine intake occurred after day 42. The differences between the control and the mean morphine self-injection rate of days 6 through 51 were significant at $P<0.01$. Recovery of the morphine self-administration frequency to control levels was relatively fast for monkeys containing naltrexone pamoate capsules.

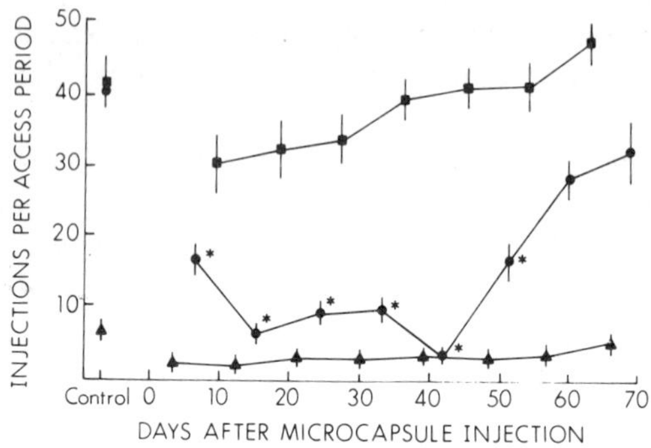

FIGURE 2 Self-administration of morphine (8 μg/kg/injection—solid circles), methamphetamine (4 μg/kg/injection—solid squares), and saline (solid triangles) following injection into monkeys of 33 mg capsules/kg. Each point represents the mean of 4 monkeys and 9 daylight access periods. Bars associated with each point represent the range of experimental values. Asterisks denote means significantly different from control ($P<0.01$).

Methamphetamine and saline self-administration were not significantly affected by the microcapsules. These results suggest that the pharmacological action of the microcapsules was specifically that of an opiate antagonist.

Figure 3 contains results of naltrexone analyses carried out on plasma samples from the same monkeys used to obtain the data in Figure 2. Naltrexone plasma levels built up slowly and peaked about 15 days after injection. The capsules did not release an initial large burst of drug, but metered it out slowly over a prolonged period. Naltrexone was not released at a constant rate since a definite peak in naltrexone plasma levels occurred. Variations in naltrexone plasma levels measured through the 51 day test period are reasonably small.

Before concluding, it is relevant to mention that no naltrexone was present in the plasma of monkeys three days after they received a 10 mg/kg dose of unencapsulated naltrexone pamoate administered in peanut oil containing 2 wt. % aluminum monostearate. This means that the prolonged action of the naltrexone pamoate is not simply due to the suspension vehicle. The injectable microcapsules meter out drug over a multi-week period *in vivo*. The rather narrow range of plasma concentrations indicates that release rates vary only over a 5–6 fold range throughout a 9 week period. The exact extent of duration of pharmacologic activity depends on the sensitivity of the response measured. The apparently less sensitive tail-flick technique

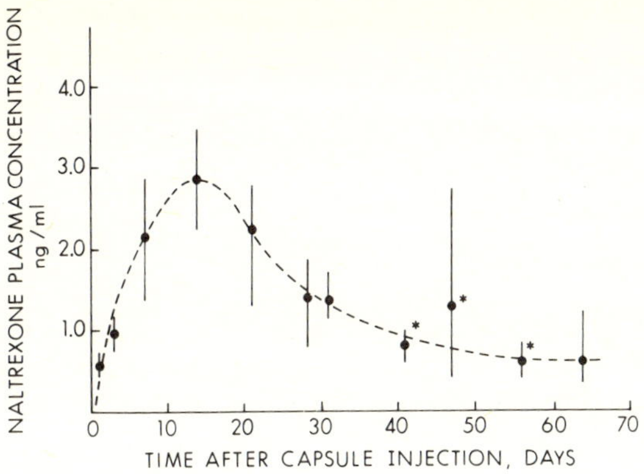

FIGURE 3 Naltrexone plasma levels in monkeys receiving 33 mg naltrexone pamoate capsules/kg. Most points represent the mean of 4 monkeys; asterisks denote mean values for 3 animals. Bars associated with each point represent the range of experimental values.

indicates about 5 weeks duration and the more sensitive monkey self-administration technique indicates about 7 weeks duration.

Acknowledgement

This work was supported by the National Institute on Drug Abuse under contracts HSM-42-72-167, HSM-42-73-182 and HSM-42-73-265. The authors acknowledge Thomas R. MacGregor, who carried out the tailflick experiments.

References

1. L. A. Luzzi, *J. Pharm. Sci.*, **59**, 1376 (1970).
2. N. Mason, C. Thies, and T. J. Cicero, *J. Pharm. Sci.*, **65**, 847 (1976).
3. R. H. Reuning, T. M. MacGregor, and C. Thies, *J. Pharm. Sci.*, manuscript in preparation.
4. W. L. Dewey, J. W. Snyder, L. S. Harris, and J. F. Howes, *J. Pharm. Pharmacol.*, **21**, 548 (1969).
5. G. Deneau, T. Yanagita, and M. H. Seevers, *Psychopharmacologia*, **16**, 30 (1969).
6. K. A. Brownless, *Statistical Theory and Methodology in Science and Engineering*, 2nd edition, John Wiley and Sons, Chapter 15 (1965).
7. V. K. Batra, R. A. Sams, R. H. Reuning, and L. Malspeis, *Acad. Pharm. Sci.*, **4**, 122 (1974).
8. L. Malspeis, M. S. Bathala, T. M. Ludden, H. B. Bhat, S. G. Frank, T. D. Sokoloski, B. E. Morrison, and R. H. Reuning, *Res. Comm. Chem. Pathol. Pharmacol.*, **12**, 43 (1975).

DISCUSSION

Dr. D. Lewis (*Southern Research Institute, Birmingham AL*): Did these microcapsules disperse well in water?

Dr. C. Thies: A wetting agent has been used to give a capsule suspension that was sufficiently stable to be injected into monkeys. The suspensions (and injections) were made at Parke-Davis, so I cannot comment further on capsule suspension ability.

Dr. D. Lewis: Have you examined poly(lactic acid) of other molecular weights?

Dr. C. Thies: The molecular weight has been varied. I prefer a specific molecular weight range, but cannot disclose what it is.

Dr. E. Goldberg (*University of Florida, Gainesville, FL*): What is the baseline control on the naltrexone in peanut oil formation?

Dr. C. Thies: Unencapsulated naltrexone is eliminated in 3–5 days.

Dr. I. Parikh (*Burroughs Wellcome Co., Research Triangle Park, NC*): One point, I suspect, that you have not considered is the baseline frequency of intake for morphine at different times. As morphine is an addictive drug, this intake frequency should increase with time. This consideration should probably lead to an even better overall intake frequency diagram than the one you presented.

Dr. C. Thies: The amount of morphine given to the monkeys is too small to have an addictive effect.

Dr. H. Hopfenberg (*North Carolina State University, Raleigh, NC*): Is there any indication that degradation of the membranes occurred?

Dr. C. Thies: The sites of the capsule injections in monkeys were not biopsied, so how much degradation of the capsule membranes occurred was not measured.

Dr. H. Hopfenberg: In one of your slides concerned with permeability and diffusivity, you mentioned flat membrane characterization. How were these characterizations performed?

Dr. C. Thies: The membranes were solvent-cast. The permeability of drug through the membranes then was established in a dialysis cell one side of which was a saturated drug solution. The other side was free of drug. Drug permeability values were calculated from the rate of drug transport through the membrane into the drug-free side of the dialysis cell.

Dr. R. Capozza (*Alza Copr., Palo Alto, CA*): A poly(lactic acid) film will generally last about one year in the body. How do you explain the shorter life-times of your films?

Dr. C. Thies: I feel that if the poly(lactic acid) is in the form of a $< 106\ \mu m$ diameter bead and the bead is 50% drug, the poly(lactic acid) will not take a year to bioabsorb. Whether or not this is the case is being examined now.

Dr. V. Stevens (*Dow Chemical Co., Midland, MI*): Would you comment on the molecular weight and molecular weight distribution of your poly(lactic acid)? Do these molecular weight factors affect the release rate?

Dr. C. Thies: As I mentioned earlier we favor a range of molecular weights. That is all I am willing to say now.

Dr. T. Roseman (*The Upjohn Co., Kalamazoo, MI*): The UV method of assay does not appear to be very specific. Have you verified independently that the material is indeed stable and not subject to degradation?

Dr. C. Thies: We've taken the pamoate and the free base and dissolved them in a buffer. These were run in our extractor just like we run our capsules. After 700 h of extraction we get less than 10% change in the 288 nm absorption maximum. This indicates major degradation effects are not detected by the spectroscopic assay. We tried to run extractions and isolate the free base from the pamoate in order to obtain a better assay. However, it appears that gas chromatography must be used to obtain reliable naltrexone assays and such assays are being carried out.

Dr. R. Capozza: You mentioned early in your talk that there is a possibility that pamoate was preferentially diffusing through the membrane. If that is the case, then you must bear in mind that the UV assay you are using will give you misleading data. In addition, naltrexone free base is oxidatively unstable and its oxidation products have a higher extinction coefficient. This also can result in misleading drug release data.

Dr. C. Thies: We have used UV assays as a means of monitoring how well our capsules retain drug under *in vitro* conditions. The assays have served well as a means of helping us develop long-lasting capsules. More rigorous assays of drug release rate are being obtained by gas chromatographic analyses. In sealed containers, we see no evidence from the spectroscopic assays that naltrexone free base (or naltrexone pamoate) is degrading to a significant degree.

Dr. A. Trouet (*International Institute of Cellular and Molecular Pathology, Brussels, Belgium*): What is the smallest capsule size that you can produce? Have you attempted intravenous injection of these small capsules?

Dr. C. Thies: We have obtained a significant number which are less than 63 μm in diameter. Most of these smaller capsules fall in the 30 μm range. It's difficult to get capsules below 10 μm.

We have not tried intravenous injection of our capsules. We're working under the constraint that the capsules must last 30 days. The smaller the capsules become the harder it is to achieve 30 day drug release.

Drug Release from and Properties of Poly(styrene) Microcapsules

YASUO NOZAWA and FUKUJI HIGASHIDE

Shizuoka College of Pharmacy, 2-2-1 Oshika, Shizuoka, Japan

Drug release from and properties of poly(styrene) (PS) microcapsules (capsules) have been studied by microscopic observation. To investigate in detail properties and permeabilities of capsule membranes, PS capsules containing α-amylase or sodium salicylate (Sal) were prepared by the method of evaporating methylene chloride from an aqueous system. The shape of the capsules was spherical and their size was in the range of 100–400 micrometers in diameter for both the dried and the water-containing capsules. The presence of some fine bubbles and some pockmark-like holes was observed on the surface of the capsules by the microscopic studies. Existence of holes in the capsule membranes was obviously confirmed by studies of the enzymatic activity of the encapsulated α-amylase.

For drug release studies, PS capsules containing Sal and enteric coating materials such as MPM-06 or dialdehyde starch were prepared. The release of Sal was affected by the pH of the medium and was higher from the dried capsules than from the water-containing capsules.

From these facts, the capsules prepared in our studies may be used in further applications such as for the sustained release of drugs, fodder and agricultural medicine and fertilizers.

Various methods for microencapsulation of miscellaneous materials have been extensively studied by many investigators. One of these methods, evaporating an organic solvent from an aqueous system, has frequently been employed for microencapsulation of labile agents such as drugs and enzymes. In a series of recent studies on microencpasulated enzymes, Chang *et al.*[1] have pointed out the semipermeability of microcapsule (capsule) membranes and attempted to apply an encapsulated enzyme to clinical pharmacology. Kitajima *et al.*[2] also investigated the properties of polystyrene (PS) and ethyl cellulose capsules containing urease or catalase. Kondo *et al.*[3] studied the semipermeability of capsule membranes in detail on poly(urethane) and poly(phthalamide) capsules prepared by the method of interfacial polycondensation reactions. In order to further study the permeabilities of capsule membranes, the drug release from and properties of PS capsules are discussed herein.

EXPERIMENTAL

Poly(styrene) capsules containing α-amylase or sodium salicylate (Sal) were prepared by the following procedures:[4]

1) A mixture of 10 parts of 10% methylene chloride solution of PS and one part of an 8% aqueous gelatin solution containing 2.5% α-amylase or 30% Sal was thoroughly emulsified by a homogenizer in the range of 500–9000 rpm. A water in oil type emulsion (W/O) was obtained.

2) The emulsion (W/O) was carefully added to a large amount of a 1% aqueous gelatin solution with stirring at 37°C. Then, the methylene chloride was slowly evaporated from the water system yielding the complex emulsion [(W/O)/W]. After 5 h the dispersed water-containing PS capsules are produced in the water system.

3) The drying of the capsules was performed in a fluidized bed at 60°C.

The apparent enzymatic activity of the encapsulated α-amylase was spectrophotometrically determined at 700 nm[5] by reacting soluble starch with a $KI–I_2$ solution. The quantities of released Sal were measured by the use of ferric chloride solution at 520 nm.[6] Scanning electron micrographs were prepared using a JSM-50A type apparatus (JEOL Co. Ltd.). A thin film of gold was evaporated onto the specimens. Enteric coating materials used here were MPM-06 (Tanabe Seiyaku Co. Ltd., methacrylate–methyl methacrylic acid–methyl methacrylate copolymer) and dialdehyde starch[7] (DAS, defatted corn-starch oxidized with periodic acid, degree of oxidation: 75%[8]).

RESULTS AND DISCUSSION

An optical micrograph of the capsules is shown in Figure 1. The shape of the capsules was spherical, and besides, the presence of some fine bubbles was observed on the surface of the capsules. The size distribution of the capsules was in the range of 50–500 micrometers in diameter for both the dried and the water-containing capsules. A scanning electron micrograph is given in Figure 2. Some pockmark-like holes (1–10 μm in diameter) were clearly observed on the surface of the capsules. This fact is a new and interesting observation found in our studies. The formation of the holes will be discussed later in this paper.

To verify the existence of the holes in the capsule membranes, release of

FIGURE 1 An optical micrograph of poly(styrene) capsules.

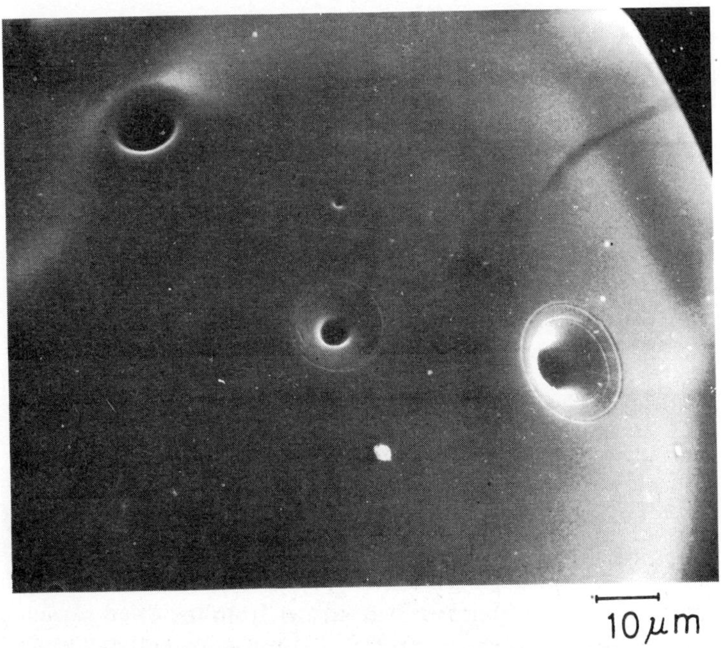

FIGURE 2 A scanning electron micrograph of poly(styrene) capsules.

α-amylase enclosed in the capsules was investigated. The time behavior of the enzymatic reaction of soluble starch with encapsulated α-amylase is shown in Figure 3. The open circular symbols in the figure represent the point at which the capsules were removed from the reaction vessel (after 1 h). The transmittance curve continuously increased, and the absorbance linearly decreased even after the removal of the capsules. The relationship between the relative enzyme activity of the encapsulated α-amylase and the number of repeated runs is shown in Figure 4 for capsules of different sizes. The enzymatic activity was estimated from Eq. (1), assuming that the enzymatic reaction obeys a first order rate law.

$$\text{Log } D = \text{Log } D_0 - kt/2.303. \tag{1}$$

In this equation D is the observed absorbance at 700 nm and k is the rate constant of the enzyme reaction (apparent enzymatic activity). The apparent activity gradually decreased as the number of repeated determinations increased to 5. The most marked decrease was in the case of the largest capsules. These facts indicate that the enclosed α-amylase was slowly released from the capsules which suggests the presence of "holes" in the capsules' membranes.

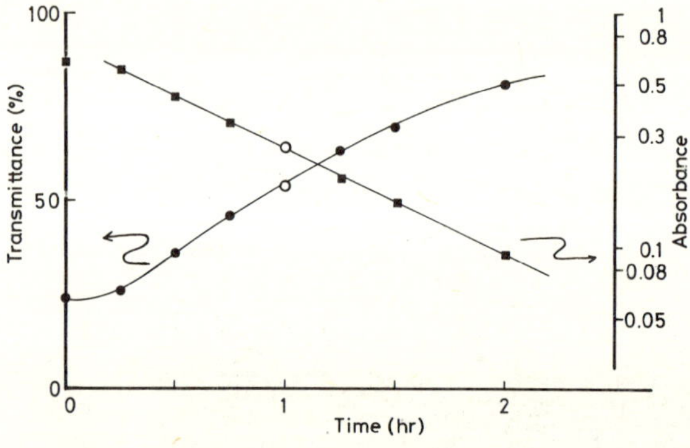

FIGURE 3 Transmittance and absorbance of starch solution containing encapsulated α-amylase as a function of time.

Further studies on drug release involved the release of Sal from PS capsules containing enteric coating materials. The percent of Sal released as a function of time from both dried and water-containing capsules are shown in Figure 5. The percent released was greater from the dried capsules than that from the water-containing capsules and was apparently affected by the pH of the medium. An enhanced release of Sal from the dried capsules

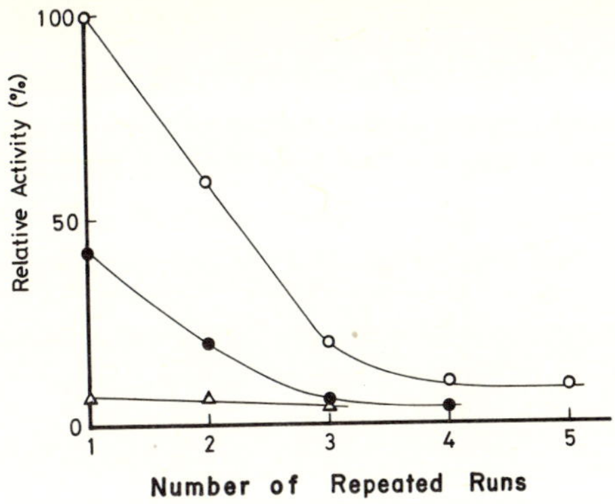

FIGURE 4 Effect of the capsule size on relation between relative activity and number of repeated runs. (△) 50–100 μm; (●) 100–200 μm; (○) 200–500 μm.

compared to that of the water-containing capsules might be explained by migration of Sal to the surface of the capsules during the drying process. The phenomena of Sal migration has been studied by the use of fluidized bed drying methods on a ball made of gypsum.[9]

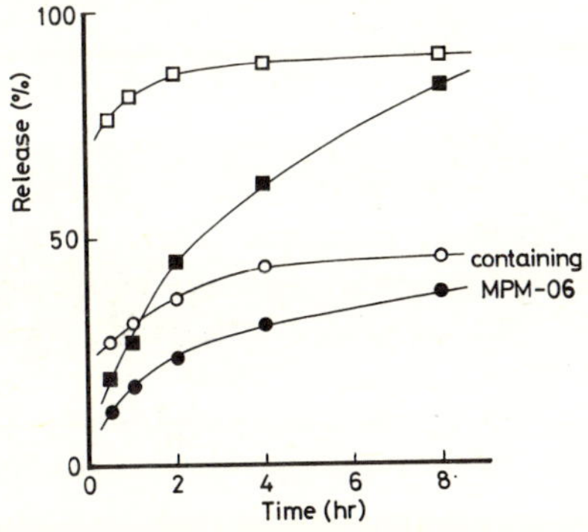

FIGURE 5 Release of Sal from dried (D) and water-containing (W) capsules. (□) pH 7.6 (D), (■) pH 7.3 (W), (○) pH 2.0 (D), (●) pH 2.1 (W).

The percent of Sal released *vs.* time for dried capsules containing DAS or MPM-06 is given in Figure 6. The pH of the medium was changed from 2 to 10 after 2 h. The curves discontinuously increased at 2 h for both capsules. This indicates that Sal molecules also permeate to the media through the holes, which the enteric coating materials filled, consequently, the release was controlled by the pH of the medium. The properties of DAS used as an enteric coating material have been discussed in the case of drug permeation in poly(vinyl alcohol) membranes containing DAS.[10]

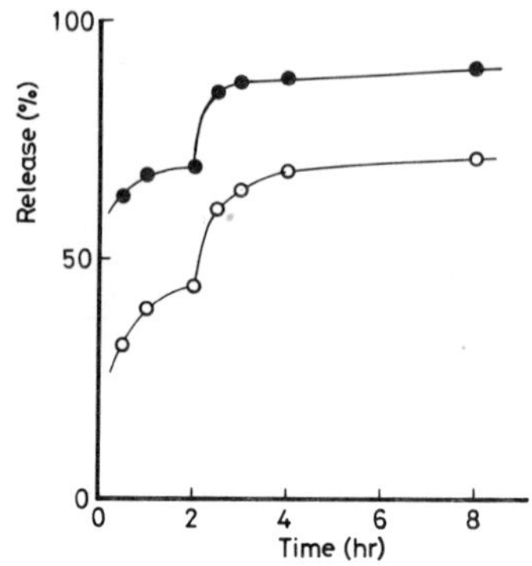

FIGURE 6 Release of Sal from dried capsules containing DAS (●) or MPM-06 (○). The pH of the medium was changed from 2–10 after 2 h.

Thus, the above presented experimental facts confirm the existence of "holes" in the PS capsules.

To further explore the formation of the holes in capsule membranes, two kinds of capsules were prepared from water in oil type emulsions agitated at either slow (500 rpm) or rapid (9000 rpm) stirring speeds. Optical micrographs of these 100–300 μm diameter capsules are shown in Figure 7. A large number of the fine bubbles, as pointed out before, were clearly observed on the surface of the capsules produced from the emulsion agitated with rapid stirring, while in the slow stirring case the bubbles on the surface are practically nonexistent. The scanning electron micrographs of these capsules are given in Figure 8. The existence of distinct holes on the surface of capsules prepared by rapid stirring is clearly evident (Figure 8b) while no holes could be found in the case of slow stirring (Figure 8a). From these microscopic observations, it may be considered that the significant factor of hole formation was mostly related to the particle size in the (W/O) type emulsion, in particular, the fine particles existing in the emulsion predominantly contributed to produce the holes in the membranes of the capsules.

On the basis of these drug release studies and microscopic observations, the existence and formation of holes in PS capsules were clearly elucidated. The existence of the holes is considered to have a significant role in the release of encapsulated core materials. Therefore, the capsules prepared in this study may be extensively used in further applications such as for sustained drug release, agricultural medicine, and fertilizer.

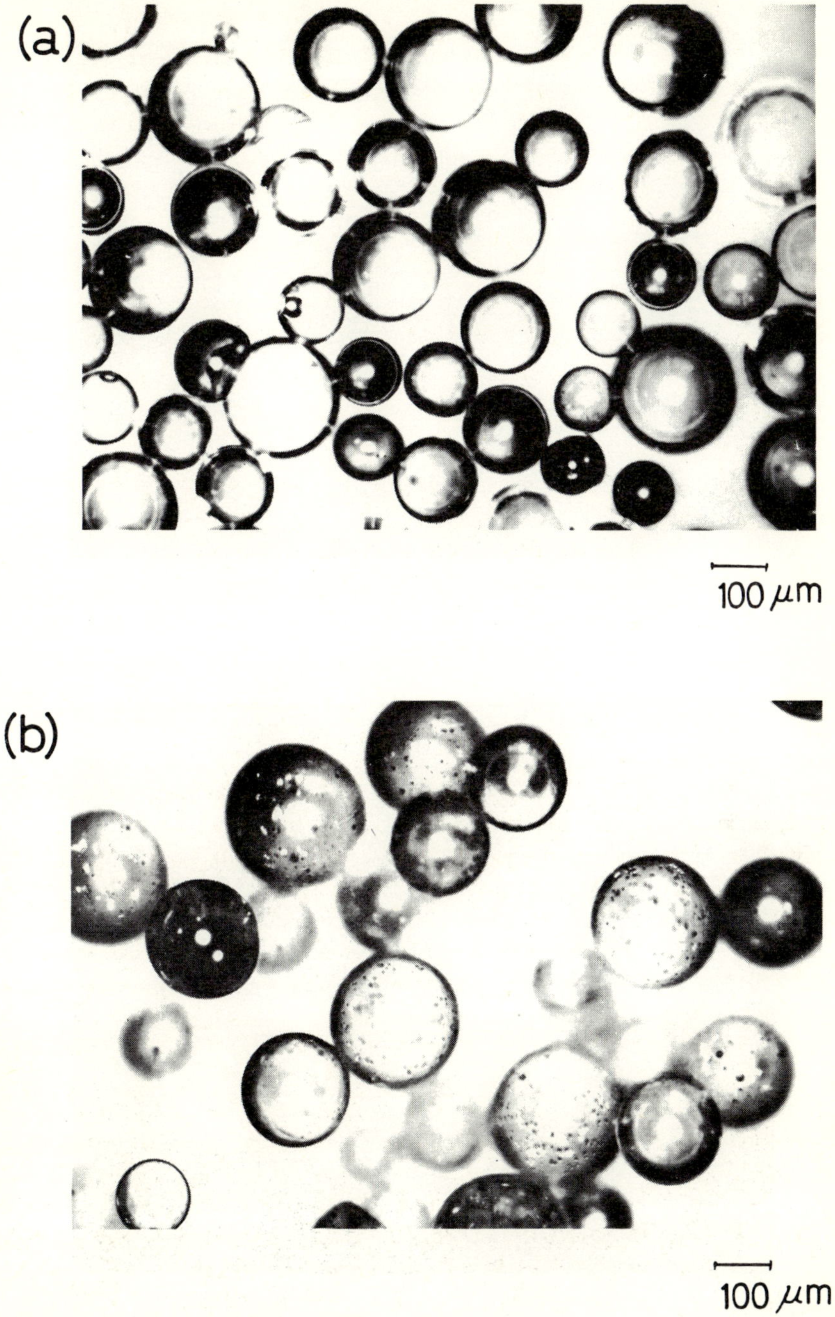

FIGURE 7 Optical micrographs of capsules of 100–300 μm diameter produced by (a) slow stirring (500 rpm) or (b) rapid stirring (900 rpm).

(a)

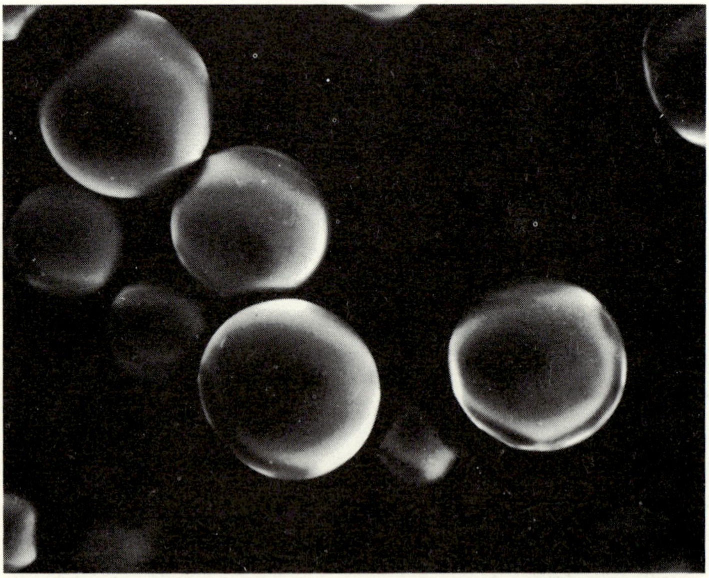

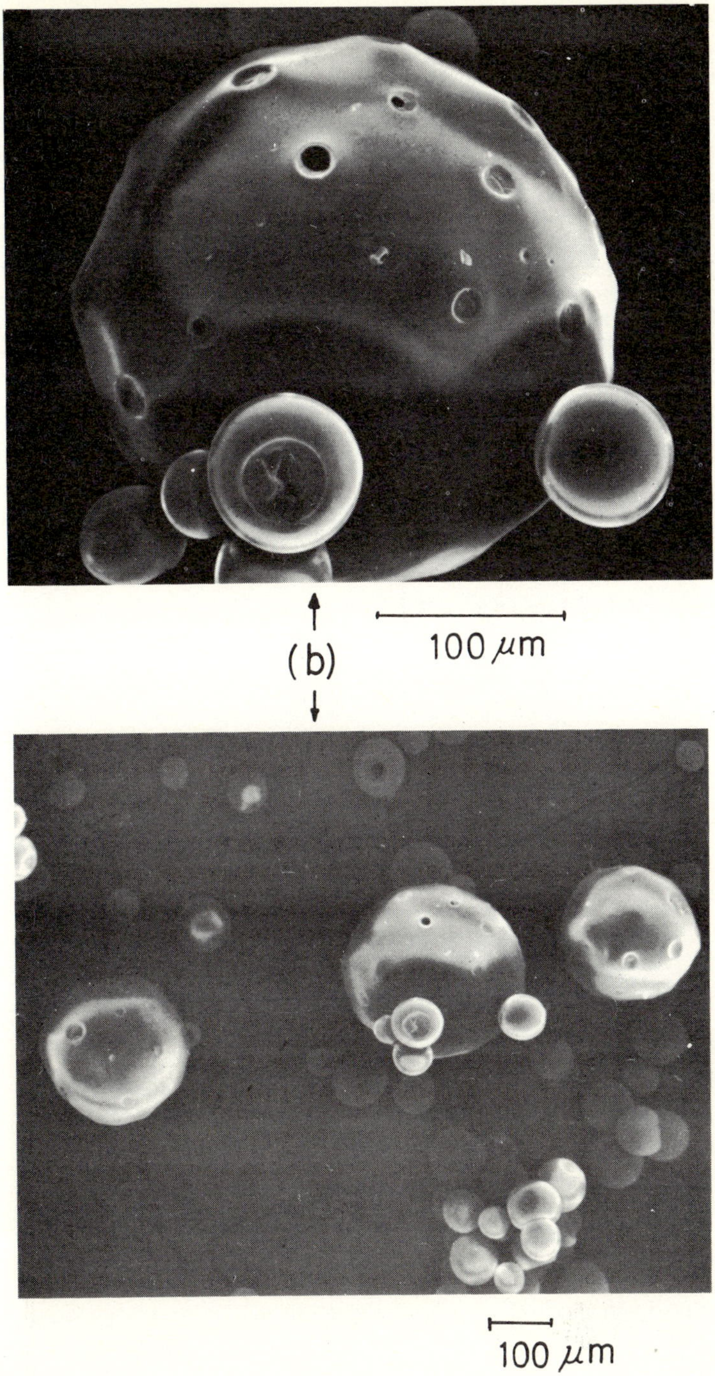

FIGURE 8 Scanning electron micrographs of capsules produced by (a) slow stirring and (b) rapid stirring.

References

1. T. M. S. Chang, *Science*, **146**, 524 (1964). T. M. S. Chang, F. C. MacIntosh, and S. G. Mason, *Canad. J. Physiol. and Pharmacol.*, **44**, 115 (1966).
2. M. Kitajima, S. Miyano, and A. Kondo, *Kogyo Kagaku Zasshi* (*J. Chem. Soc. Japan, Ind. Chem. Sect.*), **72**, 493 (1969).
3. S. Suzuki, T. Kondo, and S. G. Mason, *Chem. Pharm. Bull.*, **16**, 1639 (1968). Y. Shigeri and T. Kondo, *Chem. Pharm. Bull.*, **17**, 1073 (1969). M. Koishi, N. Fukuhara, and T. Kondo, *Chem. Pharm. Bull.*, **17**, 804 (1969).
4. F. Higashide and T. Ushikawa, *Yakuzaigaku* (*The Archives of Practical Pharmacy*), **34**, 72 (1974).
5. H. Fuwa, *J. Biochem.*, **41**, 538 (1954).
6. W. J. Keller, *Am. J. Clin. Pathol.*, **27**, 415 (1949).
7. E. L. Jackson and C. S. Hadson, *J. Am. Chem. Soc.*, **59**, 2049 (1973).
8. J. C. Rankin and C. L. Mehltretter, *Anal. Chem.*, **28**, 1012 (1956).
9. Y. Nozawa, T. Ushikawa, T. Yonezawa, and F. Higashide, *Yakuzaigaku* (*The Archives of Practical Pharmacy*), **36**, 7 (1976).
10. Y. Nozawa, F. Higashide, and T. Kanamoto, *J. Polym. Sci.*, part C, **13**, 433 (1975).

Controlled Drug Release from Small Particles Encapsulated with Acrylic Resins

KLAUS O. R. LEHMANN, HEIDE M. BÖSSLER and DIETER K. DREHER

Pharmaceutical Laboratory of Röhm, GmbH, Kirschenalle, D-6100 Darmstadt, Germany

By using fluidized bed coating machines it was possible to encapsulate spherical pellets 0.5–1.0 mm, compact granules 0.2–0.5 mm and crystals of 0.1–0.3 mm particle size with acrylic resins to control drug release. When operating conditions were optimized and with some conspergents such as talc or magnesium stearate suspended in the coating solution, it was possible to remarkably avoid agglomeration of the particles. Scanning microphotographs showed coherent film layers covering all compact structures of particle surfaces including sharp edges and corners of crystals. Encapsulated particles with the desired drug release were performed by application of 5–30%, preferably 10–20%, of dry coating material. From these experiences we conclude that spraying of acrylic resins in a fluidized bed on drug particles ranging from 0.1–1.0 mm particle size is a very practicable method of encapsulation for pharmaceutical purposes. It is of high effectiveness in respect to process data, formula evaluation and amount of coating material. In this way the well known control of drug release by film coating can be transformed to smaller particles that mean safer and more uniform drug therapy.

In comparison to coated tablets there are some important advantages in coating small drug particles and in formulating them as capsule preparations or disintegrating tablets: (1) The coated particles will distribute over the stomach and the intestine shortly after ingestion so that individual variations and local milieu conditions (pH, amount of fluids, agitation) are randomized resulting in reduced side effects and more uniform drug release. (2) One drug with different coatings can be applied to get a more effective release pattern. (3) Different active ingredients in drug combinations can be coated individually. (4) The danger of incomplete release of active material is reduced.

Based on the correlation between particle size and surface area the amount of the coating substance was calculated for a 10 μm film thickness with 1 mg per cm^2. Nearly spherical pellets in the range of 0.6–1.2 mm in diameter are expected to need 5–10% of dry coating material. A very promising working area for smaller particles could be between 0.2–0.6 mm with 10–30% of coating.

Acrylic resins with specific dissolution and permeability properties under physiological conditions were used to get the desired release pattern.[1,3] Gastrosoluble films are copolymers of methacrylic esters with dimethylamino-ethylmethacrylate (EUDRAGIT E). They give taste-proof preparations with fast drug release in the stomach by salt formation at the tertiary amino groups under acid conditions. As enteric coatings copolymers of methacrylic acid and methacrylic esters (EUDRAGIT L/S) were used to get gastro-resistant particles with fast or sustained release by dissolution of the films in the intestine under neutral to weak alkaline conditions. To achieve slow drug release–time controlled by diffusion and independent of pH–acrylic resins containing low amounts of hydrophilic quarternary amino groups (EUDRAGIT retard) were used. These coatings are insoluble but swell and are permeable to some extent. The desired diffusion rate can be obtained by mixing two types, one of high and another of low permeability. The acrylic resins were applied as organic solutions preferably in isopropyl alcohol and acetone but also in the form of aqueous dispersions obtained from emulsion polymerization.[2]

By using fluidized bed coating machines of the systems WURSTER and GLATT, it was possible to encapsulate spherical pellets 0.5–1.0 mm, compact granules 0.2–0.5 mm and crystals of 0.1–0.3 mm particle size. When operating conditions were optimized and some conspergents as talc or magnesium stearate were suspended in the coating solution, it was possible to remarkably avoid agglomeration of the particles.

Scanning microphotographs showed coherent film layers covering all of the particles' surfaces including sharp edges and corners of crystals. Encapsulated particles with the desired drug release properties were formed by application of 10–20% of coating material.

The first selected example demonstrates coating of granules with EUDRAGIT E to get a taste proof preparation. On scanning microphotographs, Figures 1 and 2, we can see that the formulation consists of compact granules of minor porosity and that the coated particles are enveloped by coherent film layers.

Another example of taste proof preparations with EUDRAGIT E was worked out with much more difficult particles: crystals in form of long columns. These crystals were broken to some extent during the coating process in the fluidized bed, but in Figures 3 and 4 we see that the broken crystals were fixed on the larger ones and coated all together.

It was much easier of course to coat compact crystals such as macro-crystalline Furadantin. With relatively low amounts of EUDRAGIT L the

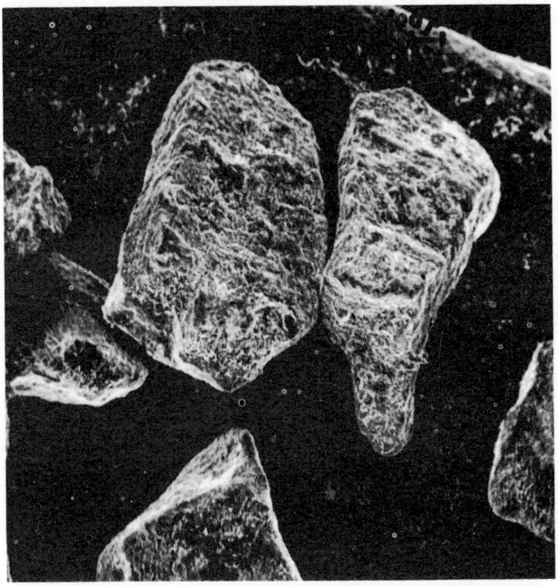

FIGURE 1 Uncoated propoxyphen granules.

FIGURE 2 Propoxyphen granules coated with EUDRAGIT E.

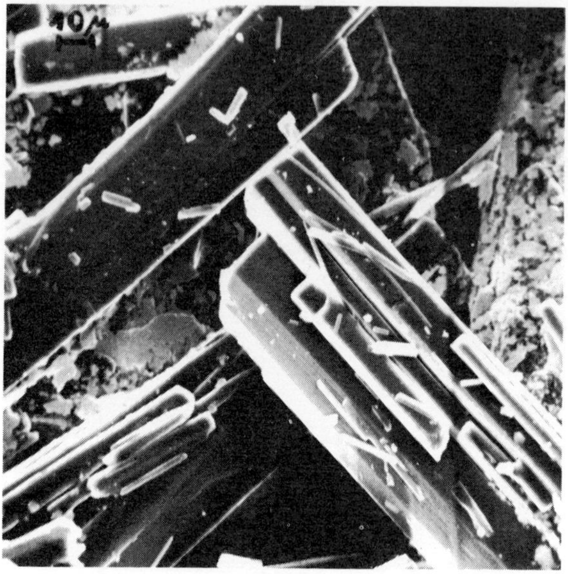

FIGURE 3 Uncoated chinidine-sulfate crystals.

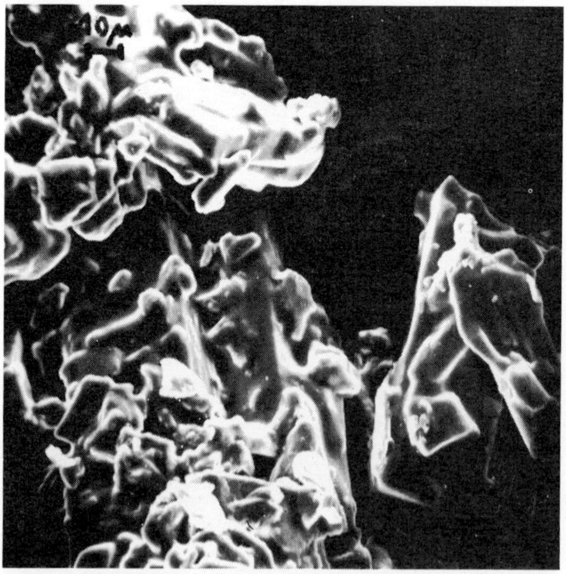

FIGURE 4 Chinidine–sulfate crystals coated with EUDRAGIT E.

film is not thick enough to give an absolute resistance against gastric juice but the coating is sufficient to give a socalled "repeated action" preparation with an initial dose in the stomach and a second dose in the intestine. If it is necessary it is of course possible to prepare enteric coated particles with a high degree of resistance against gastric juices by application of increasing amounts of EUDRAGIT L or S. Normally the desired effect is obtained with *ca.* 15–20% of dry coating material.

Another example demonstrates the function of a permeable coating on phenylpropanolamine pellets (0.8–1.2 mm dia.). In Figure 5 the release rate *in vitro* was compared with the hypothetical release *in vivo* calculated with a one-compartment model from the excretion curves in man.

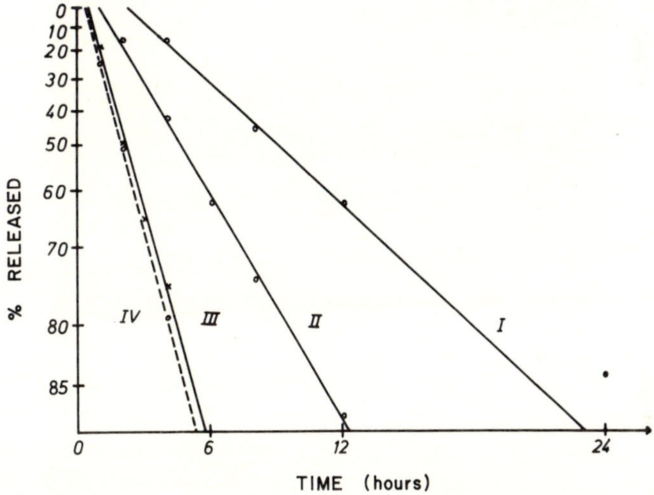

FIGURE 5 Release rate profiles for the release of phenylpropanolamine from pellets coated with EUDRAGIT retard. I. Urinary excretion from man after application of retard pellets. II. Excretion after application of uncoated drug. III. Release rate *in vitro*, half change method. IV. Calculation of hypothetical release rate *in vivo* from curves I and II with one-compartment model.

Some more extended studies were made with methaqualon crystals. They are shown uncoated and coated with EUDRAGIT retard in Figures 6 and 7. The release rate of coated crystals in correlation to the amount of coating applied was studied in the range of 5, 10 and 15%. The release rate decreases rapidly from 5–10% of coating when the majority of the particles were enveloped. In the range from 10–15% the release rate is nearly linear and influenced mainly by the permeability of the coating, releasing the drug continuously during *ca.* 4 h. So it is important to vary the permeability of the coating by mixing both EUDRAGIT retard types RL and RS until the release rate is in the desired range. Minor modifications especially during the production process are possible by variation of the amount of coating.

FIGURE 6 Uncoated methaqualone crystals.

Coated particles can be filled into hard gelatine capsules, and by some special precautions it is possible to compress them into tablets. When compression is made with high pressure and low amounts of fillers, a slow or non-distintegrating tablet is formed. We have the well known principle of a matrix tablet: the active ingredients are leached out of a porous polymer matrix. For such matrix tablets it is *not* necessary to prepare excellent coated particles. This can be achieved in a more simple way by the compression of impregnated powders, granules or mixtures of active ingredients and a polymer powder.[4] The more attractive concept is to prepare a disintegrating tablet which will release coated particles with a predetermined release pattern.

This was achieved with coated methaqualone crystals as is shown in Figure 8. The tablets were formed by compression with 50% of fillers such as starch and microcrystalline cellulose to protect the coated particles and to achieve the short disintegration time of 45 s. The redispersed particles showed a release curve very similar to that of the original coated crystals.

An increase in drug release of only 10% can be explained as coming from

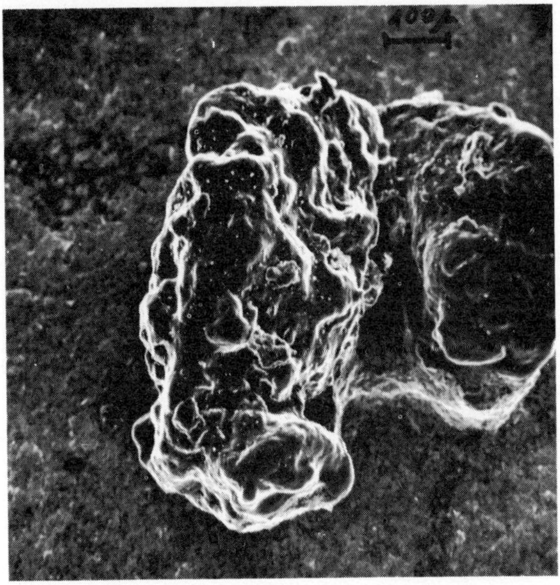

FIGURE 7 Methaqualone crystals coated with EUDRAGIT retard.

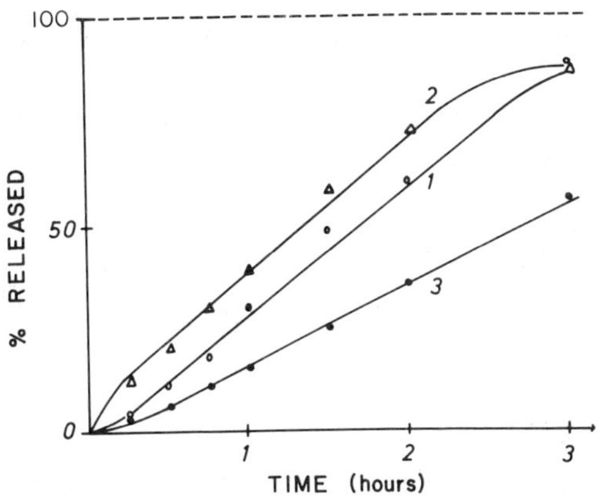

FIGURE 8 *In vitro* release rate profiles for the release of methaqualone from compressed crystals (tablets) coated with EUDRAGIT retard. 1. Crystals coated with 10% dry coating material. 2. Tablets from coated crystals disintegrating within 45 s. 3. Matrix-tablets from coated crystals disintegrating within 4 h.

limited burst of coatings during compression especially at the surface of the tablets. The parallel run of both curves demonstrates the constant release rate of the other intact particles and the unchanged permeability of the coating.

From these experiences we conclude that spraying of acrylic resins in a fluidized bed onto drug particles ranging from 0.1–1.0 mm particle size is a very practicable method of microencapsulation for pharmaceutical purposes. It is of high effectiveness in respect to process data, formula evaluation and amount of coating material which is normally in the range of 10–20% dry substance. In this way the well known control of drug release by film coating can be transformed to smaller particles that mean safer and more uniform drug therapy.

References

1. K. Lehmann, *Manufacturing Chemist and Aerosol News*, May–June 1973, pp. 36–41.
2. K. Lehmann and D. Dreher, *Drugs Made in Germany*, vol. XVI, pp. 126–136 (1973); *Pharm. Ind.*, **34**, 894–899 (1972).
3. K. Lehmann, *Manufacturing Chemist and Aerosol News*, May 1974, pp. 48–50.
4. D. Dreher, *Pharma International*, **1/2**, 3 (1975).
5. K. Lehmann, *Acta Pharmaceutica Technologica*, **21**, 255 (1975).

DISCUSSION

Dr. W. Leonard (*Dynapol, Palo Alto, CA*): Have you looked at the molecular weight distribution of the resin which you employ in the coating process?

Dr. K. Lehmann: Yes. The molecular weight of the resins used are in the range of 100,000 to 200,000. This gives us a low viscosity in our coating solution and, we have also a low viscosity in the aqueous dispersions due to the behavior of the latex.

Dr. W. Leonard: Are you concerned that the low molecular weight components of the encapsulating polymer might be transported by the G.I. system and, thus, raise separate questions regarding their toxicology?

Dr. K. Lehmann: Yes. In the aqueous dispersions the low molecular weight species or residual monomer is in the range of 0.02%. This was also the limit of detection in the film by our gas chromatograph. So we are sure that there is no remarkable amount of monomer in the coatings. We have also performed some metabolic studies with radioactively marked material. We have found only traces of radioactivity in the urine, and 98 to 104% in the feces. The traces found in the urine are excreted very rapidly, so we think this may come from other impurities from the preparation of the radioactively labelled material.

Dr. R. Capozza (*Alza Corporation, Palo Alto, CA*): What solvents do you use for placing these coatings on the particles?

Dr. K. Lehmann: The normal solvent is isopropyl alcohol. Sometimes we use acetone. One can also use chlorinated hydrocarbons mixed with the acetone, but that is not necessary. From the standpoint of air pollution and flammability of organic solvents, we think that isopropyl alcohol is the best compromise. However, we are now trying to use aqueous systems. One can also use aqueous solutions containing 60% isopropyl alcohol and 40% water. This will reduce the flammability of the solution. This mixture is not flammable by normal definitions. It is like whiskey.

Dr. G. Banker (*Purdue Univ., West Lafayette, IN*): Have you studied any differences in film properties between latex applied and organic solvent applied films?

Dr. K. Lehmann: Yes. We have studied this. The film forming mechanism from aqueous dispersion is quite different. The latex particles first link together, and then there is a confluence to a homogeneous film. Sometimes one can find a structure of deformed latex particles in the film. This depends on the drying temperature and can lead to a higher permeability of the film.

Dr. G. Banker: Are your RS and RL membranes plasticized or not?

Dr. K. Lehmann: They are normally not plasticized. The films are flexible enough that a plasticizer is not required. When it is necessary to add some plasticizer, we usually add carbowax. With the enteric coatings, we always add a plasticizer. In these cases we also use carbowax or other accepted additives.

V. Stevens (*Dow Chemical Co., Midland, MI*): The backbone of your polymers should be very stable. Would you comment on the stability of the side chains?

Dr. K. Lehmann: Yes, the backbone of the acrylics are not biodegradable. They are very stable in the body and against microorganisms. The ester groups are also very stable and are not hydrolyzed to any remarkable extent in the body. The extent of hydrolysis may be below 1%.

Results on Biodegradable Cylindrical Subdermal Implants for Fertility Control

DONALD L. WISE, JOHN B. GREGORY, PAUL M. NEWBERNE,† LESTER C. BARTHOLOW,†‡ and JOHN B. STANBURY†

Dynatech R/D Company, Cambridge, MA02139, U.S.A.

Biodegradable cylindrical subdermal implants for fertility control were evaluated in beagle dogs. L(+)-lactic, *dl*-lactic and glycolic acids were synthesized to polymers with weight ratios of 75*dl*/25G, 75L(+)/25G, 90*dl*/10G, 50*dl*/50L(+), 100*dl*, and 100L(+) and at selected molecular weights from 56,000 to 260,000. The hormonal steroid *d*-norgestrel was incorporated into these polymers as a physical matrix at several weight ratios of polymer/steroid and extruded into cylinders of 1.5 mm dia. These cylinders were inserted subcutaneously by trochar in the area of the right femoral groove of each dog. After mild edema the cylinders were well tolerated by all the dogs. Radioactive assay of urine, as well as some feces, were used to evaluate the steroid release and, to a lesser extent, polymer hydrolysis. The release of steroid appeared to follow directly the polymer susceptibility to degradation. The approximate ranking of polymers in order of most rapid biodegradation, was 75*dl*/25G, 75L(+)/25G, 90*dl*/10G, 50*dl*/50L(+), 100*dl*, and 100L(+). The 100*dl* and 50*dl*/50L(+) type materials appear to have potential for a delivery system with a lifetime of 6 months to 1 year; for this purpose the other materials may be excluded from further investigation. In general, lower molecular weight polymers and higher steroid loadings of the implant appeared to result in faster steroid release. Overcoating of cylinders with pure polymer reduced and delayed initial higher steroid release. Ratios of excreted vs. implanted ^{3}H (polymer) and ^{14}C (steroid) suggested that some lactide oligomer was possibly bonded to *d*-norgestrel fragments. All of the dogs were examined post mortem and none showed any tissue reaction at the implant site. Necropsy and histopathological studies showed no abnormalities attributable to the implants.

INTRODUCTION

Most publications on controlled release of contraceptive steroids were based on work using cylindrical subdermal implants of poly(dimethylsiloxane).[1-8] The application of poly(-L(+)-lactic acid) as a biodegradable carrier was shown earlier by testing in rats thin films of this polymer containing

† Massachusetts Institute of Technology, Cambridge, Massachusetts.
‡ Now at the University of Youngstown, Youngstown, Ohio.

d-norgestrel. In the thin film form, the release rates were found to be uniform and the steroid was released at a rate greater than the polymer hydrolyzed. The object of the present program was to evaluate cylindrical subdermal implants of polymers and copolymers of lactic and glycolic acids with *d*–norgestrel as the contraceptive steroid and to attempt to provide for uniform release of this steroid for a period of six months.

MATERIALS AND METHODS

Polymer preparation

Polymer synthesis was carried out from L(+)-lactic acid (Fluka AG), glycolic acid, and *dl*-lactic acid (both Fisher Scientific) as was reported earlier.[9-11] Labeled polymer was synthesized from lactic acid with tritium on the No. 2 carbon atom (New England Nuclear). Copolymers were prepared from various weight ratios of the respective lactides and glycolide. Molecular weights of the polymers ranged from 56,000 to 260,000. These molecular weights, $\bar{M}\eta$, were calculated by measuring the intrinsic viscosity, η, of the polymers in tetrahydrofuran at 37°C and using the equation $\eta = KM^a$. The constants $K = 1.04 \times 10^{-4}$ and $a = 0.75$ were determined for selected polymers using gel permeation chromatography, membrane osmometry, and light scattering.

Contraceptive steroid

As with the earlier study,[9] *d*-norgestrel, 13-β-ethyl-17-α-ethynyl-17-β-hydroxygon-4-en-3-one, was used (from Wyeth). ^{14}C *d*-norgestrel was labeled in the ethynyl group (synthesized by ICN Corp. under license from Wyeth). Background on selection of the *d* form of norgestrel,[12] dosage levels for fertility control,[13-15] and subdermal release from a non-biodegradable implant[16] has been presented. Also, the solubility of *d*-norgestrel in water at 37°C was reported to be 5 μg/cm^3;[17] experiments carried out during the present work, in which the steroid was equilibrated in pH 7 phosphate buffer at 37°C between Millipore® filters,[18] indicated the solubility to be 2 μg/cm^3.

Implant preparation

Thin films of the several weight ratios of polymer/steroid were prepared following the earlier procedure[9] of solvent blending of the steroid and polymer, casting films, and removing residual solvent. These films of polymer/steroid were extruded in specially designed equipment to cylindrical rods 1·5 mm in diameter.[18,19] Some rods were overcoated[19] with polymer of the same composition as the matrix. The rods were cut into 3 cm lengths for implanting.

In vivo procedure

Purebred female dogs about one year of age and sexually mature were obtained from Ridglan Research Farms (Mt. Horeb, Wisconsin) and housed in metal metabolism cages. They were fed a standard canned food twice daily and provided with water *ad libitum*. Prior to implantation the medial surface of the right hind leg in the area of the femoral groove was shaved and a local anesthetic (xylocaine hydrochloride) was injected at the implant site. Three or more cylinders were inserted subcutaneously by trochar (12 Stubbs gauge) in the area of the right femoral groove of each dog. Sutures were used following implantation for the first few dogs but were found to be unnecessary and were omitted later. After it was apparent from radioactive measurements of feces and urine that the initial cylinders were exhausted of steroid and metabolically degraded, the dogs were reimplanted. Mild edema was visible at the implant site for approximately 7 days after implantation after which it subsided. The cylinders were well tolerated by all the experimental animals and in no case were attempts made by the animals to dislodge them.

Urine was collected and measured daily [9 cm^3 of Aquasol (New England Nuclear/1 cm^3 urine counted for 20 min (Unilux II liquid scintillation counter)]. Background counts were determined using an aliquot of urine from a control animal which had been implanted with a similar cylinder of unlabeled polymer/steroid. Feces were periodically collected, weighed, and dried in a 65°C oven for 4 h; combustion analysis was then performed by New England Nuclear, Boston, Mass., following the method of Gupta.[20] Although the polymers used in this study were labeled with tritium, the combination of color quenching and low specific activity prevented detection of urinary or serum ^{3}H-radioactivity; this activity in the feces was determined, however. Also, whenever serum, feces, or urinary ^{14}C radioactivity is mentioned in the text, the ^{14}C label from *d*-norgestrel and all metabolites is implied.

RESULTS

The method for evaluating the pattern of release of steroid from implants was to plot the average quantity of drug found in samples of urine or feces. The pattern may be appreciated most directly by chronological plots of the amount recovered. The results were grouped as shown in Table I in descending order of polymer susceptibility to degradation. This ranking is approximate, but is consistent with the over-all results.

TABLE I

Ranking of over-all results by polymer composition

Polymer composition (% by weight)			Figure presenting data	Polymer molecular weight	Fraction *d*-norgestrel in implant (by weight)	Dog identification number
dl-lactic	L(+)-lactic	glycolic				
75		25	1	56,000	27.5	3
			2	78,000	27.5	5
	75	25	3	56,000	50.0	4
90		10	4	59,000	50.0	2
50	50		5	145,000	20 and 33	no coat: 13 @ 20%/14 @ 33% coat: 9 @ 20%/10 @ 33%
			6	240,000	20.0	5B
100			7	69,000	50.0	6
			8	135,000	20.0	no coat: 11 coat: 7
	100		9	260,000	20.0	4B

In Figures 1 to 3 the early burst of steroid release from the implants is shown when readily biodegradable material was used to form the matrix. The behavior shown in Figure 2 is typical in respect to showing an early burst, but unique in respect to the extreme height of the burst. At the conclusion of 90 days testing 34% of the implanted drug radioactivity had been accounted for through urinary excretion (Figure 2).

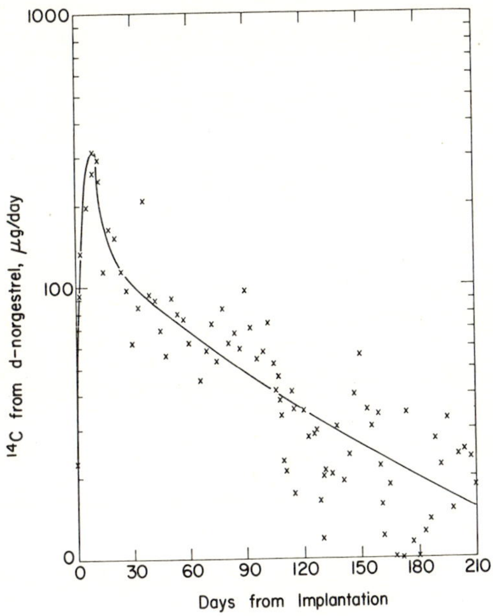

FIGURE 1 Urinary ^{14}C recovery, dog 3, 75*dl*/25G, 56,000 $\bar{M}\eta$, 27.5% *d*-norgestrel, implant area 5.6 cm^2.

An interesting double burst is shown in Figure 4 when 90*dl*/10G copolymer was used. As is seen with all results, an initial burst is noted; however, the second burst of drug release at 50–70 days shown in Figure 4 is interpreted to indicate that the relatively susceptible polymer began to degrade significantly at this time. Here the very high urinary release led to elimination of about 1/2 of the implanted radioactivity by 90 days. Taking into account probable loss of radioactive label by other routes, it may be concluded that substantial drug exhaustion had occurred following the second burst.

In Figure 5 are shown the results of testing 4 dogs with implants using 50*dl*/50L(+) copolymer. Here the initial urinary burst was observed, but release was steady out to 90 days rather than trending upward with time. When the proportion of drug in the implant was raised to 33% (Dog 14), the initial peak was enhanced and the following release showed a gradual decline. Implants overcoated with polymer of the same composition as the matrix were tested. When the overcoat was composed of a relatively hydrolysis resistant polymer, such as a 50*dl*/50L(+) copolymer (Figure 5) the overcoat technique reduced the early burst effectively.

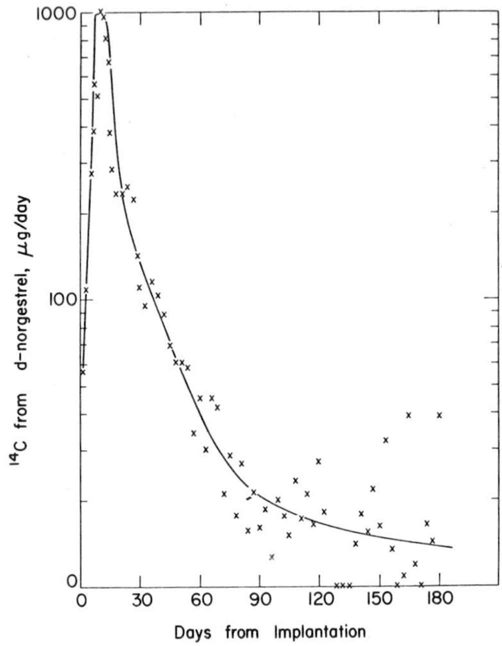

FIGURE 2 Urinary ^{14}C recovery, dog 5, 75*dl*/25G, 78,000 $\bar{M}\eta$, 27.5% *d*-norgestrel, implant area 6.7 cm^2.

A 50*dl*/50L(+) copolymer was also tested using a much higher molecular weight material. Here the steroid release for the results shown in Figure 6 was followed for a duration of 240 days. An initial urinary and fecal burst was encountered but this was of relatively modest magnitude. No later urinary burst occurred although there was a slow rate increase with time. A later burst was noted in the feces at 272 days that may indicate degradation of the polymer matrix at this point. Only 0·2% of the original drug was found at the implant site when the site was excised at necropsy. Only 9·3% of the drug was recovered in the urine. It appears that 90% of the C–14 labeled steroid was excreted in the feces. This is a higher fecal to urinary recovery ratio than that indicated in Figure 6 for the first 250 days but is consistent with the final high levels of recovery in the feces.

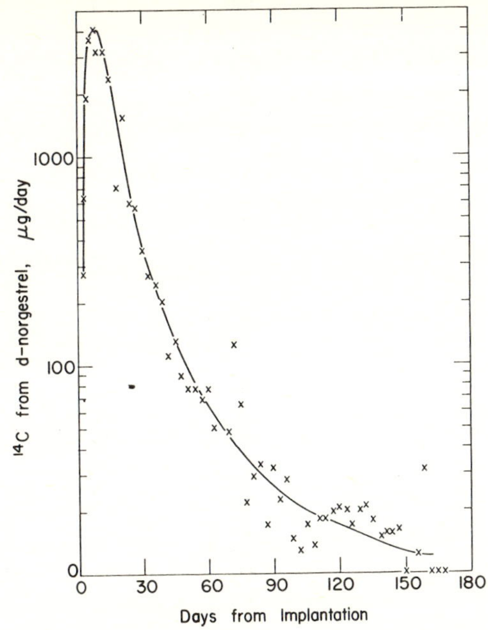

FIGURE 3 Urinary ^{14}C recovery, dog 4, 75L(+)/25G, 56,000 $\bar{M}\eta$, 50% *d*-norgestrel, implant area 5.9 cm^2.

Implants prepared from 100*dl* polymer of two molecular weights were tested in dogs as shown in Figures 7 and 8. At a 50% drug loading (Figure 7), a substantial late burst was observed. However, the animal implanted with formulations containing 20% drug (Figure 8) exhibited no late burst; the pattern was that of initial burst followed by declining release. When a *dl* polymer of 135,000 molecular weight was used for the overcoat (Figure 8) the early burst was eliminated but, apparently after the polymer overcoat hydrolyzed, a later burst occured.

The implant prepared with 100L(+) polymer showed an initial burst and continued steady release of steroid at a low level as seen in Figure 9. The 100L(+) polymer was judged the most resistant to degradation of all the matrices utilized. This may be because the 100L(+) polymer is crystalline[21] and the chain segments have restricted mobility. It appears that the diffusion rate of drug may be lower in 100L(+) polymers than in similar but non-crystalline *dl* homopolymers and glycolide/lactide copolymers. The available fecal data were plotted in the various figures. While the scope of these data was limited, the major excretion path for *d*-norgestrel was via the feces. The fecal urinary excretion ratio was about 3. Also, the fecal to urinary ratio of *d*-norgestrel metabolites varied widely before and after any burst of measured urinary release.

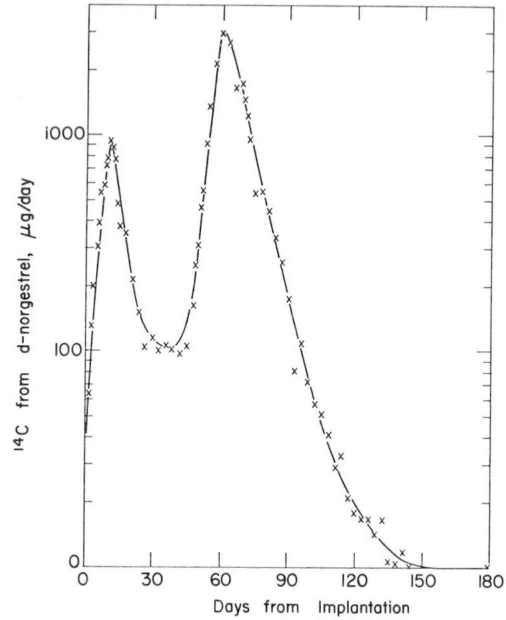

FIGURE 4 Urinary ^{14}C recovery, dog 2, 90*dl*/10G, 59,000 $\bar{M}\eta$, 50% *d*-norgestrel, implant area 5.0 cm^2.

Comparison of excreted versus implanted ^{3}H and ^{14}C

The feces of dogs 5B (Figure 6) and 6 (Figure 7) were dried and analyzed for ^{3}H and ^{14}C radioactivity by combustion analysis. The ratio of polymer and drug metabolites in the feces was obtained as seen in Table II. The ratios for the two labels are almost the same and the quantity of polymer excreted in the feces appears to be related to the proportion of drug in the implant. It is reasonable to anticipate that implanted polylactic acid becomes hydrolyzed initially to lactic acid, which would then be metabolized to carbon dioxide and water and, hence, would not appear in the feces. As evidence that this may occur, Brady, *et al.*, report that 92% of the ^{14}C label of polylactic acid appears as carbon dioxide and only 8% in the urine and feces.[22]

The detection of large amounts of tritium in the feces, plus the reasonably constant ratio of excreted vs. implanted ^{3}H and ^{14}C, suggest that some of the lactic acid polymer was not undergoing complete degradation but was excreted in the feces as lactic acid oligomer, possibly bonded to *d*-norgestrel fragments.

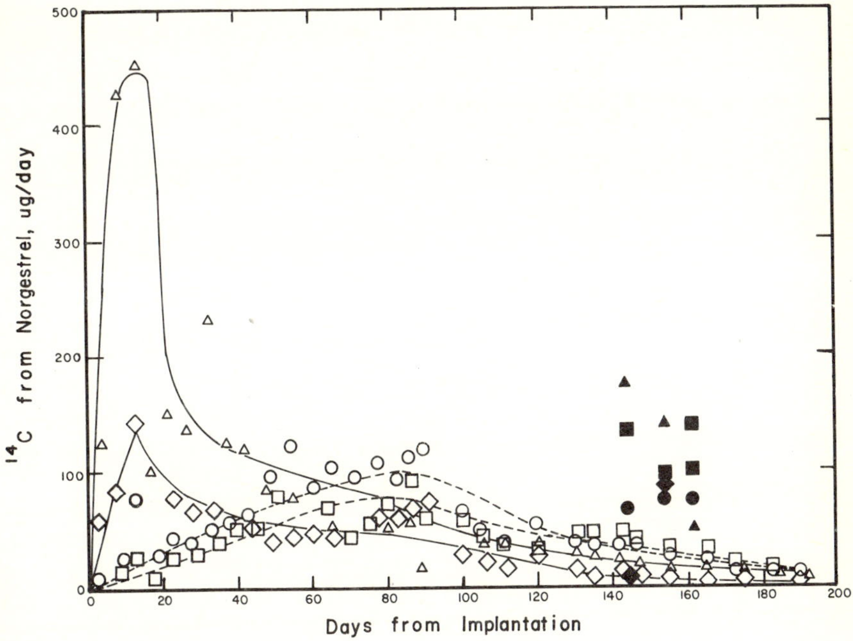

FIGURE 5 Recovery of excreted ^{14}C from dogs with implants of 50*dl*/50L(+) copolymer of 45,000 $\bar{M}\eta$. Implant area was 5.3 cm^2. (Fecal recovery—solid black symbols.)

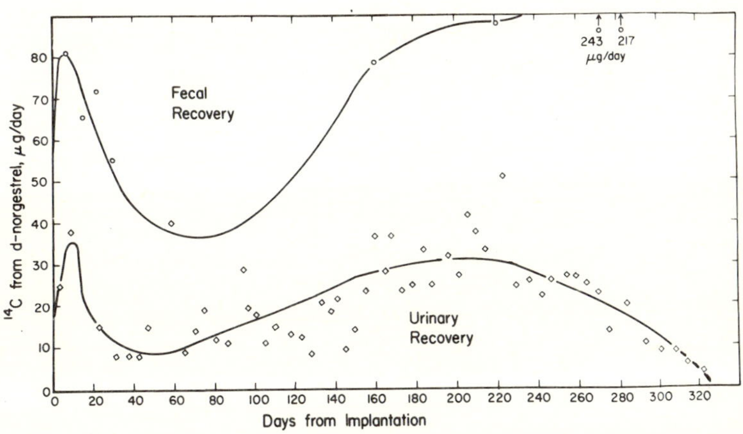

FIGURE 6 Recovery of excreted ^{14}C from dog 5B with implants of 50*dl*/50L(+) copolymer of 240,000 $\bar{M}\eta$ and with 20% *d*-norgestrel. Implant area was 6.6 cm^2.

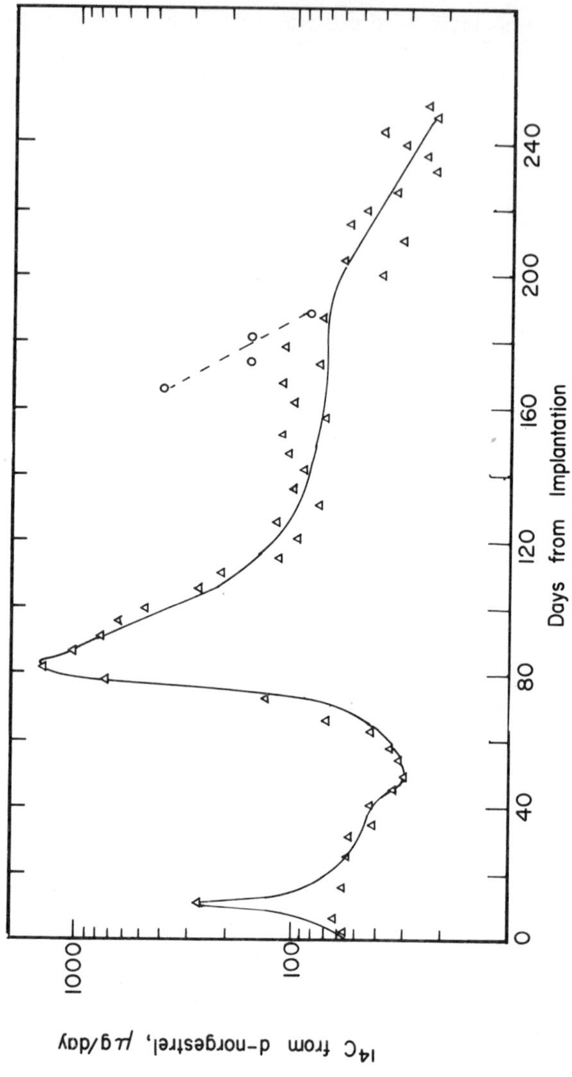

FIGURE 7 Recovery of excreted ^{14}C from dog 6 with implants of 100*dl* polymer of 69,000 $\bar{M}\eta$ and with 50% *d*-norgestrel. Implant area was 4.8 cm².

TABLE II

Comparison of excreted vs. implanted ^{3}H and ^{14}C

Dog	Fecal lactic acid/ fecal norgestrel	Implanted lactic acid/ implanted norgestrel	Excreted/ implanted
5B	4.89 : 1	4 : 1	1 : 1
6	0.91 : 1	1 : 1	1 : 1

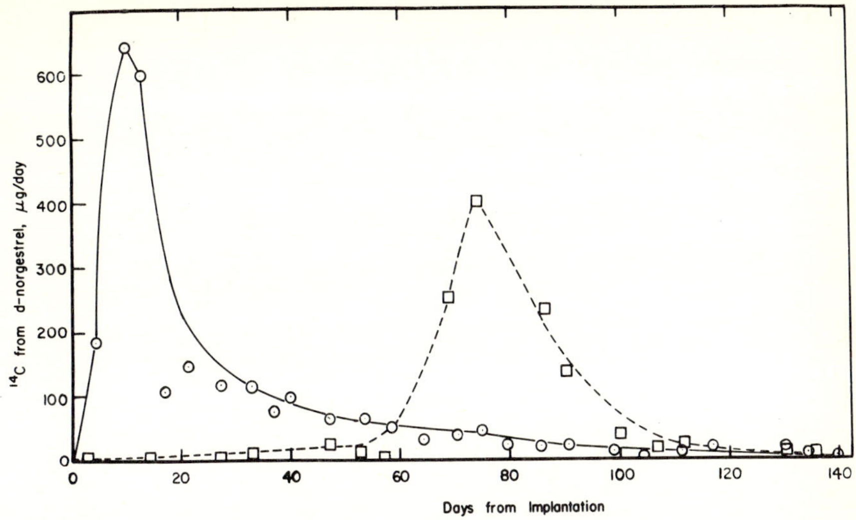

FIGURE 8 Recovery of excreted ^{14}C from dogs with implants of 100*dl* polymer of 135,000 $\bar{M}\eta$ and 20% *d*-norestrel. Implant area in each dog was 4.9 cm^2. (Fecal recovery–solid symbols.)

Results of post mortem examinations

The necropsy of Dog 6 (Figure 7) permitted observation of the distribution of polymer and *d*-norgestrel residues among the organs. The results of the necropsy are presented in Table III and show that little polymer remained in the implant site, but 4% of the drug remained.

TABLE III

Necropsy studies on dog 6 residue of implant found, μg/g

Site	Polymer	Drug	Ratio polymer drug
Uterus	none	none	—
Stem of uterus	148	13.0	11/1
Liver	2	0.3	6/1
Lung	49	1.3	40/1
Ovary	none	none	—
Jejunum	none	none	—
Mesenteric lymph node	38	2.9	13/1
Colon	49	9.0	5/1
Kidney	22	1.7	11/1
Adrenal	57	2.7	21/1

Residue at implant site	Polymer	Drug
μg	203.0	4672.0
As % of material implanted	0.2	4.2

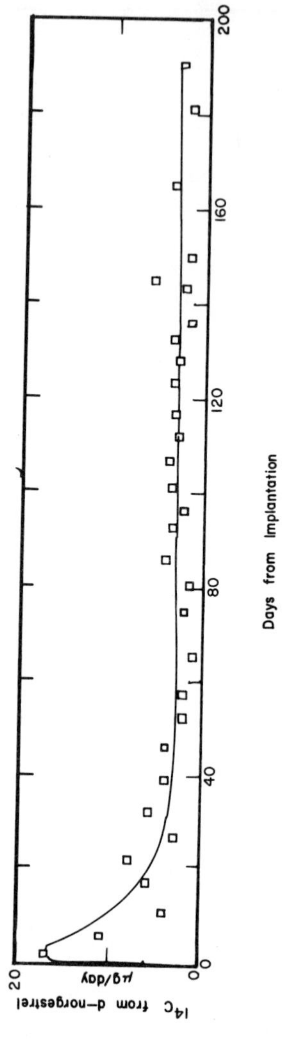

FIGURE 9 Recovery of excreted ^{14}C from dog 4B with implants of 100L(+) polymer of 260,000 $\bar{M}\eta$ and 20% *d*-norgestrel. Implant area was 5.0 cm^2.

Dog 6 was sacrificed 253 days after implantation with 222 mg of ^3H labeled *dl* poly(lactic acid) cylinder containing 50% ^{14}C labeled *d*-norgestrel. An initial burst of drug in the urine was noted shortly after implantation (Figure 7). This initial burst peaked at about 10 days with a release of 800 μg/day. A second burst occurred starting about 60 days after implantation. The second burst peaked at 83 days with a release of over 1000 μg/day. From 120 days to the date of sacrifice the release rate steadily dropped from about 100 μg/day to 25 μg/day. The total recovery of *d*-norgestrel in the urine was 37% of the amount implanted. Fecal recovery between day 166 and day 190 was 2·3 ± 1 times the urine recovery. If it is assumed that fecal recovery is twice the urine recovery then the total recovered equals 37%+74%+4%=115%; the experimental error on the feces/urine recovery ratio probably accounts for this being greater than 100%. No large proportions of drug or polymer residues were observed anywhere in the animal's body. Very little drug was stored in the tissues; the only significant reservoirs were in the stem of the uterus and the colon. Polymer residues or metabolites were detected throughout the body. Gross observation at necropsy was normal except for two small cysts. Such cysts are not infrequent in the bitch. Histologic studies showed the ovaries to be within normal range with numerous developing follicles. The uterus was in a state of anestrus. All other organs and tissues were without significant lesions. Drug and polymer were not associated as shown in Table II. This may not be surprising since most of the polymer was apparently hydrolyzed out of the implant leaving almost pure drug behind.

Histopathological studies on Dogs 2 (Figure 4) and 5 (Figure 2) were normal except for Dog 2. Note that Dogs 2 and 5 had two implants. When data was collected for the second implants, the dogs were designated Dogs 2B and 5B. In Dog 5 there was no significant histopathological changes. Dog 2 had cystic endometrial hyperplasia which is not uncommon in females but usually come at a late age. It may be that this case was enhanced by the presence of low levels of steroid. Dog 2 had the second implant 7 months after the first. This dog became sick (diarrhea) one month after the second implantation so that fecal and urinary *d*-norgestrel recovery was not obtained after that date. This dog was sacrificed six months after the second implant was administered. The remaining dogs were also sacrificed and examination of the implant sites at necropsy showed no indication of tissue reaction.

It is to be noted for completeness that blood was drawn twice monthly from the dogs and the radioactivity of the serum measured. The serum radioactivity was significantly above background only just prior to or closely following those days when very large amounts of C–14 from *d*-norgestrel were measured in the urine. Cholesterol determinations (following Levine and Zak[23]) revealed nothing significant and are, therefore, not reported. Triglycerides were determined by a procedure analogous to cholesterol determinations (following Benagiano *et al.*[5]). There was a suggestion for a decrease in serum triglyceride concentrations but this, too, was not statistically significant. Similarly, serum uric acid levels (following Henry[24]) did not change appreciably and determinations were discontinued after several months of the study. It is to be noted that the body weights of the dogs showed the expected increase in weight with age.

DISCUSSION AND CONCLUSIONS

In the earlier study[9] in which thin films of *d*-norgestrel in the relatively nonbiodegradable polymer of L(+) lactic acid were used, it was postulated that those findings might relate to cylindrically shaped implants. That postulation related those results to a biodegradable implant lasting for several years and releasing about 50 μg/day of *d*-norgestrel at a rate of about 4 μg/day·cm^2. Because of the lower rate of polymer hydrolysis with respect to steroid release in that study, the present investigation was carried out using polymers with greater susceptibility to degradation. Further, it appeared practical to consider implants that might have an effective lifetime of six months.

The results of the present study with cylindrical implants using dogs indicate that polymer from 100L(+)-lactic released *d*-norgestrel at a uniform rate of about 4 μg/day·cm^2, as was found with films. However, this apparent diffusion controlled steroid release system appears to have a lifetime much greater than the desired six months. On the other hand, the copolymers tested of 75*dl*/25G, 75L(+)/25G, and 90*dl*/10G appear to be so readily biodegradable that an implantable cylinder prepared from those materials for six months continuous delivery of *d*-norgestrel may not be possible.

From the present study, polymers of 100*dl* and 50*dl*/50L(+) may have the potential to meet the desired goals. The 1.5 mm diameter cylinders prepared from these materials provided for rather erratic release of the *d*-norgestrel but did show the potential for maintaining system integrity for at least six months while not appearing to have an impractically long polymer lifetime. An implant of the 100*dl* type polymer provided for a nominal *d*-norgestrel release of about 100 μg/day·cm^2 for a period of less than one year (data shown in Figure 7 from Dog 6). For this implant system, however, the initial burst 10 days post-implantation of about 800 μg/day and the later burst of over 1000 μg/day after 83 days appear to make this system unsatisfactory in its present formulation.

The application of an overcoat of a relatively hydrolysis resistant polymer to the cylinder appears to reduce the initial burst (Figure 5). Results, however, for some polymers, indicate that this may only delay a sudden release of steroid (Figure 8).

The potential for these types of biodegradable implants, however, may be seen in the results upon necropsy in which very little polymer or drug was found to remain at the implant site. Moreover, histologic studies revealed no reservoir of steroid or polymer.

The results of testing a 50*dl*/50L(+) type copolymer show certain potential to meet the desired criteria (data shown in Figure 6 for Dog 5B). In this case the implant system had a lifetime of almost one year and a nominal *d*-norgestrel release rate was about 20 μg/day·cm^2. This system was noted for a rather modest initial burst of steroid compared with the other systems tested; a somewhat swelling form of final burst of steroid was seen as opposed to the more dramic release patterns observed.

While all results have been ranked with respect to polymer composition, it is well to call attention to both polymer molecular weight and *d*-norgestrel loading in the implant. Although the 100L(+) polymer system tested is excluded as not being within the range of desired results, it is to be noted that this system had a polymer of high molecular weight and a low steroid loading relative to the other systems. Likewise, compared with the other systems tested, the more promising 50*dl*/50L(+) type system was also prepared from high molecular weight polymer and had a low loading of steroid. On the other hand, the 100*dl* system which appeared to have an over-all equivalence with the one 50*dl*/50L(+) system (data shown in Figure 7 for Dog 6), had polymer with one of the lowest molecular weights tested and was among one of the highest steroid loadings tested.

While these somewhat conflicting characteristics for the 100*dl* and 50*dl*/50L(+) systems may not be immediately answerable, the results do tend to support the exclusion of the other systems on the basis of polymer composition. That is, the 75*dl*/25G, 75L(+)/25G, and 90*dl*/10G systems, albeit all from lower molecular weight polymers, may be excluded from consideration as a six month delivery system. Also, the 100L(+) polymer system appears to provide for steroid release over a period outside the desired time frame. Based on the meritorious biological results of all systems, however, it is concluded that select biodegradable cylindrical subdermal implants have potential for purposes of fertility control. Those that may be selected for further investigation would appear to be those synthesized from 100*dl*-lactide and 50*dl*/50L(+) lacide.

Acknowledgement

This work supported by National Institute of Child Health and Human Development Contract No. NO1–HD–3–2736.

References

1. G. Benagiano, M. Ermini, C. C. Chang, K. Sundaram, and F. A. Kincl, *Acta Endocrinol.*, **63,** 29 (1970).
2. E. M. Coutinho, C. E. R. Mattos, A. R. S. Santanna, J. A. Filho, M. C. Silva, and J. H. Tatum, *Contraception*, **2**, 313 (1970).
3. H. B. Croxatto, S. Diaz, P. Atria, S. Cheviakoff, S. Rosatii, and H. Oddo, *Contraception*, **4**, 155 (1971).
4. M. C. Chang, J. H. Casas, and D. M. Hunt, *Fertil. and Steril.*, **22**, 383 (1971).

5. G. Benagiano, M. Ermini, L. Carenza, and G. Rolfini, *Acta Endocrinol.*, **73**, 335 (1973).
6. G. Benagiano, M. Ermini, L. Carenza, and P. Donini, *Acta Endocrinol.*, **73**, 347 (1973).
7. M. Ermini, F. Carpino, M. Russo, and G. Benagiano, *Acta Endocrinol.*, **73**, 360 (1973).
8. E. M. Coutinho and A. R. daDilva, *Fertil. and Steril.*, **25**, 170 (1974).
9. T. M. Jackanicz, H. A. Nash, D. L. Wise, and J. B. Gregory, *Contraception*, **8**, 227 (1973).
10. R. K. Kulkarni, E. G. Moore, A. F. Hegyeli, and F. Leonard, *J. Biomed. Mater. Res.*, **5**, 169 (1971).
11. R. K. Kulkarni, K. C. Pani, C. Neuman, and F. Leonard, *Arch. Surg.*, **93**, 839 (1966).
12. S. F. Sisenwine, H. B. Kimmel, A. L. Liu and H. W. Ruelius, *Drug Metab. Dispos.*, **2**, 65 (1974).
13. A. Ferrari, J. C. Meyrelles, J. H. Sartoretto, and A. Soares-Filho, *Int. J. Fertil.*, **18**, 133 (1973).
14. R. Apelo and I. Veloso, *Fertil. Steril.*, **24**, 191 (1973).
15. F. Garmendia, E. Kesseru, and L. A. Llerena, *Horm. Metab. Res.*, **5**, 134 (1973).
16. J. Zbuzkova and F. A. Kincl, *Endocrinol. Exp.*, **4**, 215 (1970).
17. K. Sundaram and F. A. Kincl, *Steroids*, **12**, 517 (1968).
18. L. C. Anderson, D. L. Wise, and J. F. Howes, *Contraception*, **13**, 375 (1976).
19. A. D. Schwope, D. L. Wise, and J. F. Howes, *Life Sci.*, **17**, 1877 (1975).
20. G. N. Gupta, *Microchem. J.*, **13**, 4 (1968).
21. R. K. Kulkarni, E. G. Moore, A. F. Hegyeli, and F. Leonard, *J. Biomed. Mater. Res.*, **5**, 169 (1971).
22. J. M. Brady, D. E. Cutright, R. A. Miller, G. C. Battistone, and E. E. Hunsuck, *J. Biomed. Mater. Res.*, **7**, 155 (1973).
23. J. Levine and B. Zak, *Clin. Chim. Acta*, **19**, 381 (1964).
24. R. J. Henry, *Clinical Chemistry: Principles and Techniques*, New York, Harper and Row, p. 278 (1964).

DISCUSSION

Dr. R. Nevin (*Eli Lilly and Co., Indianapolis, IN*): You have referred in your paper today and in your paper the other day to the polymerization of DL lactide. Is this, in fact, DL lactide?

Dr. D. Wise: Yes.

Dr. R. Nevin: What bothers me about this is that there are two asymmetric centers in lactide. Thus, theoretically there should be four geometrical isomers, two *cis* and two *trans*. I'm convinced the two *cis* isomers constitute a D and L pair. As hard as I try, I cannot rationalize a D and L pair from the *trans* isomers. I think they constitute a *meso* form. So the fact that your working with an optically inactive lactide means that it could be a DL or *meso* form. If I'm wrong, then there is another DL pair. Which DL pair are we talking about? The question is relevant because it reflects on the stereochemistry of the polymer and, thus, would affect the biodegradability of the polymer. I know the literature refers to the lactide which is commonly available as the DL lactide, but I don't believe it is actually DL lactide.

Dr. D. Wise: I'm not sure if I can see the practical merit of other than the racemic DL mixture or L(+) crystalline form.

Dr. R. Nevin: That's really what determines the difference in biodegradability of the so called DL and the L. One is highly crystalline and the other is not so crystalline. One is much less biodegradable than the other. This will be so even in the copolymers. But I think that since so many people have worked with this system, it would be nice if they knew exactly what material they were working with.

The difference between the *trans* and *cis* forms is between 8 and 10 kilocalories per mol which would favor the formation of the *trans* isomer during synthesis. You should really know whether you are working with a racemic mixture or the *meso* form.

The other point is that you can't have one DL without having two DL's.

Dr. D. Wise: I'm a little lost in your explanation, because when we work with these materials in synthesis, we use what is termed an L(+) lactic acid, which is a product of fermentation of lactose. As I understand it, this is the conventional way to get the L(+) form of lactic acid. We drive the water off, and after recrystallization get a sharp melting point for that lactide. On the other hand, the racemic mixture of lactic acid, which is made by Monsanto as food grade lactic acid, is referred to as DL-lactic acid. Following a similar procedure, we obtain DL-lactide.

Dr. R. Nevin: That is racemic, but is the lactide derived from that racemic or *meso*? It seems that all you know is that it is optically inactive. The NMR spectra of the L(+) vs. the DL should tell you what you have.

Dr. D. Wise: We've used a polarimeter to determine whether we have the D or the L form and to check the purity of our L lactide. This is mostly a quality control technique, and it doesn't distinguish between the (−) or (+) form.

Dr. D. Meier (*Midland Macromolecular Institute, Midland, MI*): Is the mechanism of release of drug from your poly(lactic acid) system the same as that from the Alzo Chronomer® material, namely surface erosion?

Dr. D. Wise: No, it isn't. We find that in no case does this material change its shape. The center of an implanted cylinder will, after some time, begin to get soft while the outside apparently gets harder than it was at the beginning. The device does not change its initial dimensions.

Dr. L. Bangs (*Dow Chemical Company, Midland, MI*): I thought you showed that these disappeared after a time?

Dr. D. Wise: That's true. But the way this appears to happen is not by a simplistic "erosion" mechanism.

Diffusion of Water Soluble Drugs from Initially Dry Hydrogels

WILLIAM R. GOOD

Research Services Division, Central Research, CIBA-GEIGY Corporation, Ardsley, NY10502, U.S.A.

A mathematical model for diffusion of water soluble drugs from hydrogel materials is developed. The model is derived from Fick's second law of diffusion utilizing a time dependent diffusion coefficient to take into account the simultaneous influx of solvent and eflux of drug which occurs when a dry hydrogel containing the drug is brought into contact with the extracting medium. The equations are used to fit experimental diffusion data for tripelennamine·HCl release from hydrogels of crosslinked poly(2-hydroxyethylmethacrylate) and the effect of initial drug concentration in the hydrogel on solute diffusion coefficient and solvent diffusion coefficient is discussed.

INTRODUCTION

Presently, the use of polymeric materials as carriers for drugs is an intense area of research in the pharmaceutical industry. A large variety of synthetic polymers used in controlled release applications has been reported in the literature. Among the most common types of polymers investigated are poly(dimethylsiloxane),[1–5] poly(lactic acid),[6] poly(amides),[5] poly(ethylene),[1,5] and various types of hydrogel polymers.[7,8] The release mechanism considered to be most prevalent in all these polymers is pure diffusion with the exception of poly(lactic acid) which is a biodegradable polymer. In this case the rate of drug release is controlled by a combination of diffusion and polymer hydrolysis. Polymers which do not absorb significant amounts of solvent are thought to release enclosed drug via a pure diffusion mechanism involving a thin diffusion layer near the polymeric surface.

Hydrogel materials, on the other hand, are unique with respect to their ability to release an enclosed drug in aqueous medium. Since these materials are capable of absorbing water while simultaneously releasing an enclosed drug, the release rate of the drug is modified by the degree of hydration of the drug/polymer device, and the nature and concentration of the drug modifies the rate and extent of hydration. When the drug is water soluble and its release rate is characterized by diffusion coefficients of the same order of magnitude as the solvent, this mutual perturbation of transport properties becomes quite significant. If, *in vivo*, the bioavailability of the drug hinges on the rate of delivery to the target area, any perturbation in the release from the dose form ultimately effects the physiologic response, and accurate correlations of release properties with *in vivo* response requires a knowledge of this transient period of modified release rate. With these considerations in mind, a modified solution to Fick's second law of diffusion to accommodate simultaneous solute diffusion and solvent infusion has been derived and evaluated experimentally, *in vitro*.

THEORY

In one dimension, Fick's second law of diffusion is,

$$\partial C/\partial t = \partial^2(DC)/\partial x^2, \tag{1}$$

where x is the space coordinate, C is concentration of diffusing material within the membrane, t is time and D, the diffusion coefficient is here expressed in its most general form. The units of D are area/time, so we speak of a rate law per unit surface area.

Consider a flat sheet of hydrogel having its boundary faces at $x=0$ and $x=1$. The gel initially is free of solvent and has homogenously distributed within it the solute at concentration C_0. At time zero the flat sheet is brought into contact with the solvent and the processes of solvent absorption and solute diffusion begin simultaneously. If we neglect any changes in thickness of the device due to solvent swelling the diffusion coefficient for the solute can be reasonably expected to be proportional to the degree of solvent absorption. That is,

$$D = kQ^*. \tag{2}$$

In addition, assuming that only that fraction of the solute which is solubilized by the solvent medium has significant mobility, the solute fraction closest to the surfaces of the device are characterized by a larger diffusion coefficient than the fractions near the center of the device, at least during the initial solvent absorption stage of the process. Since the degree of solvent absorption can be expressed as a function of time, we may characterize the solute diffusion coefficient as a simple function of time, i.e.,

$$D(t) = D_s(Q^*/Q_\infty), \tag{3}$$

where D_s represents the solute diffusion coefficient at equilibrium solvent absorption and (Q^*/Q_∞) is the fraction of solvent absorbed with respect to its equilbrium solvent content at time t.

Returning now to Eq. (1), we may rearrange to produce,

$$\partial C/\partial t = D(t)\, \partial^2 C/\partial x^2 \tag{4}$$

and introduce the auxiliary variable, T in such a way that by definition,

$$dT/dt = D(t)/l^2 \tag{5}$$

and

$$T = \int_0^t [D(t)/l^2]dt. \tag{6}$$

By the chain rule,

$$dC/dt = dC/dT\, dT/dt \tag{7}$$

substitution of which into Eq. (4) generates the linear differential equation,

$$\partial C/\partial T = l^2\, \partial^2 C/\partial x^2. \tag{8}$$

Equation (8) is simply Fick's second law restated with the constant parameter l^2 replacing the time dependent diffusion coefficient and the function T replacing the time.

The solution of Eq. (8) is easily obtained by the method of separation of variables using the boundary conditions,

$$C = 0 \quad \begin{matrix} x = 0 \\ \\ x = l \end{matrix} \quad t > 0 \tag{9}$$

and initial conditions

$$C = C_0, 0 < x < l\}t = 0. \tag{10}$$

The solution to equation (8) desired is,

$$C = \frac{4C_0}{\pi} \sum_0^\infty \frac{\exp[-(2n+1)^2\pi^2 T]}{(2n+1)} \sin(2n+1)\frac{\pi x}{l}. \tag{11}$$

The fraction of solute diffused at time t may be obtained by integrating Eq. (11) over the space variable for arbitrary time and is given by

$$Q_t = Q_\infty \left[1 - \frac{8}{\pi^2} \sum_0^\infty \exp[-(2n+1)^2\pi^2 T]/(2n+1)^2\right], \tag{12}$$

where Q_t is the amount of solute diffused at time t and Q_∞ is the amount of solute diffused at infinite time.

Our task now becomes one of defining the function T in terms of real time, or more exactly, defining the fraction of solvent absorption as a function of time. Just as a concentration gradient in solute exists across the polymer boundaries, a solvent concentration gradient exists across the same boundaries, although in the opposite direction. The absorption of solvent is also governed by Fick's second law but differs from the solute expression in that the diffusion coefficient for solvent is assumed to be constant for a given initial solute concentration. The boundary and initial conditions for solvent absorption are,

$$C^* = C_0^* \quad \left.\begin{matrix} x = 0 \\ \\ x = l \end{matrix}\right\} t > 0$$

$$C^* = 0 \quad 0 < x < l\} t = 0 \tag{13}$$

and the solution of the differential equation for solvent absorption is,

$$C^* = \frac{4C_0^*}{\pi} \sum_0^\infty \frac{(-1)^n \exp[-(2n+1)^2\pi^2 D^* t/l^2]}{(2n+1)} \cos (2n+1) \frac{\pi x}{l}. \tag{14}$$

If Q^* represents the amount of solvent which has entered the sheet at time t and Q_∞ represents the corresponding quantity after infinite time, then

$$Q^* = Q_\infty \left[1 - \frac{8}{\pi^2} \sum_0^\infty \exp[-(2n+1)^2\pi^2 D^* t/l^2]/(2n+1)^2 \right], \tag{15}$$

where D^* is the diffusion coefficient for solvent. Substitution of this expression into Eq. (3) produces the desired time dependent diffusion coefficient for solute in terms of real time and the diffusion coefficient for the solvent,

$$D(t) = D_s \left[1 - \frac{8}{\pi^2} \sum_0^\infty \exp[-(2n+1)^2\pi^2 D^* t/l^2]/(2n+1)^2 \right]. \tag{16}$$

Further substitution of Eq. (16) into the expression for the function T given by Eq. (6) and with subsequent integration generates our complete time dependence for T,

$$T = \frac{D_s t}{l^2} + \frac{8D_s}{\pi^4 D^*} \sum_0^\infty \frac{\exp[-(2n+1)^2\pi^2 D^* t/l^2]}{(2n+1)^4} - \frac{8D_s}{\pi^4 D^*} \sum_0^\infty \frac{1}{(2n+1)^4}. \tag{17}$$

The infinite series in the last term of Eq. (17) converges to $\pi^4/96$, and we finally obtain the function T in terms of time, solvent diffusion coefficient and solute diffusion coefficient at equilibrium solvent content,

$$T = \frac{D_s t}{l^2} + \frac{8D_s}{\pi^4 D^*} \sum_0^\infty \frac{\exp[-(2n+1)^2\pi^2 D^* t/l^2]}{(2n+1)^4} - \frac{D_s}{12D^*}. \tag{18}$$

Equation (18) exhibits some interesting characteristics including linear time dependence, exponential time dependence and a constant term. For even moderately short times the infinite series converges very rapidly giving rise to a single exponential term. For large values of D^*t/l^2 the second term of Eq. (18) becomes insignificant producing

$$T \simeq \frac{D_s t}{l^2} - \frac{D_s}{12D^*}, \tag{19}$$

a purely linear relationship between T and t. Therefore, once the linear approximation is realized, it is possible to calculate not only the steady state diffusion coefficient for the solute, but the diffusion coefficient for the solvent as well. Using hypothetical values for D_s, D^* and l the graphical representation of Eq. (18) is shown in Figure 1. By extrapolating the linear portion

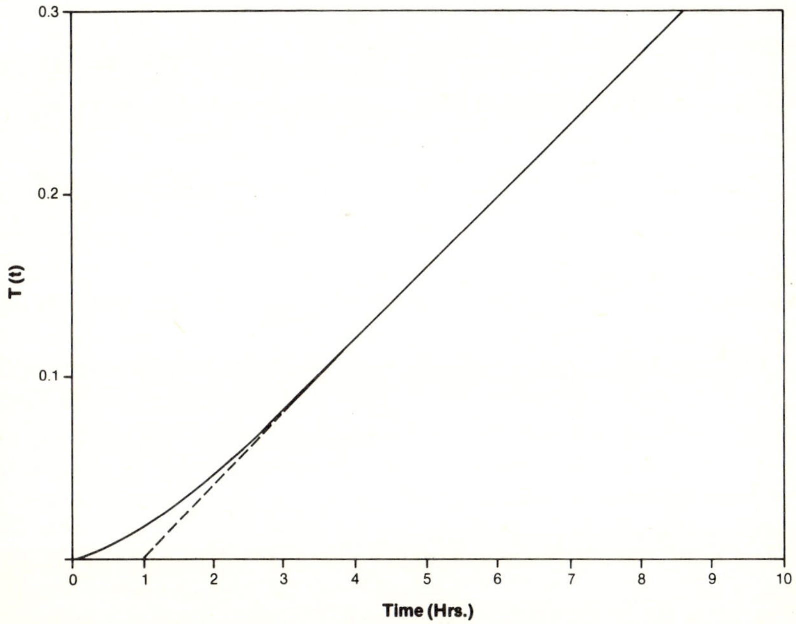

FIGURE 1 The function $T(t)$ plotted vs. time for the hypothetical values of $D_s = 1.11 \times 10^{-7}$ cm^2 s^{-1}, $D^* = 2.33 \times 10^{-7}$ cm^2 s^{-1}, and thickness of 0.100 cm.

of the curve to the time axis the time lag associated with the transient solvent absorption phase of the diffusion process may be obtained, and having calculated D_s from the slope of this portion of the curve, the value of D^*

may be calculated. For these hypothetical values it is found that the linear portion of the curve representing a constant diffusion coefficient for the solute characterizes only the last 30% of the release profile. The initial 70% of the release profile falls into the transient region of simultaneous diffusion of solute and infusion of solvent. The application of the modified diffusion Eq. (12) together with the mathematical definition given by Eq. (16) is, therefore, an important part of characterizing the release profile of solutes from initially dry hydrogels.

EXPERIMENTAL

Chemicals

2-Hydroxyethyl methacrylate (HEMA) of high purity was obtained from Alcolac Corp. and tetraethyleneglycol dimethacrylate was purchased from Rohm and Haas. The initiator used in hydrogel synthesis was *t*-butylperoctoate obtained from Lucidol Division of Wallace and Tiernan Corp.

The drug used in this work was tripelennamine hydrochloride. This compound was chosen as being representative of extremely water soluble drugs and was obtained from CIBA-GEIGY Corporation.

Hydrogels

Insoluble hydrogels consisting of lightly crosslinked poly(2-hydroxyethyl methacrylate) have been known for some time[10] and their preparation either in the absence of water or in the presence of water has been well documented.[11] The hydrogels prepared for this study were polymerized at 80°C for 3 h using *t*-butylperoctoate (0.5%) as a free radical initiator. Flat sheets were cast of a mixture of HEMA and 1.5% tetraethyleneglycol dimethacrylate in a mold consisting of two Mylar sheets with a silicone O-ring material of the desired thickness serving as spacer. The Mylar sheets were backed with plate glass, and the entire assembly was clamped to produce a mold sealed on three sides. The monomer solution was injected into the mold cavity with a needle and syringe. Although conversion of monomer to polymer in the anhydrous state is generally better than 98%, it was necessary to remove the last traces of residual HEMA by washing the sheets in a large volume of water at 50°C for 3 days as HEMA absorbs ultraviolet radiation strongly at a wavelength of 212 nm. After allowing the hydrogel to cool to 37°C a specimen of the material was dried at 80°C *in vacuo* for 24 h and the equilibrium water content was found to be 36.05%. The remaining sheets were then maintained in water at 37°C until the drug was incorporated into them.

Active dose forms

Since tripelennamine·HCl is such a highly water soluble drug it affords the opportunity to incorporate in the hydrogel a wide range of concentrations. The amount of drug taken up per unit volume of hydrogel is a function of solubility of the drug in the polymer network and the concentration of the drug in the aqueous phase. Aqueous concentrations of tripelennamine·HCl chosen were 5, 10 and 20% w/w.

Hydrogels of thickness ranging from approximately 0.5 to 1.5 mm were allowed to equilibrate in these solutions for a sufficient time period so as to be sure that no significant concentration gradient existed through the polymer sheets. Circular specimens were then cut from the sheets with a cork borer of 2 cm dia. and dried *in vacuo* at 80°C to constant weight. Specimen weights were monitored during the entire process to determine the amount of drug absorbed by the hydrogel, although these data were used primarily in a qualitative way and quantitative drug assay was performed spectrophotometrically as described in the next section.

Tripelennamine·HCl assay

The ultraviolet spectrum of tripelennamine·HCl exhibits a maximum absorption at 315 nm (pH=2.1). A stock solution of the drug was prepared at a concentration of 482 mg/l of HCl solution, pH=2.15, and six quantitative dilutions were used to measure the absorbance at 315 nm. The Beer–Lambert law was applied and a least squares fit of concentration *vs.* absorbance gave a correlation coefficient of 0.9999 over the concentration range of 4.82 to 57.84 mg/l. The molar absorption coefficient at 315 nm is 8181.6 and the concentration may be expressed by,

$$C(\text{mg/l}) = 35.6690\, A_{315} \pm 0.01. \tag{20}$$

A specimen of each of the loaded hydrogels was pulverized and extracted four days in a known volume of HCl solution and the amount of drug determined by the previously described UV method. The instrumentation used was a Beckman Acta CIII UV/Visible spectrophotometer with automatic sampler.

Diffusion experiments

All drug release experiments were conducted at 37°C in a sufficient volume of 0.007 N HCl to maintain sink conditions. Cumulative release curves were obtained using the automatic sampler together with flow through UV cells to which the extracting solution was pumped from the diffusion flask at a rate of 1 l per h through medical grade silicone tubing. The chart speed and interval between absorbance readings at 315 nm were adjusted to produce a release curve in which the time span was sufficiently expanded to generate at least 10 in. of chart on the time scale for no less than 80% of the absorbance scale with an absorbance reading every tenth inch of the time scale. This procedure produces an almost continuous release curve and allows very accurate visual interpolation between absorbance readings.

RESULTS

Solute absorption and specimen assay

Each solution used for absorption was continuously mixed to avoid concentration polarization at the hydrogel surfaces and it was found that after several days the gels had come to equilibrium with their respective solutions. The equilibrated gels remained transparent even after drying, indicating that the absorped drug was completely dissolved within the polymeric network. A rather interesting phenomenon occurred upon placing some of the dose forms into fresh HCl solution. The thickest of the specimens (*ca.* 2 mm) which had been swollen in the highest concentration of tripelennamine·HCl cracked and finally separated longitudinally into two sheets of half the original thickness. Generally, this occurred within a short period of time and prohibited the use of these high concentration, thick specimens for quantitative diffusion study. It is apparent that excessive pressure within the specimens due to extreme solvent absorption was sufficient to surpass the mechanical strength of the polymer.

Tripelennamine·HCl assay was performed at room temperature in 0.007 N HCl under sink conditions using finely pulverized dry material obtained by grinding the excess material of each sheet after the diffusion specimens had been cut. Table I presents the absorption data obtained. These results agreed well with total drug calculated from final concentration readings during diffusion runs.

TABLE I

Tripelennamine·HCl loadings achieved by soaking poly(2-hydroxyethylmethacrylate) hydrogels in various aqueous solutions of the drugs. Drug loadings are given as percent (w/w) based on the dry specimens

Solution concentration, %	Drug loading, %
5.0	12.54
10.0	23.17
20.0	35.59

Diffusion experiments

Data was obtained for six diffusion runs in which both initial drug concentration and thickness were varied. The disc thicknesses were measured prior to placement into the diffusion cell and after completion of the experiment. Table II summarizes the initial and final conditions of the diffusion specimens. All of these specimens remained transparent throughout the release experiments, and no cracking or separation of the specimens occurred.

TABLE II

Comparison of thickness of flat sheet specimens in the dry state with thickness of the same specimen after all dissolved drug has been diffused and the specimen has been equilibrated in an aqueous environment

Specimen	Dry loading, %	Dry thickness, cm	Wet thickness, cm
1	12.54	0.0699	0.0673
2	12.54	0.1344	0.1374
3	23.17	0.0699	0.0673
4	23.17	0.1443	0.1458
5	35.59	0.0711	0.0676
6	35.59	0.1529	0.1448

The release of the drug was recorded automatically and a continuous time *vs.* absorbance curve was drawn directly onto the chart paper for each run. Using a standard set of values for the function T, Eq. (12) was solved to produce a set of values for the fraction of drug released corresponding to the set of T values. This set of numbers representing the fraction of drug released was then marked off on the release curve for each experimental diffusion run and the corresponding time was then recorded. Figures 2, 3 and 4 are release curves which were redrawn from this data for the three different initial concentrations used in this study. The effect of thickness variation for each concentration is obvious from the figures and the higher rate of release from devices with higher initial concentrations is also quite apparent.

In order to determine the values of D_s and D^* for a particular specimen directly, it is necessary to fit the experimental data to Eq. (18). This was done graphically by plotting T *vs.* time as in Figure 5 and fitting the long time portion of the curves to Eq. (19). This linear approximation was fitted by a least square method and the values of D_s, D^* and the time lag were computed from the slope and intercept of the approximation. In all cases the fit of this linear approximation to Eq. (19) was excellent, the worst correlation coefficients being 0.9993 based on a perfect fit having correlation coefficient of 1.0. The results of this treatment are presented in Table III.

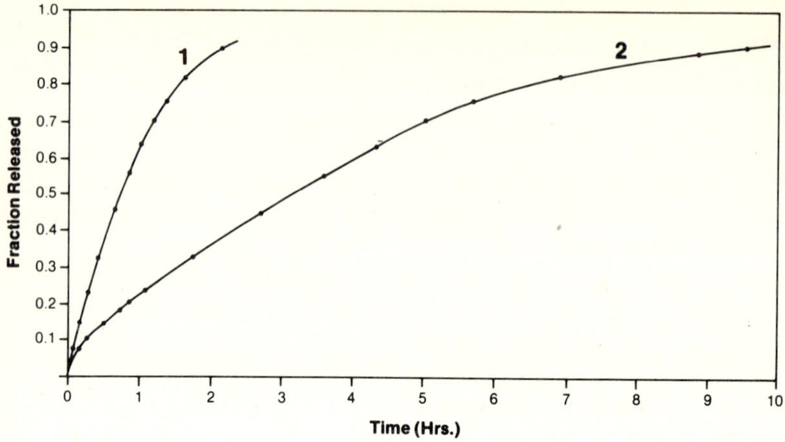

FIGURE 2 Release profiles for tripelennamine·HCl from poly(2-HEMA) sheets having initial dry loading of the drug of 12.54% w/w. Profile 1 results from a sheet of thickness 0.0699 cm and Profile 2 results from a sheet of thickness 0.1344 cm.

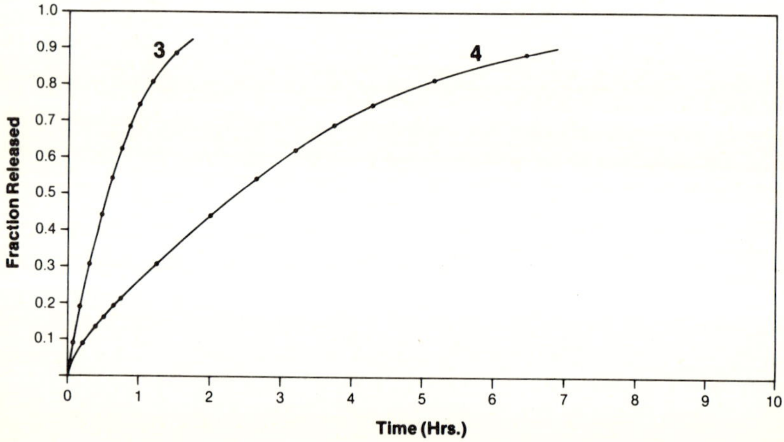

FIGURE 3 Release profiles for tripelennamine·HCl from poly(2-HEMA) sheets having initial dry loading of 23.17% w/w drug. Profile 3 represents diffusion from a sheet of thickness 0.0699 cm and Profile 4 shows diffusion from a sheet of thickness 0.1443 cm.

DISCUSSION

Time lag

The time associated with the approach of the diffusion coefficient to its steady state value is given by,

$$t^* = l^2/12D^*. \tag{21}$$

This parameter is independent of the solute diffusion coefficient but heavily dependent upon thickness and solvent diffusion coefficient. An independent verification of Eq. (21) is shown in Figure 6 where the time lag determined by extrapolating to the time axis the linear portion of the T *vs.* time curve is

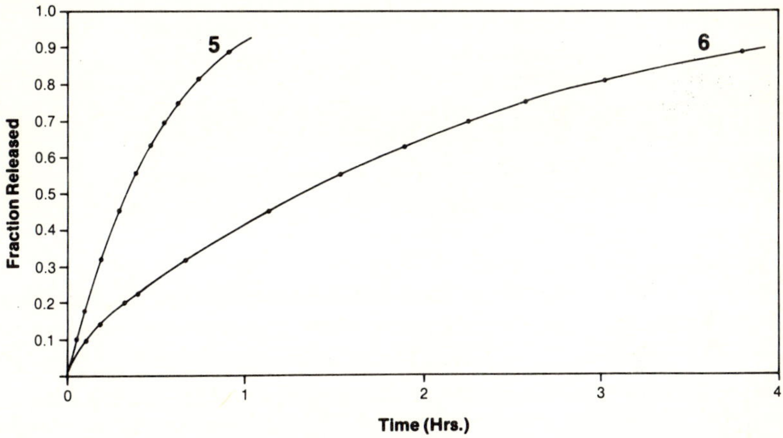

FIGURE 4 Release profiles for tripelennamine·HCl from poly(2-HEMA) sheets having initial dry loading of 35.59% w/w drug. Profile 5 represents diffusion from a sheet of thickness 0.0711 cm while Profile 6 shows diffusion from a sheet of thickness 0.1529 cm.

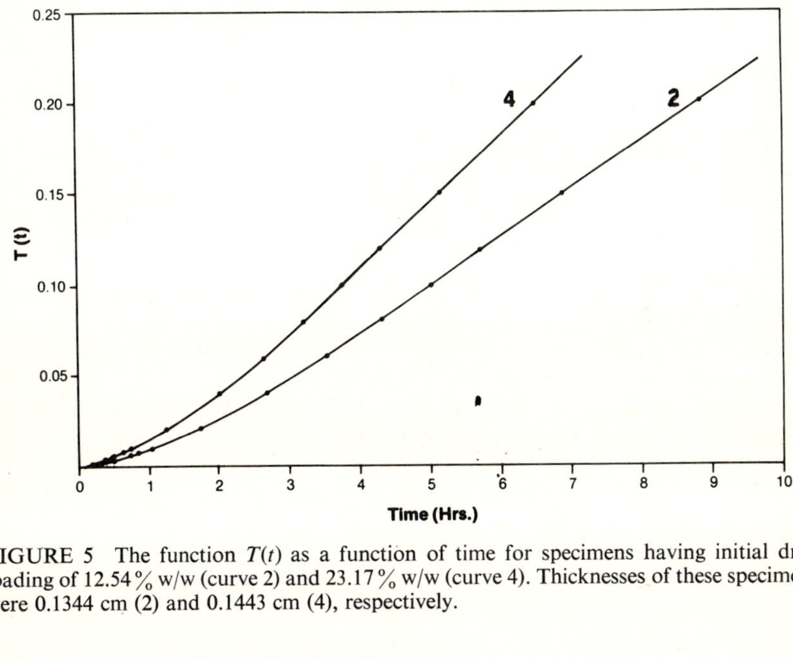

FIGURE 5 The function $T(t)$ as a function of time for specimens having initial drug loading of 12.54% w/w (curve 2) and 23.17% w/w (curve 4). Thicknesses of these specimens were 0.1344 cm (2) and 0.1443 cm (4), respectively.

TABLE III

Steady state diffusion coefficient (D_s) for tripelennamine·HCl, water diffusion coefficient (D^*) and time lag (t^*) for diffusion specimens having variable thickness and initial drug loading

Drug loading, %	Thickness, cm	D_s, (cm^2 s^{-1})$\times 10^7$	D^*, (cm^2 s^{-1})$\times 10^7$	t^*, (h)
12.54	0.0699	1.61	3.20	0.353
12.54	0.1344	1.34	3.25	1.287
23.17	0.0699	2.20	4.18	0.270
23.17	0.1443	2.12	4.70	1.026
35.59	0.0711	3.77	6.76	0.173
35.59	0.1529	4.07	8.82	0.613

plotted *vs.* the square of the thickness for the specimens at each initial concentration. The slopes of these lines can be used to determine the corresponding solvent diffusion coefficients. The dependency of the solvent diffusion coefficient upon the initial drug loading is quite marked, increasing as the initial drug load is increased. This is to be expected since at these high initial loadings of a water soluble drug, the hydrophilicity of the system is increased. Conversely, one might expect that the presence of a relatively water insoluble drug would cause a corresponding decrease in D^*.

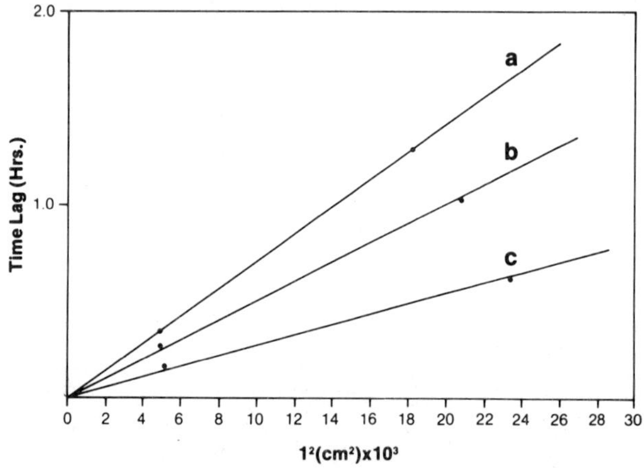

FIGURE 6 Time lag for the onset of steady state diffusion coefficient as a function of the square of the thickness. The three lines represent the relationship for the three initial concentrations used in this study, 12.54% (a), 23.17% (b) and 35.59% (c).

Solute steady state diffusion coefficients

In order to determine an average solute diffusion coefficient, the values of D^* for the appropriate initial concentrations were determined from the slopes of the lines in Figure 6 and the experimental data refitted using the averaged time lags in the least square calculation. The solute steady state diffusion coefficients calculated in this way are presented in Table IV. These data compare favorably with the values of D_s determined from the slopes of the linear portions of the *T vs.* time curves which did not include the time lag in the fitting procedure. The effect of higher initial concentration on the steady state diffusion coefficient is also quite marked.

TABLE IV

Average steady state diffusion coefficient (D_s) for tripelennamine·HCl and water diffusion coefficient (D^*) for specimens having variable initial drug loading

Drug loading, %	D_s, $(cm^2\ s^{-1}) \times 10^7$	D^*, $(cm^2\ s^{-1}) \times 10^7$
12.54	1.43 ± 0.10	3.25
23.17	2.14 ± 0.02	4.67
35.59	3.82 ± 0.20	9.03

Prediction of diffusion patterns

Since the solvent diffusion coefficient and steady state solute diffusion coefficient were determined from the steady state conditions, these constants may be used to check the validity of the present model during the transient period in which solvent infusion and solute diffusion are occurring simultaneously. The values for D^* and D_s shown in Table IV were used to generate data from Eqs. (12) and (18). For this purpose a Fortran IV computer program was written which included a subroutine which carried out the infinite series calculations indicated to an accuracy of 8 decimal places. Figure 7 illustrates the comparison of the calculated release patterns for the thin specimens for each concentration (solid lines) with the experimentally determined data points. The conformity of the experimental data to the predicted curves is excellent throughout the transient time period and is even more dramatically illustrated in Figure 8 where we have plotted the function *T vs.* time for the same three specimens. Again in Figure 8 the curves are calculated while the points are experimentally determined.

SUMMARY

A solution to Fick's second law of diffusion was obtained which takes into account the simultaneous influx of solvent and efflux of solute from a solid polymeric dose form. This phenomenon is handled by treating the solute

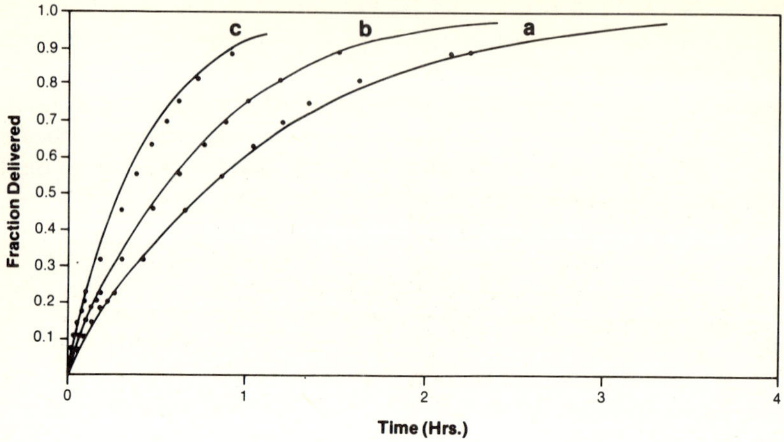

FIGURE 7 Comparison of release profiles calculated (solid lines) for specimens having initial drug concentrations of 12.54% (a), 23.17% (b) and 35.59% (c). Experimental values appear as data points. These curves were generated for the thinnest specimens studied in each case.

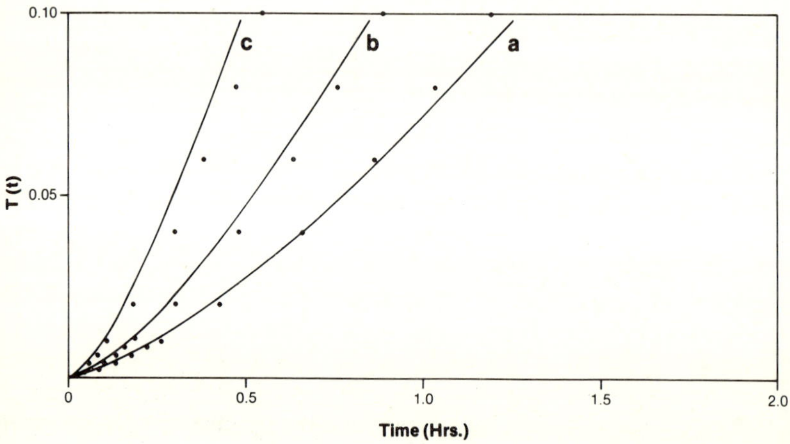

FIGURE 8 Comparison of calculated vs. experimental values of the relationship between the function $T(t)$ and time. The values of D_s and D^* were averaged for the two thicknesses studied at each concentration in order to generate the theoretical curves (solid lines). As in Figure 7, the curves were calculated for the thinnest specimens at initial loadings of 12.54% (a), 23.17% (b) and 35.59% (c).

diffusion coefficient as a time dependent variable. A series of polymeric dose forms composed of crosslinked poly(2-hydroxyethylmethacrylate) containing a water soluble drug in various initial concentrations was used to verify the validity of the derived solution to Fick's second law and illustrate the importance of taking into account the water influx in attempting to predict the release pattern of the drug from a water insoluble hydrophilic polymer.

References

1. Y. W. Chen, H. J. Lambert, and D. E. Grant, *J. Pharm. Sci.*, **63**, 365 (1974).
2. E. G. Lovering and D. B. Black, *J. Pharm. Sci.*, **62**, 602 (1973).
3. E. G. Lovering and K. B. McRae, *J. Pharm. Sci.*, **62**, 971 (1973).
4. D. M. Long, L. R. Sehgal, M. E. DeRios, M. V. Rios, P. B. Szanto, and R. Forrest, *Rev. Surg.*, **30**, 229 (1973).
5. E. R. Garrett and P. B. Chemburkar, *J. Pharm. Sci.*, **57**, 1401 (1968).
6. J. H. R. Woodland and S. Yolles, *J. Pharm. Sci.*, **16**, 897 (1973).
7. H. Yasuda, L. D. Ikenberry, and C. E. Lamaze, *Die Makromol. Chemie*, **125**, 108 (1969).
8. M. Arlen, J. W. Scherrer, J. Cukierman, F. Nagin, and B. S. Levowitz, *Arch. Surg.*, **105**, 100 (1972).
9. J. Crank, *Mathematics of Diffusion*, Oxford University Press, London, 1970.
10. K. Kliment, J. Vacik, Z. Ott, and V. Majkus, U.S. Patent 3,551,556, 1970.
11. M. F. Refojo and H. Yasuda, *J. Appl. Polym. Sci.*, **9**, 2425 (1965).

DISCUSSION

Dr. H. Hopfenberg (*North Carolina State Univ., Raleigh, NC*): What is the magnitude of the thickness change as a consequence of swelling?

Mr. W. Good: There wasn't any because the loading was such that the diffusion of drug out almost matched the absorption of water so that there was virtually no change in thickness over the course of these experiments. This is one of the things I should have mentioned. If the drug is moving out as fast as the water is moving in, one should get virtually no change in the thickness. If one has the case in which the diffusion coefficient of water is much larger than that of the drug, then one finds that the thickness will increase to a large extent during the experiment. So this kind of a derivation is most applicable for fast diffusion of drugs.

Dr. D. Meier (*Midland Macromolecular Institute, Midland, MI*): Will your equations then apply if the system swells?

Mr. W. Good: Swells from what state?

Dr. D. Meier: From some initial thickness to some larger thickness. It should not turn out that the equations are inapplicable in that case.

Mr. W. Good: Correct. One would have to take into account the change in thickness or build that into the time dependence of the diffusion coefficients. In other words the maximum swelling would have to be taken into account. One can do this empirically of course. You could define your function, Φ, such that it gives the absorption curve the way it really happens. That's an empirical approach and will probably have to be done in many cases.

Dr. R. Capozza (*Alza Corporation, Palo Alto, CA*): I was fascinated by the fact that the hydrochloride form of the drug results in clear films.

Mr. W. Good: Yes, we were too.

Dr. R. Capozza: I wonder if what's happening in this case is that the free base is diffusing into the polymer. The free base would be soluble in the polymer but the hydrochloride would not be soluble in that matrix once it's dry.

Mr. W. Good: The free base happens to be a brilliant yellow colour. These gels were colorless. I believe it's the hydrochloride in the gels. The loading was done in distilled water so the concentration of HCl has to be quite low. There has to be at least some hydrochloride in the device. Frankly, we didn't look for chloride in the experiment. Perhaps we should have.

Dr. R. Capozza: I still suspect it's the free base.

Mr. W. Good: It could be the free base but as I said the free base is a brilliant yellow color, and as you saw, these gels were colorless. Unless the amine is associated with hydroxyl groups one will get color.

V. Stevens (*Dow Chemical Company, Midland, MI*): You mentioned that the polymer you have is poly(hydroxyethylmethacrylate) and that it is crosslinked with the diacrylate of ethylene glycol so that what you have is a pendant hydroxyethyl group along with the backbone.

Mr. W. Good: That's correct.

V. Stevens: Do you have any problem with the release of ethylene glycol in the presence of free base. Ethylene glycol itself is a toxic substance.

Mr. W. Good: No. However, poly(hydroxyethylmethacrylate) is not the polymer CIBA-GEIGY is anticipating using, because it is already patented. We have looked at the stability of poly(hydroxyethylmethacrylate) in acid and base over long periods of time and, amazingly, hydrolysis is virtually negligible even though the degree of swelling is quite high.

V. Stevens: Under what pH conditions?

Mr. W. Good: pH of 2, 60°C for several weeks and pH of 8 at the same temperature conditions. It's an amazingly stable side chain. It takes a very concentrated acid to hydrolyze it.

V. Stevens: That's interesting. The drug you're using is an amine. I was wondering whether there is any catalytic action caused by the drug.

Mr. W. Good: Not that we've noticed.

Dr. H. Hopfenberg: Did these composites take on different physical characteristics over the drug concentration range which you studied?

Mr. W. Good: No, they didn't. They were all totally transparent and apparently soluble.

Dr. H. Hopfenberg: And all rubbery?

Mr. W. Good: Not all rubbery. At lower concentrations, 12% for example, the composite was still glassy and brittle. In grinding up the specimens for doing the assays, the lower concentrations were very easy to grind and could be done with a mortar and pestle. The specimens with higher drug concentrations had to be frozen with liquid nitrogen before grinding.

Dr. H. Ringsdorf (*Univ. of Mainz, Mainz, Germany*): We have discussed many different drug delivery systems during the last few days. Would you try to rate or compare your method with the other ones? Would you discuss the advantages and disadvantages?

Mr. W. Good: I think that as far as the pharmaceutical industry is concerned, the oral delivery systems are preferred. Quite frankly, the monolithic device for delivery in the GI tract, in my opinion, is one of the best systems to consider simply because there is a certain residence time in the GI tract that has to be considered as the ultimate rate determining step. One can't have a drug delivered over 20 days if the residence time in the GI tract is only 18 h. We've done some very revealing theoretical calculations using a three compartment pharmacokinetic model showing that zero order release in the GI tract has advantages in only very specific instances over first order controlled release. So I feel that if the monolithic devices can be loaded at sufficiently high concentrations, these systems will be as effective as, say, capsular dosage forms.

Dr. H. Ringsdorf: Would you compare the polymer you used with similar descriptions of acrylamides which basically have been the starting materials for this type of application?

Mr. W. Good: Are you talking about poly(hydroxyethylmethacrylate)?

Dr. H. Ringsdorf: Yes. Compare that to acrylamides.

Mr. W. Good: Acrylamides unfortunately swell to too high a degree to be practical because the particle size will become too large to serve as a practical oral dosage form. The polymers which we are interested in ourselves are proprietary materials. These materials do not release drugs as fast as poly (hydroxyethylmethacrylate).

Dr. H. Ringsdorf: But you could control the release rate by crosslinking or by just adding another comonomer unit.

Mr. W. Good: Sure, one can add another comonomer, but you may run into problems with the FDA. It's important to know the toxicity of any added monomer and how much added monomer will be allowed. What we're interested in is satisfying all the requirements for a controlled release oral dose form. That means living under the conditions set forth by the FDA. We're talking about 5–10 parts per million of allowable residual monomer so we try to work with materials which are least likely to cause problems in this respect. So we stay away from acrylates, for example, which are very toxic. Also we don't work with comonomers because the more components one puts in, the more one has to prove are safe. For that reason our systems are not based on a whole variety of materials but just on a couple or three materials.

Carriers for Bioactive Materials

ANDRÉ TROUET

Institute of Cellular and Molecular Pathology, University of Louvain, 75, avenue Hippocrate, B-1200 Bruxelles, Belgium

The advantages and limitations conferred to bioactive agents by their association to carriers are discussed. The various requirements for the carrier and carrier-agent linkages are defined according to the different sites of pharmacological and biological action (extra-, peri-, intracellular). The different levels of selectivity that can be imparted to the active compounds are considered as well as the possible selective carriers. Two practical examples of carrier-drug complexes in the field of cancer and trypanosomiasis therapy are finally given.

Many efforts have been made for years to decrease the toxicity and to increase the therapeutic efficiency of bioactive agents by associating or linking them to carrier molecules. Without going into an exhaustive review of the literature we can enumerate as follows the papers describing the association of:

antitumoral agents and—antibodies[1–12]
—proteins[13–19]
—DNA[20,21]
—liposomes[22–29]
—lectin like compounds[30,31]
Adamantadine derivatives and—a dipeptide[32]
—synthetic polymers[33]
Catecholamines and synthetic polymers[34]
Naloxone and polysaccharides[35]
Mescaline and polyvinylpyrrolidone[36]
Diphtheria toxin and antibodies[37]
Enzymes and liposomes[38–40]
Antibodies and liposomes[41]
Varia[42–46]

On the basis of these data, of our results and of the most recent developments in the field of cell biology we will try to state the carrier-drug problem as clearly as possible and try to draw some guide-lines for further research in this field.

For the sake of simplification we will use the term "agent" to designate any drug or material possessing bioactive properties. By "carrier" we mean any molecule or assembly of molecules that can serve to bind agents and that by itself or combined to agents is unable to permeate through cellular plasma membranes.

ADVANTAGES OF CARRIER LINKING

The most common benefit sought by the use of carriers is to provide therapeutic agents with a higher specificity towards their targets. Transforming unspecific bioactive compounds into target seeking guided missiles is indeed one of the most appealing if not the most difficult goal to achieve in drug design. It seems logical to combine an active but unspecific compound with a target specific carrier since it appears very difficult, if not impossible, to manipulate the molecular structure of an active compound in such a way that it keeps its pharmacodynamic properties while acquiring target recognizing and cell penetration abilities.

The association with a carrier can provide the agent with several other advantages besides an increased specificity. Most interestingly the carrier can modify the mechanism of intracellular penetration and allow the cellular uptake of compounds that are unable to enter cells because of a native or acquired impermeability of the cell membrane.

The pharmacokinetics of the carrier linked agent will also be profoundly altered and this modification will be more pronounced the greater the molecular size of the carrier. The rate of excretion and intratissular distribution of the carrier linked compound will be modified and this property can be used to increase the plasma level of the agents, to maintain it for a longer time and to increase their therapeutic effectiveness.

Finally, linked to a suitable carrier an agent can be slowly released from the carrier in an active form providing in this way a depot formulation with delayed action.

Limitations of carrier linking

The linking to a carrier will limit very severely, if not completely, the absorption from the gastrointestinal tract of compounds given by oral route.

After parenteral administration the tissular distribution of the active compounds will be restricted by the ability of the carrier to cross the capillary membranes that separate the site of injection from the targets. The physico-chemical properties and molecular size of the carrier will play a determinant role in this respect as well as the anatomical localization of the target. The effectiveness of the capillary barrier towards various macromolecules varies indeed very much from one organ to another. Targets localized in the bone marrow, spleen or liver, for example, will be much more accessible to the carrier-agent complexes than those located in the muscles or the central nervous system. In the former case the capillary barrier is not very tight since it is constituted by an incomplete basal membrane and by noncontiguous endothelial cells; in the latter case on the contrary, the basal membrane and the endothelial lining are complete in such a way that the carrier-agent complexes have to cross both the cells and the membrane.

The carrier itself must be devoid of any toxic effects. Immunogenic molecules, like foreign proteins, should be excluded as carriers. Great care should also be taken to avoid allergic reactions elicited by agents acting as haptens when linked to a macromolecular carrier even if the latter is nonimmunogenic by itself.

Carriers should also be biodegradable in order to avoid toxic effects due to intracellular uptake and overloading as observed already after clinical use of poorly or nondegradable polymers like dextran or polyvinylpyrrolidone.[47–49]

CARRIERS AND CARRIER-AGENT LINKS

The carrier and the type of carrier-agent link to be used will depend not only on the considerations mentioned above but even more on the site of action of the bioactive agents.

EXTRACELLULAR SITE OF ACTION

Without being exhaustive we can classify in this category drugs like: (a) antibiotics acting on extracellular bacteria or parasites; (b) inhibitors that block the deleterious effects of enzymes released in the extracellular medium during pathological processes like inflammation, shock or rheumatoid arthritis; (c) enzymes like asparaginase and urease; (d) anticoagulants. The main advantage of carrier complexes is, in this case, to provide the active compounds with a slow release mechanism or to maintain them at a higher level and for a longer time in the extracellular space or cavities (bloodstream, peritoneum, pleural cavity, joint cavity, cerebrospinal fluid). The carrier can also shield the agent in order to decrease its immunogenicity and decrease its reactivity with free or cell bound antibodies.[24,50,53]

The carrier should (a) have sufficient molecular size to slow down the passage of the active compound through the capillary barriers; (b) not be taken up at a high rate by endocytizing cells namely the cells of the reticulum endothelial system.

The link between the carrier and the agent should on the other hand not interfere with the biological activity of the latter or the link should allow the agent to be dissociated from the carrier and activated by enzymatic or physico-chemical processes prevailing in the extracellular space.

PERICELLULAR SITE OF ACTION

In this category fall all the compounds the pharmacological action of which is mediated by an interaction with the plasma membrane of the target cell and does not require an intracellular penetration.

By linking these agents to carriers in such a way that they remain active, the advantage will be mainly to confer to the agents slow release properties and to restrict their intratissular distribution, according to the carrier type.[34–36]

The carrier can also provide the active compounds with a higher selectivity against certain cell types provided the agent is inactive when linked and that it is released from the carrier in an active form after contact between the target cell and the carrier-agent complex. This latter condition can be fulfilled if the agent is released from the carrier by an enzyme of the cellular plasma membrane or by an enzyme secreted by the target cell and possessing a short range of action (Figure 1). One could think for example of anti-inflammatory or anti-rheumatoid agents which would be activated specifically by the neutral proteases or collagenases secreted by the cells involved in the initiation and development of inflammation or rheumatoid arthritis.

INTRACELLULAR SITE OF ACTION

In this category we find the great majority of drugs and nearly all of them penetrate inside the cell by permeation through the plasma membrane. This passage occurs either by passive diffusion or by an active transport mechanism.

The intracellular penetration of bioactive compounds linked to a carrier can occur according to two distinct pathways: (a) endocytosis; (b) transport through the cellular membrane. These two pathways will be discussed separately since they are characterized by different therapeutic possibilities and very different carrier requirements.

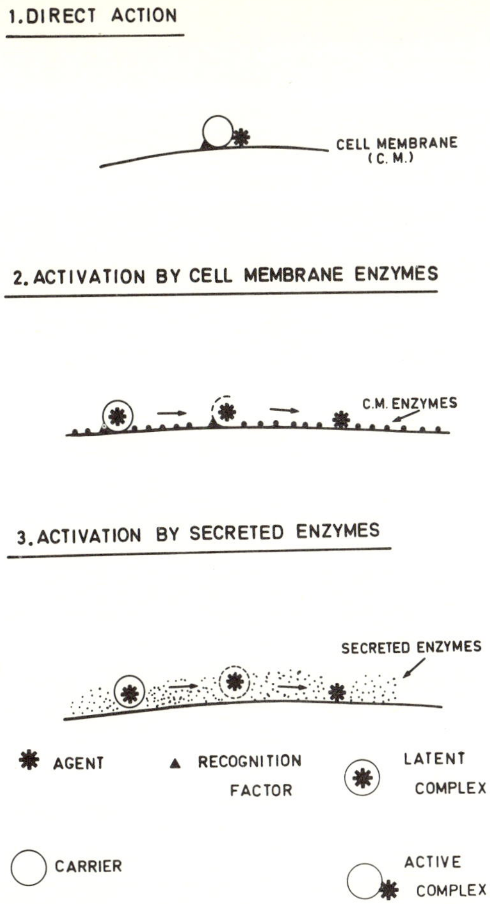

FIGURE 1 Modes of action of "carrier-agent" complexes acting at the cell periphery.

a) Uptake by endocytosis (lysosomotropic carrier-agents)

The most obvious mode of cellular entry for carrier-agent complexes is endocytosis. During this process the complexes will reach the lysosomal compartment after fusion of the endocytic vacuole with the lysosomes[51] (Figure 2).

This type of intracellular transport is of course the most appealing for all the bioactive materials acting inside the lysosomes. In this case the complexes will have one or more of the following advantages over the free agent: (a) the active compound will be selectively concentrated at its intracellular site of action; (b) agents normally not taken up by endocytosis and unable to

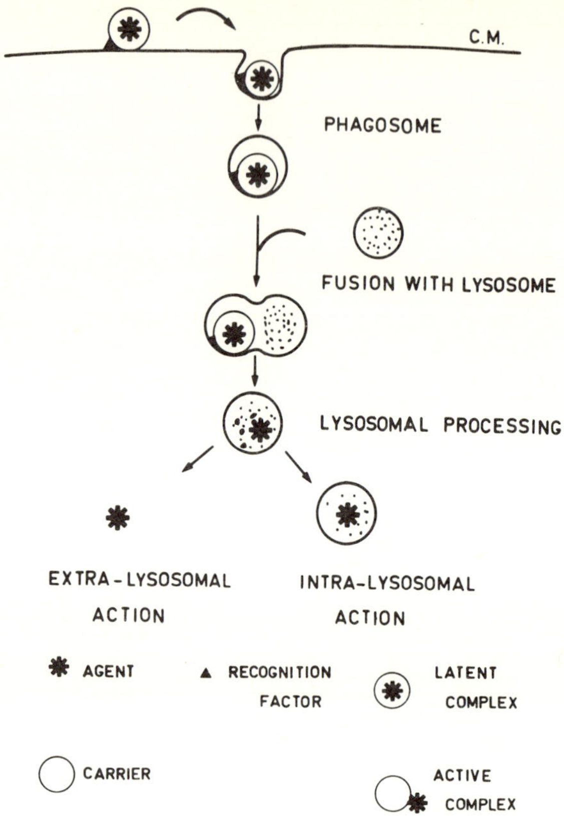

FIGURE 2 Mechanism of entry and mode of action of "carrier-agent" complexes taken up by endocytosis.

cross the plasma membrane will reach the lysosomes; (c) since endocytosis occurs at different rates for different carrier molecules and different cells, the cell selectivity of a drug can be increased in this manner.

In this class we find antibiotics acting on intralysosomal bacteria and parasites[52] as well as lysosomal enzymes themselves which could be used in the treatment of lysosomal storage diseases.[38–40] We find also drugs, the likely site of action of which are the lysosomal enzymes or the lysosomal membranes themselves, like gold salts, chloroquine and so-called "membrane stabilizing agents."

All these compounds should be resistant to the lysosomal hydrolases and their liberation from the carrier must occur in an active form after digestion of the carrier or hydrolysis of the carrier-agent link. This hydrolysis can occur by an enzymatic process or be due to the acid pH prevailing inside the lysosomal compartment. The carrier itself should be endocytozable and degradable by the lysosomal enzymes.

The usefulness of lysosomotropic carrier-agents can be extended to drugs acting on the extralysosomal compartments. After being released from the carrier, the drug can indeed permeate out of the lysosomes and act on the nucleus or other cytoplasmic organelles. The active agent must in this case be able to cross the lysosomal membrane. The advantage of such a carrier complex is that an increased cell selectivity can be obtained by making use of the varying endocytic properties of different cell types or of the cell selecting properties of the carrier.

b) Uptake through the cell membrane

Recent developments in the field of unilamellar lipid vesicles open complete new perspectives for drug design. The indications are that unilamellar vesicles can fuse with the cell membrane and that in this way the content of these vesicles can be introduced directly inside the cytoplasm[54,55] (Figure 3). These carriers provide us with the possibility of introducing inside the cytoplasm (a) macromolecules that in this way escape endocytosis and degradation inside the lysosomes; (b) compounds that cannot diffuse or be transported through the cell membrane.

The field of genetic engineering, and enzyme replacement therapy in mammalian cells lays thus wide open as well as the bypassing of the cell membrane by any drugs towards which these membranes are impermeable.[56] All these benefits can be combined moreover with a cell specific type of interaction mediated by the carrier.

CELL SELECTIVITY MEDIATED BY CARRIERS

As we have seen, one of the most interesting possibilities arising from the use of carrier linked agents is the obtention of agents that are highly selective for one or several cell types. We will discuss these possibilities in more detail with regard to the compounds acting peri- or intracellularly.

When an agent is linked to a carrier, we can distinguish a gradation in three levels of increasing selectivity. At the first level the carrier only prevents the active compound from interacting with one or a few nontarget cell types. At the second level a higher concentration of the compound is achieved at the target while its concentration at the nontarget sites is equal or less than that obtained with the free agents. The third and highest degree of selectivity is the ideal condition in which only the target sites or cells come in contact with the bioactive agents.

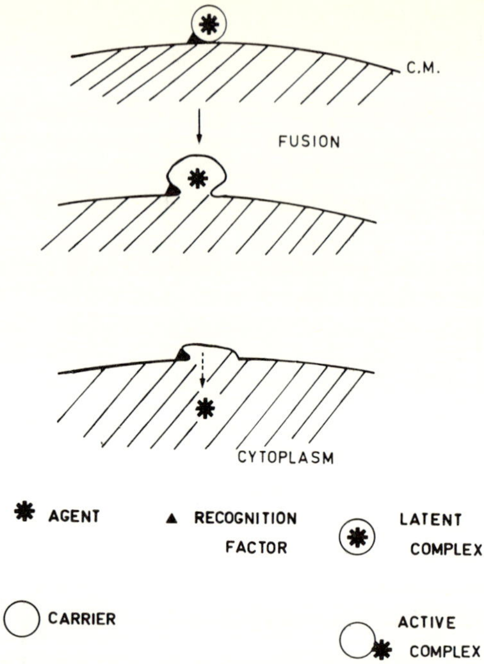

FIGURE 3 Mechanism of entry of "carrier-agent" complexes taken up through the cellular membrane (C.M.).

First level of selectivity

The first and second levels of selectivity can only be obtained with agents acting at the intracellular level. Almost all the compounds of this type penetrate after parenteral or oral administration into almost all the cells of the body by diffusion or active transport through the cell membranes. The linking to an endocytozable carrier or to monolamellar liposomes will modify the mode of penetration of the agents. They will be prevented from entering nonendocytizing cells and/or of coming in contact with cells separated from the site of injection by capillary or anatomical barriers impermeable to the carrier. In order to reach this level of selectivity it is sufficient that the target cells can come in contact with the carrier and endocytize it or fuse with it. Limiting toxic site effects on nontarget cells will be eliminated if these cells are unable to interact with the carrier-agent complexes.

Second level of selectivity

The second level of increased selectivity can only be obtained with endocytozable carriers if the target cells display a higher endocytic activity towards the carrier molecule than the nontarget cells. In these conditions the intracellular concentration of the active compound can reach higher levels than if the agent was given alone.

Third level of selectivity

An optimal degree of selectivity can be reached when a compound is linked to a carrier that can recognize a single cell type and allows a specific interaction of the therapeutic agent with this cell.

The recognition process can only occur at the plasma membrane level and after interaction between this membrane and the carrier one of the following events can happen: (a) the carrier can remain at the cell surface; (b) the carrier can be taken up by endocytosis; (c) the carrier (only for unilamellar vesicles) can fuse with the plasma membrane.

Whatever their recognition properties these carriers should not be absorbed and taken up in a random way by the cells with which they first contact after parenteral administration. These cells are mainly the endothelial and reticuloendothelial cells. The recognition determinants of the carrier will indeed only play a significant role if they are not spoiled by unspecific fixation phenomena. The only exception to this rule is of course the case in which the target cells are precisely the endothelial or reticuloendothelial cells.

As possible recognition molecules one thinks first of all antibodies directed against cell specific antigens. A second possible group is constituted of the various peptidic hormones that, like insulin, vasopressin, thyrotropin, etc. . . , interact with the membranes of some cell types. A third group consists of viral envelopes or surface components whose interaction with cell surfaces constitutes the first step of the viral infectious process. Cells seem to also have specific endocytic receptors; this has been shown conclusively for hepatocytes that endocytize very specifically desialated glycopeptides with terminal galactosyl residues on their side chains. Lectins as well as polypeptidic fragments of toxins (dipheteric, abrin, ricin) responsible for the cellular fixation and penetration of the toxins can also be considered as possible specific carriers.

After recognition of the specific cell surface, the properties of the carrier should, moreover, be such that their subsequent handling by the cell leads the bioactive compound to its cellular site of action. Very little is known today about the factors that determine the fate of a molecule after its fixation to the cell surface membrane. Using antibodies of various specificities and rat embryo fibroblast, Y. J. Schneider has shown in our laboratory that the type of antigens with which an antibody reacts play a crucial role in its uptake by endocytosis. Antibodies directed against plasma membrane specific antigens remain for a much longer time at the cell surface than antibodies directed against lysosomal type antigens. The latter type of antibodies react indeed first with the plasma membrane but are very rapidly endocytized and taken up inside the lysosomes. The cellular fate of IgG molecules (devoid of any known immunological specificity) can also be modulated by the fixation of fluorescein isothiocyanate. While normal IgG is endocytized very slowly and to a very small extent by cells in culture, IgG molecules labelled with fluorescein are taken up very rapidly and are concentrated specifically inside the lysosomal system.

A similar problem arises with regard to the unilamellar vesicles the nature and physico-chemical characteristics of which determine the cellular uptake by endocytosis or the fusion with the plasma membrane.[55]

The requirements for the ideal specific carrier are so multiple and complex that a great amount of further basic research remains to be performed. It will most probably be very difficult to find a molecule possessing all the needed properties. We think that it will be necessary to make use of a transport molecule to which the agents can be bound in a proper way and that possesses the characteristics required for the planned cellular handling. A recognition element will then be finally linked to it to make it cell specific. Attempts in this direction have already been made using liposomes associated with immunoglobulin,[57] viral components[58] or sialoglycoproteins.[59]

EXAMPLES OF CARRIER-DRUG COMPLEXES WITH INCREASED SELECTIVITY

To illustrate the changes in specificity that can be obtained by linking agents to carriers, we will briefly report our results obtained in the field of cancer and trypanosomiasis chemotherapy.

Cancer chemotherapy

Daunorubicin (DNR) and Adriamycin (AM) are very powerful but also very toxic antitumoral drugs. Their toxicity results from their deleterious effects not only on normal dividing cells but also on the heart muscle. Given by intravenous route they tend to diffuse and to distribute themselves more or less uniformly into most cells of the body. In collaboration with D. Deprez-De Campeneere, we have made use of the DNA binding properties of DNR and AM to prepare DNR–DNA and AM–DNA complexes (DNA is extracted from calf thymus or herring sperm DNA and has a molecular weight above 1,000,000). When bound to DNA, DNR and AM can only penetrate inside cells by endocytosis. After reaching the lysosomes the DNA is digested by the lysosomal deoxyribonucleases while DNR and AM, resistant to the lysosomal enzymes, become free to diffuse out of the lysosomal granules and reach the other cellular compartments.[20,21,60]

Animal studies[61] and clinical trials performed on more than 300 patients indicate that DNR–DNA and AM–DNA have a much lower cardiac toxicity than the free drugs. This decreased cardiac toxicity is very probably due to the fact that AM and DNR, when linked to DNA, accumulate to a much lesser extent in the heart muscle cells which are separated from the bloodstream by a complete capillary barrier and probably possess low endocytic properties. Animal experiments have also shown that the DNA complexes of AM are less toxic for the hematopoietic stem cells and for the gastric epithelium.

The chemotherapic efficiency of the complex drugs is, moreover, significantly higher than that of the free compounds suggesting that the uptake of AM as a DNA complex by endocytosis is more efficient than the uptake by permeation of the free drug (Table I). These results indicate that in clinical tests the selectivity of the drugs has been increased by one level at least and that in L1210 leukemic mice this increase is of two levels.

TABLE I

Chemotherapeutic activity of adriamycin and adriamycin–DNA

L1210 leukemic cells Number	Drug Form	Drug Doses mg/kg/day	Median Survival time	% increase in life span	Long term survivors (30 days)	Total number of mice
10^4	AM	2.5	13	62	0	15
10^4	AM–DNA	2.5	15	87	0	15
10^4	AM	4	15.5	94	0	13
10^4	AM–DNA	4	18.5	131	2	14
10^4	AM	5	14.5	81	0	14
10^4	AM–DNA	5	23.5	194	4	15
10^4	AM	6	8	0	0	13
10^4	AM–DNA	6	60	650	8	12
10^4	—	—	8	—	—	29
10^5	AM	5	16	167	0	10
10^5	AM–DNA	5	18.5	208	0	8
10^5	AM	6	16	167	0	11
10^5	AM–DNA	6	30.5	408	5	10
10^5	—	—	6	—	—	10

Drugs (AM: adriamycin; AM–DNA: adriamycin–DNA) were adminstered intravenously on 5 consecutive days starting on day 1. DBA/2 mice were inoculated on day 0 with leukemic cells harvested from a 6 to 8 days old ascitic form of L1210 leukemia.

Trypanosomiasis chemotherapy

In collaboration with J. M. Jadin and F. Van Hoof we have shown that complexes formed between ethidium bromide (EB) and DNA have a very much increased chemotherapic efficiency when administered intraperitoneally to mice infected with *Trypanosoma Cruzi* (Figure 4). This can be due to a higher uptake of the complex drug by the endocytizing parasites during their stay in the bloodstream. More probably, however, the complex drugs are concentrated specifically inside the lysosomes of the infected host cells which precisely shelter the parasites during their prevalent intracellular stage.

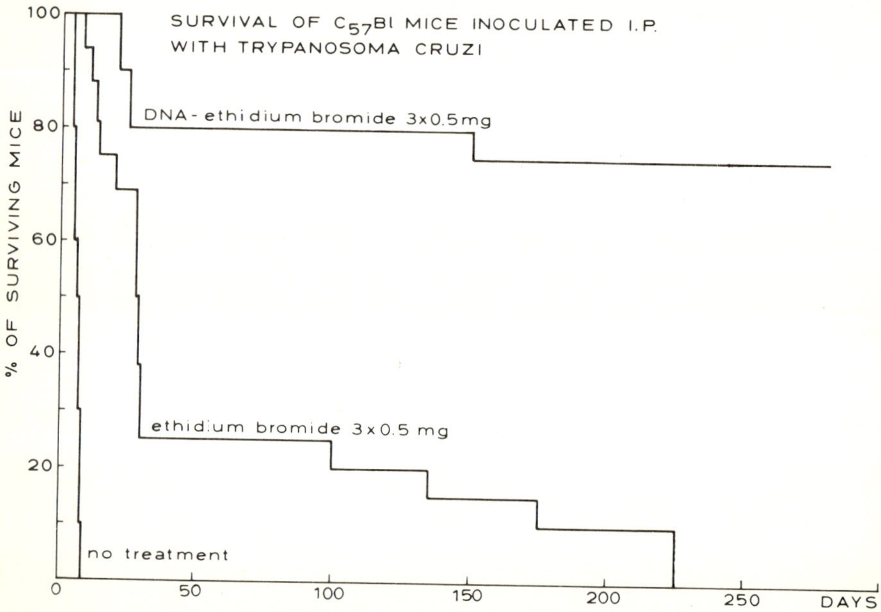

FIGURE 4 Percent of surviving mice after intraperitoneal inoculation of about 100 *Trypanosoma Cruzi*. The treated animals received 3 injections of EB or EB–DNA on days 2, 3 and 4

CONCLUSION

A great amount of basic research is still required before the "carrier-bioactive agent" concept will be put to practical use with the greatest benefit. Various molecules will have to be purified from biological materials or synthetized and their properties carefully analyzed in view of their possible use as carriers. We can enumerate as follows the possible molecules or molecular complexes to be studied:

1) Plasma proteins—albumin
—immunoglobulins
—fibrinogen, etc.
2) Nucleic acids (ribo- and deoxy-).
3) Synthetic polynucleotides (ribo- and deoxy-).
4) Synthetic polypeptides of various molecular weight and aminoacid composition covering a large spectrum of physicochemical characteristics.
5) Polysaccharides.
6) Liposomes.
7) Unilamellar lipid vesicles.
8) Biodegradable synthetic microcapsules.
9) Cell specific antibodies.
10) Lectins.
11) Peptidic fragments of diphtheria toxin, ricin and abrin.
12) Peptidic hormones.

These compounds should be studied alone or in combination in order to analyze their suitability as drug carriers. This will of course vary according to the pathological state one wants to cure but the following properties should almost certainly be looked for:

1) Plasmatic half life and mode of excretion,
2) Uptake by the reticulo-endothelial system,
3) Passage through various capillary barrier membranes,
4) Tissular distribution and cell specificities,
5) Cellular handling with special emphasis on
   a: Plasma membrane fixation.
   b: Endocytosis and lysosomal processing.
   c: Fusion with plasma membrane (for the unilamellar vesicles),
6) Catabolism (plasmatic as well as intracellular),
7) Immunogenicity, and
8) Acute and chronic toxicity.

Although not new, the idea of vectoring bioactive material selectively to their targets by the use of carriers appears to become more and more liable to practical applications. The recent developments in cellular and molecular biology as well as the latest expertise in biological and organic chemistry will without any doubt play an important role in bringing the concept to its ultimate goal, namely, the design of cell selective drugs and agents so badly needed in the treatment of cancer as well as of degenerative and parasitic diseases.

References

1. W. F. Bale and I. L. Spar, *Adv. Biol. Med. Phys.*, **5**, 285 (1957).
2. G. Mathe, Tran Ba Loc, and J. Bernard, *Compt. Rend. Acad. Sci. Paris*, **246**, 1626 (1958).
3. J. D. Broome and J. G. Kidd, *J. Exp. Med.*, **120**, 467 (1964).
4. T. Ghose, M. Cerini, M. Carter, and R. C. Nairn, *Brit. Med. J.*, **1**, 90 (1967).
5. T. Ghose, M. R. C. Path, and S. P. Nigam, *Cancer*, **29**, 1398 (1972).
6. T. Ghose, S. T. Norvell, A. Guclu, D. Cameron, A. Bodurtha, and A. S. MacDonald, *Brit. Med. J.*, **3**, 495 (1972).
7. F. L. Moolten, N. J. Capparell, and S. R. Cooperband, *J. Natl. Cancer Inst.*, **49**, 1057 (1972).
8. D. A. L. Davies and G. J. O'Neill, *Brit. J. Cancer*, **28**, suppl. 285 (1973).
9. I. Flechner, *Eur. J. Cancer*, **9**, 741 (1973).
10. R. Levy, E. Hurwitz, R. Maron, R. Arnon, and M. Sela, *Cancer Res.*, **35**, 1182 (1975).
11. G. F. Rowland, G. J. O'Neill, and D. A. Davies, *Nature*, **255**, 487 (1975).
12. A. Guclu, T. Ghose, J. Tai, and M. Mammen, *Eur. J. Cancer*, **12**, 95 (1976).
13. R. Wade, M. E. Whisson, and M. Szekerke, *Nature*, **215**, 1303 (1967).
14. B. Larsen, *Eur. J. Cancer*, **2**, 63 (1967).
15. G. Magnenat, R. Schindler, and H. Isliker, *Eur. J. Cancer*, **5**, 33 (1969).
16. M. Szekerke, R. Wade, and M. E. Whisson, *Neoplasma*, **19**, 199 (1972).
17. M. Szekerke, R. Wade, and M. E. Whisson, *Neoplasma*, **19**, 211 (1972).
18. M. G. Nair and C. M. Baugh, *Biochem.*, **12**, 3923 (1973).
19. P. A. Kramer, *J. Pharm. Sci.*, **63**, 1646 (1974).
20. A. Trouet, D. Deprez-De Campeneere, and C. de Duve, *Nature New Biol.*, **239**, 110 (1972).
21. A. Trouet, D. Deprez-De Campeneere, M. De Smedt-Malengreaux, and G. Atassi, *Eur. J. Cancer*, **10**, 405 (1974).
22. J. Bartulin, M. Przybylski, H. Ringsdorf, and H. Ritter, *Makromol. Chem.*, **175**, 1007 (1974).
23. H. G. Batz, H. Ringsdorf and H. Ritter, *Makromol. Chem.*, **175**, 2229 (1974).
24. Y. Fischman and N. Citri, *FEBS Letters*, **60**, 17 (1975).
25. G. Gregoriadis, C. P. Swain, E. J. Wills, and A. S. Tavill, *The Lancet*, **1**, 1313 (1974).
26. G. Gregoriadis, *FEBS Letters*, **36**, 292 (1973).
27. G. Gregoriadis and E. D. Neerunjun, *Chem. Pathol. Pharmacol. Res. Comm.*, **10**, 351 (1975).
28. T. Kobayashi, S. Tsukagoshi, and Y. Sakurai, *Gann*, **66**, 719 (1975).
29. Y. E. Rahman, W. E. Kisielski, E. M. Buess, and E. A. Cerny, *Eur. J. Cancer*, **11**, 883 (1975).
30. M. von Ardenne, W. Kruger, O. Prokop, and S. Schnitzler, *Deut. Gesundheitsw.*, **24**, 588 (1969).
31. M. von Ardenne, *Z. Naturforsch.*, **25b**, 897 (1970).
32. H. T. Nagasawa, J. E. Elberling, and F. N. Shirota, *J. Med. Chem.*, **18**, 826 (1975).
33. H. Ringsdorf, H. Ritter, and H. Rolly, *Makromol. Chem.*, **177**, 741 (1976).
34. M. S. Verlander, J. C. Venter, M. Goodman, N. O. Kaplan, and B. Saks, *Proc. Natl. Acad. Sci. U.S.A.*, **73**, 1009 (1976).
35. G. W. Pasternak, S. C. March, I. Parikh, S. H. Snyder, and P. Cuatrecasas, *Life Sci.*, **18**, 977 (1976).
36. H. Jatzkewitz, *Z. Naturforsch.*, **10b**, 27 (1955).
37. F. L. Moolten and S. R. Cooperband, *Science, N.Y.*, **169**, 68 (1970).
38. G. Gregoriadis, P. D. Leathwood, and B. E. Ryman, *FEBS Letters*, **14**, 95 (1971).

39. G. Gregoriadis and B. E. Ryman, *Eur. J. Biochem.*, **24**, 485 (1972).
40. G. Eytan, M. J. Matheson, and E. Racker, *FEBS Letters*, **57**, 121 (1975).
41. T. de Barsy, P. Devos, and F. Van Hoof, *Lab. Invest.*, **34**, 273 (1976).
42. R. L. Baehner, D. G. Nathan, and M. L. Karnovsky, *J. Clin. Invest.*, **49**, 865 (1970).
43. G. Barbanti-Brodano and L. Fiume, *Nature New Biol.*, **244**, 281 (1973).
44. Y. E. Rahman, M. W. Rosenthal, and E. A. Cerny, *Science, N.Y.*, **180**, 300 (1973).
45. L. G. Donaruma, *Progr. Polymer Sci.*, **4**, 1 (1975).
46. H. Ringsdorf, *J. Polymer Sci.*, Symposium ed. **51**, 135 (1975).
47. J. M. Lachapelle, *Dermatologica*, **132**, 476 (1966).
48. F. Basset, L. Bocquet, J. Chabot, J. P. Paolaggi, and F. Delbarre, *Ann. Anat. Path.*, **16**, 159 (1971).
49. C. E. Hall, O. Hall, and S. Ayachi, *Lab. Invest.*, **24**, 67 (1971).
50. T. M. S. Chang, *Science, N.Y.*, **146**, 524 (1964).
51. C. de Duve, T. de Barsy, B. Poole, A. Trouet, P. Tulkens, and F. Van Hoof, *Biochem. Pharmacol.*, **23**, 2495 (1974).
52. C. de Duve and A. Trouet, in "*Non specific*" *factors influencing host resistance*, W. Braun and J. Ungar, eds., A. G. Karger, Basel, p. 336 (1973).
53. G. Gregoriadis and A. C. Allison, *FEBS Letters*, **45**, 71 (1974).
54. S. Batzri and E. D. Korn, *J. Cell Biol.*, **66**, 621 (1975).
55. G. Poste and D. Papahadjopoulos, *Proc. Natl. Acad. Sci. U.S.A.*, **73**, 1603 (1976).
56. G. Poste and D. Papahadjopoulos, *Nature*, **261**, 699 (1976).
57. G. Weissmann, D. Bloomgarden, R. Kaplan, C. Cohen, S. Hoffstein, T. Collins, A. Gotlieb, and D. Nagle, *Proc. Natl. Acad. Sci. U.S.A.*, **72**, 88 (1975).
58. A. M. Haywood, *J. Mol. Biol.*, **87**, 625 (1974).
59. R. I. MacDonald and R. C. MacDonald, *J. Biol. Chem.*, **250**, 9206 (1975).
60. G. Cornu, J. L. Michaux, G. Sokal, and A. Trouet, *Eur. J. Cancer*, **10**, 695 (1974).
61. A. Langslet, I. Øye, and S. O. Lie, *Acta Pharmacol. Toxicol.*, **35**, 379 (1974).

DISCUSSION

Dr. D. Bennett (*Midland Hospital, Midland, MI*): I was wondering if ethidium was altered with regard to toxicity when it's combined as the DNA complex?

Dr. A. Trouet: The toxicity of the ethidium bromide was decreased.

Dr. D. Bennett: Can you comment on the effect of these complexes on the immune system? Is there depression or enhancement of their action on the immune system?

Dr. A. Trouet: Daunorubicin and adriamycin, as you know, are very immunosuppressant, and we haven't seen any increase or decrease in their toxicity.

Dr. W. Walles (*Dow Chemical Company, Midland, MI*): Have you any experience with the complex with poly(vinylpyrolidone)?

Dr. A. Trouet: I have none. I would not like to use this polymer because it is nondegradable and toxic. We use DNA only because the drugs link spontaneously to DNA. DNA might not be the best carrier to use. We are now studying the linkage of daunorubicin and adriamycin to proteins by a covalent linkage.

Dr. R. Kostelnik (*Midland Macromolecular Institute, Midland, MI*): What is the nature of the linkage of these drugs to the DNA?

Dr. A. Trouet: Intercalation.

Dr. K. Doebel (*Ciba-Geigy Corporation, Summit, NJ*): In the beginning you mentioned the desirability of the carrier having a high specificity to the target site. That brings us to the still unknown problem of chemotaxis. I agree with you that if one knew how to regulate chemotaxis, then one could completely eliminate the random walk and bring the compound precisely to the target organ or receptor site. You would then probably be able to mimic the normal processes by which the organism fights a pathological condition. So if I interpret you correctly, this is truly the direction of your thinking. Unfortunately, we know very little about this.

Dr. A. Trouet: Yes, but as you say, we know very little about this.

Dr. K. Doebel: Another problem I have involves the so called intralysosomal drug activity. You mentioned hydrochloroquine and other drugs which are "anti-malarial-types." It has been demonstrated *in vitro* that these drugs have a high membrane stabilization effect for lysosomes. It is undesirable to have the lysosomal membrane rupture and have a large amount of lysosomal enzymes released because of the secondary reactions which are claimed to be part of the chronic cycle of rheumatoid arthritis. Thus, I have some difficulty with your concept of getting an active drug into an unruptured lysosomal membrane. Do I take it that what you want the drug to do is to act only inside of the lysosome?

Dr. A. Trouet: Yes. You see, for example, that the toxicity of chloroquine is very high and does not allow its use as an anti-rheumatoidal. If one could link chloroquine to a carrier so that it would not act on the retinal cells any more but still on the target, one would get the desired effect. With regard to chemotaxis, I would stress the point that if one wants to get a selective carrier, one has to look for carriers which are not taken up as most of them are now by the R.E. system. For example, injected liposomes go to the liver and they go to the spleen. What one needs are carriers which, first of all, remain in the blood stream. Then one can attach agents to the carriers to give them specificity. You cannot get specificity if the carriers get lost in the R.E. system. That's the actual situation even with DNA as the carrier.

Dr. T. Roseman (*Upjohn Company, Kalamazoo, MI*): Do you view the problem of not being able to give the drug orally as a major obstacle for acceptance?

Dr. A. Trouet: Yes, it is a problem. For example, if one thinks of parasitic diseases and treatment of patients in Africa or South America, it's difficult to envision large scale administration of drugs by intravenous injections. This is a problem, and it would be better to give the drugs by an oral route.

Dr. H. Ringsdorf (*Univ. of Mainz, Mainz, Germany*): I would like to come back to the difference in uptake by leukemic tumor and liver cells. You mentioned that toxicity in kidney and liver cells is very small although at least the liver cells have a high rate of endocytotic uptake. So then, why is there no effect on the liver cells?

Dr. A. Trout: I don't know why that is in regard to DNA. It's not enough to say that a cell has a specific uptake. One has to say with regard to what?

Dr. H. Hopfenberg (*North Carolina State Univ., Raleigh, NC*): Have you made any studies with other polymers especially with regard to their endocytotic uptake?

Dr. A. Trouet: Yes. We started to study to uptake of carriers *in vitro*. We are interested for example in the fate of antibodies. Antibodies directed against plasma membranes stay apparently on the plasma membranes of cells *in vitro* for several days. These antibodies seem not to get into the cell and should not be good lysosomotropic carriers.

Dr. H. Ringsdorf: In relation to the last question, I think the only intensive study of the uptake of synthetic polymers by endocytosis was by Ryser and his group who together with André Trouet would like to go ahead and vary the polymeric compound, i.e., change from basic to acidic units, and then to study the absorption and uptake. The question of the difference between absorption and uptake, in the case of the antibody, is more of an interaction between proteins in the membrane and of the antibody. The endocytotic rate may depend upon surface coverage and surface effects alone. But this is a very open and interesting question. I think it has to be decided independently of drug uptake or drug absorption.

Sustained Release of Macromolecules from Polymers

ROBERT LANGER and JUDAH FOLKMAN

Children's Hospital Medical Center and
Harvard Medical School, Boston, MA02115, U.S.A.

Proteins and other macromolecules can be incorporated into polymers and delivered for prolonged periods both *in vivo* and *in vitro*. The polymers studied were Hydron, poly(vinylalcohol), and ethylene–vinyl acetate copolymer. Macromolecules with molecular weights between 10^3 to 10^7 Daltons, including soybean trypsin inhibitor, lysozyme, alkaline phosphatase, catalase, insulin, heparin and DNA were examined. Poly(vinylalcohol) polymers were mildly inflammatory when implanted in corneas. Neither Hydron nor ethylene–vinyl acetate copolymer produced inflammation. A single bead of ethylene–vinyl acetate copolymer (2 mm dia.) released nearly 1 μg per day of biochemically active macromolecule for periods exceeding 100 days. Ethylene–vinyl acetate copolymer has been used to continuously deliver insulin and other biologically active molecules in animals.

INTRODUCTION

The use of implantable polymers for sustained release of drugs and other pharmaceutical agents is a young field. Although the prolonged release of oral medications has been widely used by the pharmaceutical industry since 1950, slow release depended primarily on salts of low solubility, or enteric formulations, or complexation. The sustained release of drugs from implanted polymers was first reported in 1964 by Folkman and Long.[1] Certain drugs such as tri-iodothyroine, isoproterenol, and digitoxin were found to diffuse through silicone rubber tubing. The drug in dry powder form was inserted into the lumen of a short length of silicone rubber tubing, and each end sealed with silicone cement (Figure 1). Implants of such tubing released drugs for several weeks.

The idea came from two unrelated experiments carried out at the same time in the laboratory of the Naval Medical Research Center in Bethesda. Folkman was attempting to treat experimental heart block in animals by implanting pellets of thyroid hormone and sought a sustained release vehicle

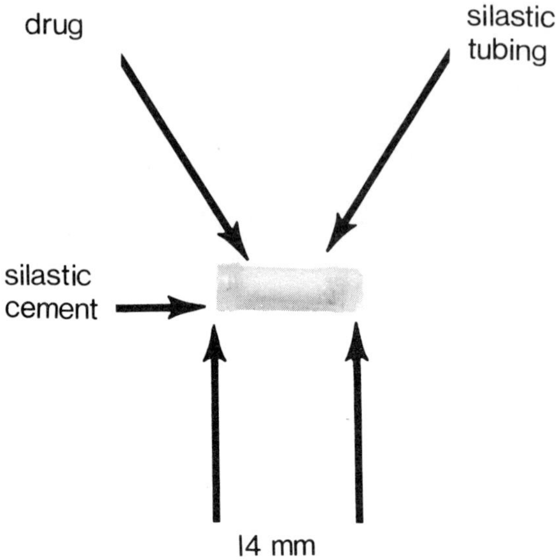

FIGURE 1 Silicone rubber tubing, 3.5 mm I.D., 0.75 mm wall thickness, sealed at each end with silastic cement. The tube contains 30 mg of tri-iodothyroine which diffuses slowly through the wall.

which was non-inflammatory. Long was studying the turbulence induced in water flowing around a heart valve made of silicone rubber. For photographic purposes, he attempted to stain the silicone rubber. It was found that certain dyes such as rhodamine or Sudan III, i.e. oil-soluble dyes, would diffuse into silicone rubber, whereas most water-soluble dyes such as chlorazol black or methylene blue, would not diffuse into the polymer (Figure 2).

It was then found that powders of rhodamine or Sudan when placed in silicone rubber tubing would diffuse through the wall for months. The next step was to try a variety of drugs. In general, it turned out that drugs of low molecular weight (under 1000) and those which were generally non-polar, diffused through silicone rubber, whereas polar compounds and large molecular weight compounds would not. In the following year, Bass *et al.* showed that atropine and histamine would diffuse through silicone rubber over a prolonged period.[2] The same year, Powers showed that certain anti-malarial drugs would also diffuse through silicone rubber.[3] Then, in 1966, Dzuik and Cook reported that the steroid hormones could be diffused through silicone rubber, and especially that small tubes of progesterone and silicone rubber implanted in cattle would release the drug for more than one year and could be used to control fertility.[4]

FIGURE 2 Three days after powdered dyes were sealed in silicone rubber tubing, Rhodamine B diffused through the wall and stained the entire tubing. By contrast, Chlorazol Black did not diffuse through the silicone rubber even after 100 days. Note the clarity of the tubing wall in the presence of Chlorazol.

In 1968, Folkman *et al.* showed that all of the general anesthetics would diffuse through silicone and could produce prolonged analgesia.[5,6]

Since 1957, other investigators have studied the permeability of silicone rubber for a variety of compounds such as caffeine barbital, etc., most of those being under 1000 molecular weight.

Some of these polymer vehicles are already in clinical use. For example, ALZA's Ocusert®, which is a membrane control device that resides in the cul-de-sac of the eye, delivers pilocarpine at a precisely controlled rate for the treatment of glaucoma. Many other types of drug polymer combinations are being tested for prolonged release in other diseases. Baker and Lonsdale[7] have reviewed the current status of this new field.

However, all of this technology has largely been limited to relatively small molecules, which could diffuse through the polymers. The sustained release of polypeptides or proteins from implanted polymers has not previously been demonstrated, except for the work of Davis,[8,9] in which polyacrylamide and poly(vinylpyrrolidone) were used as the vehicle. However, these polymers only permit brief periods of sustained release and they cause damage to some tissues.

In this report, we describe a new principle by which large molecules possessing molecular weights up to 2 million Daltons, can be released steadily from implanted polymers over periods of more than three months.

This finding grew out of work in our laboratory on tumor angiogenesis. We have previously shown[10–13] that most solid tumors secrete a diffusible material (tumor angiogenesis factor, TAF), which induces growth of new vessels from the host. These new capillaries grow toward the tumor and penetrate it, thus permitting further growth of the tumor. This factor is being purified from a pool of proteins which are released from tumor cells in culture. The rabbit cornea is used as one of the bioassays for TAF activity. The test compound is placed in a pocket in the cornea, and new vessels grow into the cornea. The rate of vessel growth correlates with the biologic activity of the fraction which is being purified. Because the more highly purified fractions of TAF are very soluble, they disappear quickly after implantation. Therefore, a sustained release vehicle was required that could provide steady diffusion of TAF into the cornea from a depot source, analogous to the way that TAF is continuously secreted from tumor cells. A further requirement was that such a polymer be absolutely inert and non-inflammatory because of the great sensitivity of the cornea to inflammation from foreign bodies implanted within its stroma.

In our early work,[14] we had tried to use poly(acrylamide) pellets for this purpose. It was necessary to mix the test protein with the poly(acrylamide) before polymerization. After polymerization, however, the small pellets were often highly inflammatory. This inflammation could be reduced by extensive washing, but it could never be eliminated completely. Furthermore, extensive washing leached out most of the test protein. Therefore, we turned to other polymers such as ethylene–vinyl acetate copolymer (Elvax 40®), obtained from ALZA Corporation. When prepared as described below, this polymer turned out to be very useful.

As a further extension of the work on tumor angiogenesis, an inhibitor of capillary proliferation was found in cartilage.[15] It was isolated and purified.[16] Purification of this compound also required the use of the cornea as a bioassay. In this system, tumor implants in the cornea were used to attract new vessels, and a polymer implant containing the test fraction of cartilage inhibitor was placed between the tumor and the vessel bed. The degree of inhibition of new vessel growth was a measure of activity of this inhibitor. The ethylene–vinyl acetate copolymer also served as an excellent slow-release vehicle for the cartilage inhibitor. Without it, the isolation and purification could probably not have been achieved. In the following sections we review methods for constructing the slow release pellets and present studies demonstrating *in vitro* and *in vivo* release of macromolecules.

POLYMERS, BIOCOMPATIBILITY, AND METHODS OF CONSTRUCTING SLOW RELEASE PELLETS

Polymers

Poly(acrylamide) and poly(vinylpyrrolidone), 20% (w/v), were made as described by Davis[9] in sterile glass petri dishes. Poly(vinylalcohol) (Elvanol®, lot 71-30), Hydron-S® (a polymer of hydroxyethyl–methacrylate, lot HL 55-34), and ethylene–vinyl acetate copolymer (40% by weight vinyl acetate–Elvax 40®) in powder or pellet form were gifts of the ALZA Corporation, Palo Alto, California. Poly(vinylalcohol) and ethylene–vinyl acetate copolymer were washed in 2.5% (w/v) reagent quality absolute pure ethyl alcohol USP at 37°C in sterile roller bottles. One hundred changes of alcohol were made with a minimum of 3 h per wash. After the final wash, the alcohol was decanted and the wet polymer was poured into a sterile glass petri dish. The remaining alcohol was removed by vacuum drying.

The alcohol wash removed various impurities which are inflammatory in normal tissues, e.g., antioxidants. This process also sterilized the polymers. The inflammatory impurities absorbed light in the ultraviolet region. The alcohol was analyzed after washing, spectrophotometrically at 230 nm, to monitor the effectiveness of the washing procedure. The absorbance of the alcohol at the end of the first wash was between 1 and 2 units. This absorbance decreased with each successive wash. After the final wash the absorbance at 230 nm was less than 0.005.

Biocompatibility

The cornea was used to study the host response to the polymers. The clarity and avascularity of the cornea permits stereomicroscopic observation of inflammatory characteristics: edema, white cell infiltration, and neovascularization.[17,18] Inflammation was graded as non-existent, mild, or significant. The cornea is the most sensitive of all indicators of inflammation.

Sterile polymer pellets ($1.5 \times 1.5 \times 0.5$ mm^3) were implanted within the corneal lamellae of 2–3 month old New Zealand white male rabbits by aseptically creating intracorneal pouches.[14] Using a Wheeler cyclodialysis spatula and fine forceps, the pellets were placed at the lower edge of the pouch within 1–2 mm of the vessels of the corneal, scleral junction.

Corneal implants were examined four times weekly over 2-month periods with a slit-lamp stereomicroscope at 6-40× magnification. The corneas were fixed in 10% buffered formalin and stained with hematoxylin and eosin for histological examination.

Only Hydron-S and alcohol washed ethylene–vinyl acetate copolymer caused no inflammation in the cornea (Table I). In 35 experiments, no inflammatory characteristics were detected stereomicroscopically (Figure 3a). Histologic sections taken 3 to 4 weeks after implantation showed no foreign cells in the region surrounding the polymer or other areas of the cornea.

By contrast, polyacrylamide and polyvinylpyrrolidone, particularly the latter (Figure 3b), were significantly inflammatory (Table I).

TABLE I

Inflammation in the rabbit cornea due to polymer implants[a]

Polymer	Number of tests	Inflammation: None %	Mild %	Significant %
Poly(acrylamide)	10	20	30	50
Poly(vinylpyrrolidone)	4	0	0	100
Poly(vinylalcohol)	20	15	70	15
Ethylene–vinyl acetate copolymer	20	40	60	0
Hydron-S	15	100	0	0
After alcohol wash:				
Ethylene–vinyl acetate copolymer	20	100	0	0
Poly(vinylalcohol)	6	67	33	0

[a] The criteria for inflammation were the presence of edema, white cell infiltration, and neovascularization. Inflammation was observed by stereomicroscopy and evaluated as mild if any one of the three characteristics was detected and significant if corneal opacity resulted.

Figures 3a and 3b appear as the frontispiece, facing the title page.

Unwashed poly(vinylalcohol) pellets made of several different initial polymer to solvent ratios also caused inflammation. Washing the polymer exhaustively with absolute alcohol greatly reduced, but did not eliminate, the inflammatory response (Table I).

Because the cornea is so extremely sensitive to inflammation, substances which cause this phenomenon in the cornea may not do so in other systems. For example, we have found that unwashed poly(vinylalcohol), which was inflammatory in the cornea, was completely non-inflammatory, as judged by stereomicroscopy, when implanted on the chorioallantoic membrane (CAM) of 10 or 11 day old chick embryos.

Construction of slow release pellets

Polymer casting solutions were made by dissolving polymers in appropriate solvents: Hydron in absolute alcohol at 37°C; ethylene–vinyl acetate copolymer in methylene chloride at 37°C; and poly(vinylalcohol) by autoclaving in distilled deionized water.

The slow release pellets were made by mixing 100 to 500 μl of casting solution with the macromolecules to be released, and placing a small amount of the resultant mixture in an appropriate mold. The molds were then dried under vacuum overnight causing the solvent to evaporate with the macromolecule trapped within the polymer matrix. The dried polymer was then rehydrated, by adding a drop of lactated Ringer's solution, and gently removed from the mold. The entire procedure was performed simply, quickly, sterilely, and at room temperature.

The apparatus used to dry the polymers consisted of a laboratory desiccator attached to wall suction (600 mm Hg) (Figure 4). The inlet of the desiccator contained a 25 mm dia., 0.45 μm Millipore filter. This allowed a constant exchange of sterile air at slightly less than atmospheric pressure.

The molds used for Hydron and poly(vinylalcohol) preparations were plastic Costar Histo-Plates containing conical shaped wells 4 mm in diameter and 2.5 mm deep. The ethylene–vinyl acetate copolymer solution reacted with plastic, therefore glass molds were made using 4 mm thick plate glass. A No. 761, 1/8 inch diameter carbide-tipped glass drill was used to bore holes 1.5 mm deep, resulting in conical shaped molds 2 mm in diameter (Figure 5).

In order to retard the diffusion of the macromolecules from the polymer matrix, "sandwiches" were sometimes made with a coat of pure polymer. Pure polymer solution was placed in the mold, allowed to dry under vacuum, and the solvent to evaporate overnight. The polymer-macromolecule solution was placed over this layer and dried under vacuum. Finally, a top layer of pure polymer was added and vacuum dried, resulting in a three-layer

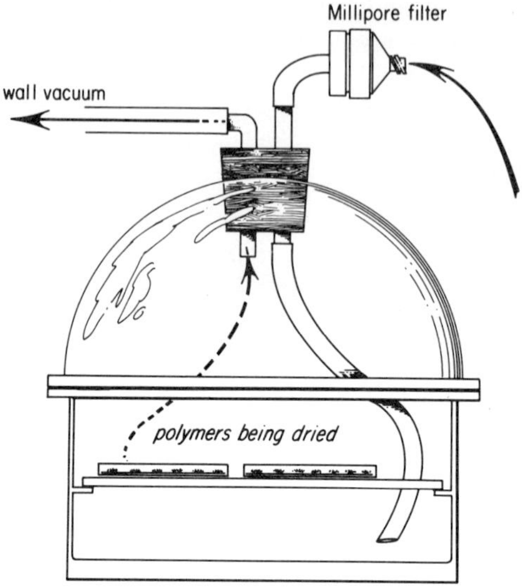

FIGURE 4 Drying unit for polymer pellets.

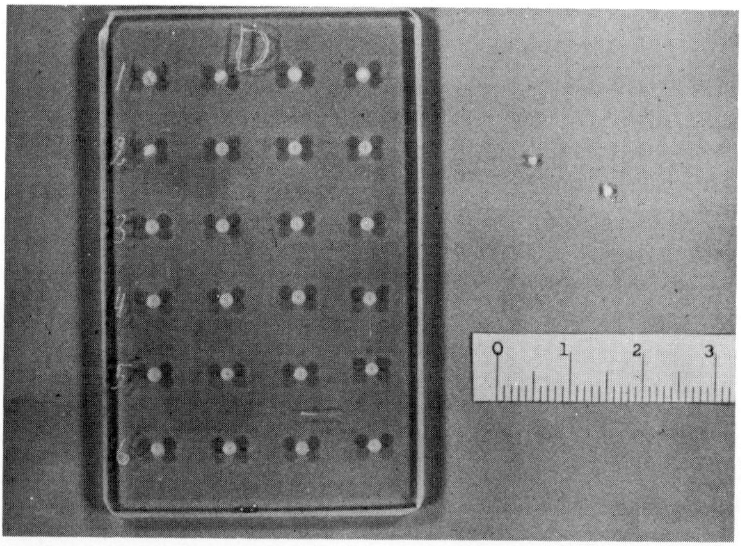

FIGURE 5 Glass mold and ethylene–vinyl acetate copolymer pellets. The scale is in mm.

polymer with the macromolecules in the middle layer. Alternatively, polymer "sandwiches" were made by removing the dry polymer containing macromolecules from its mold and suspending it in a puddle of pure polymer solution for approximately 20 s. Pure polymer adhered to the polymer-macromolecule surface, coating it and forming a "sandwich". This "sandwich" was then vacuum dried.

In the third method, macromolecules were dispersed in approximately 200 μl of polymer solution, placed in a glass petri dish, dried under vacuum, and cut into desired sizes. To make polymer "sandwiches", the cut pellets were dipped into a puddle of polymer solution as described previously.

RELEASE OF MACROMOLECULES IN VITRO

Kinetics

To examine kinetics, we incorporated different proteins into polymers and tested for the amount of protein released as a function of time.[18] This was done by incubating pellets in aqueous medium and periodically analyzing for protein released into medium using the Lowry assay.[19] All experiments were performed in duplicate or triplicate. Average error for percent released was $\pm 5\%$.

Poly(vinylalcohol), Hydron, and ethylene–vinyl acetate copolymer were examined for their ability to release soybean trypsin inhibitor (mol. wt. 21,000). The polymer to water ratio in the casting solution of poly(vinylalcohol) was important in determining release rates. The effects of various poly(vinylalcohol) concentrations in the casting solution on the rate of release of soybean trypsin inhibitor are summarized in Figure 6. Increasing the polymer concentration in the casting solution significantly decreased the initial rate at which protein was released. A threshold level was apparently reached at about 10% polyvinylalcohol, since further increases in polymer concentration did little to retard diffusion out of the pellet. Polymer concentrations above 20% results in solutions so viscous that they were technically difficult to work with.

The results in Figure 6 illustrate two additional features of these slow release systems. First, a "burst" effect was observed in which a large percentage of protein was released during the first hour or so of incubation. Second, by constructing a polymer "sandwich", the rate of diffusion was significantly retarded.

Water was also added to the Hydron–Alcohol–Protein casting solution. The greater the water to polymer ratio, the greater was the rate of protein release from the pellet.

The release of soybean trypsin inhibitor from Hydron, poly(vinyalcohol), and ethylene–vinyl acetate copolymer "sandwiches" is shown in Figure 7. Casting solutions were 12% Hydron, 10% poly(vinylalcohol), and 10% ethylene–vinyl acetate copolymer. These concentrations minimized initial

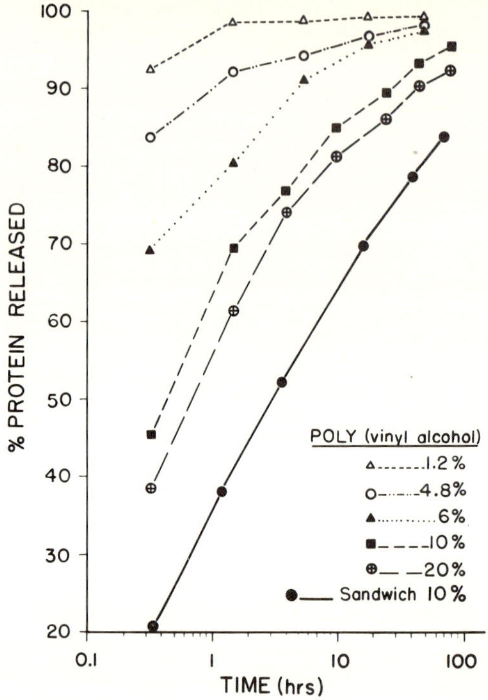

FIGURE 6 Release of soybean trypsin inhibitor from poly(vinylalcohol). The concentration of protein in the casting solution was 12 mg/cm³.

rates of release from each of the individual polymer systems and still permitted viscosities low enough so that the solutions were easy to work with.

Protein diffused out of Hydron relatively rapidly, somewhat more slowly from poly(vinylalcohol), and least rapidly from ethylene–vinyl acetate copolymer. The "burst" was evident for all three slow release systems.

Release profiles for four different proteins from ethylene–vinyl acetate copolymer "sandwiches" are shown in Figure 8. Release rates approach zero order kinetics between 20 and 100 days.

Molecular weight appears to be a weak but significant parameter in determining release profiles, particularly the "burst" effect. Lysozyme (mol. wt. 14,400) showed the greatest percent protein escaped at any given time, while catalase (mol. wt. 250,000) showed the least. Soybean trypsin inhibitor (mol. wt. 21,000) and alkaline phosphatase (mol. wt. 88,000) fell in between. We have found, however, that molecular weight is not the only parameter affecting release profiles. When blue dextran (mol. wt. 2,000,000, Pharmacia) was incorporated into pellets identical to these above, over 90% was released in a 3-day period.

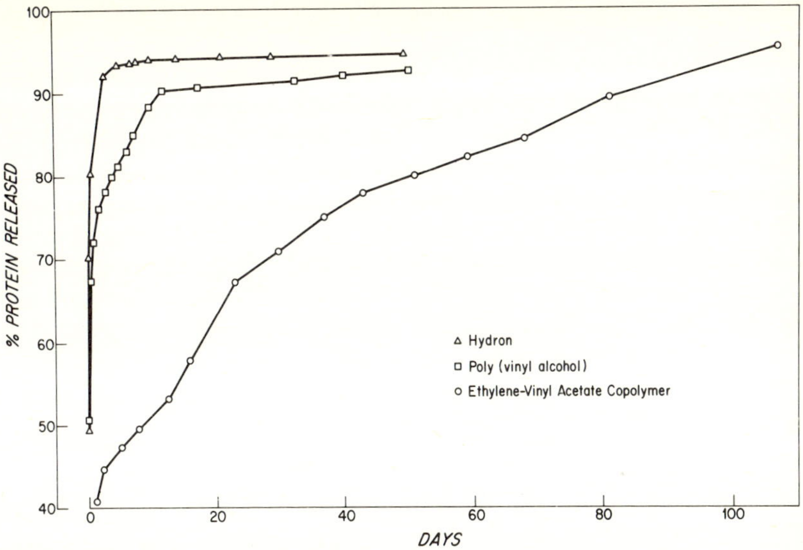

FIGURE 7 Release of soybean trypsin inhibitor from polymer "sandwiches". Protein concentration in casting solution was 50 mg/cm³.

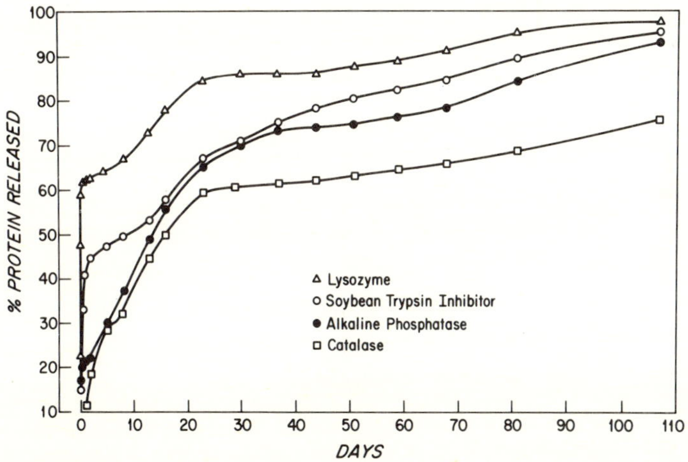

FIGURE 8 Release of proteins from ethylene–vinyl acetate copolymer "sandwiches". Protein concentration in casting solution was 50 mg/cm³.

Analysis of slow release agents for biochemical activities

We have also conducted studies to determine whether the polymers were releasing biochemically active material. The Lowry assay measures only the presence of protein. Therefore, biochemical activity was tested using agar diffusion gels.[18]

Test polymers containing 300 μg of either soybean trypsin inhibitor, lysozyme, or alkaline phosphatase were tested as follows: Polymers were incubated in 20 cm^3 volumes of lactated Ringer's solution at 37°C. The solution was changed 5 times during the first day of incubation, daily each of the next 9 days, and every 2 days during subsequent incubation. Before each change, the polymer was blotted dry on an absorbent tissue to remove, in part, adherent solution, and then washed with additional Ringer's. Polymers were removed from incubation periodically, washed with Ringer's, and placed into wells on agar slides as shown in Figure 9. More Ringer's was used to fill the remaining space in each well.

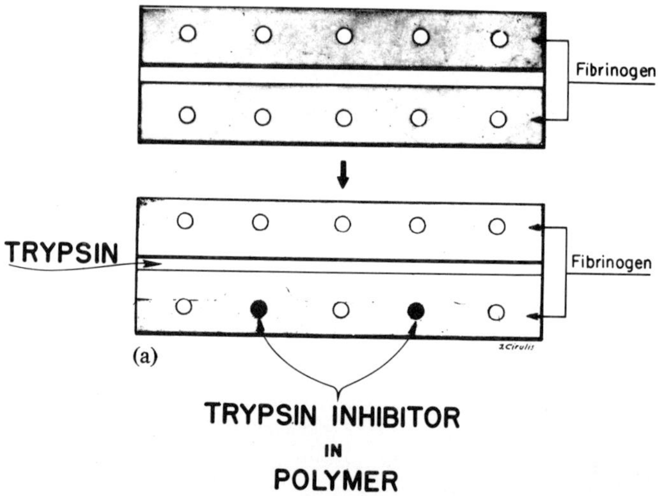

FIGURE 9 Agar slides used to assay polymers: (a) Trypsin inhibitor—the slide contains 1% agar and 0.1% fibrinogen in 0.1 M Tris buffer, pH 8. Trypsin (1 mg, Sigma, Type III, trypsin per cm^3 of Tris buffer, pH 8) is loaded into the center trough and digests the opaque fibrinogen. When the trypsin approaches a well containing a polymer releasing inhibitor, a zone is formed. (b) Lysozyme—the slide contains 1% agar, 0.03% Micrococcus lysodeikiticus in 0.05 M phosphate buffer, pH 7. Polymers releasing lysozyme digest the opaque bacteria forming a clear zone around the polymer. (c) Alkaline phosphatase—the slide contains 1% agar, 0.08% Napthol AS-GR phosphate in 0.1 M Tris buffer, pH 8. Polymers releasing alkaline phosphatase cause the formation of zones which appear as orange precipitates, due to the formation of Napthol AS-GR, on a translucent yellow slide.

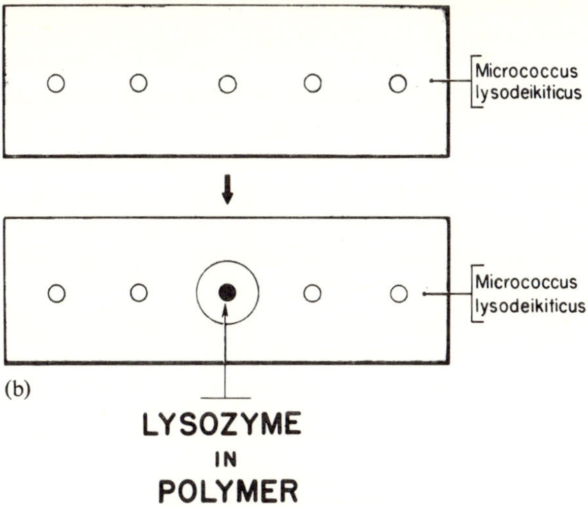

(b)

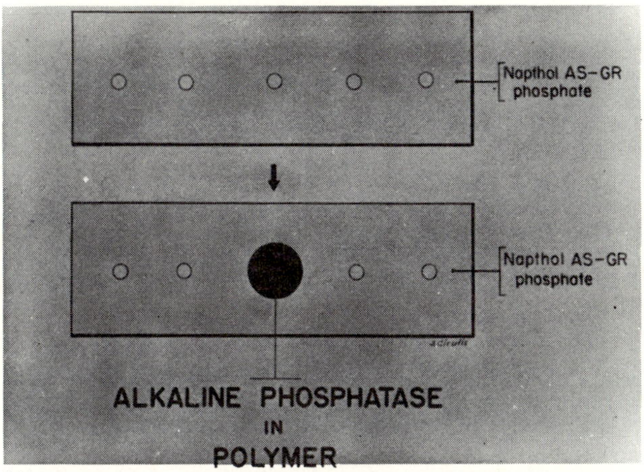

(c)

Order of magnitude titrations showed that zones were observed during a 24-h period in each of the assays, if nearly 1 μg or more of biochemically active protein was present in a well. Control polymers containing no proteins, or proteins not detected by these assays, did not produce zones.

"Sandwich" polymers released detectable amounts of soybean trypsin inhibitor for periods significantly longer than uncoated polymers. For example, when casting solutions contained 10% poly(vinylalcohol) and 50 mg/cm^3 soybean trypsin inhibitor, uncoated polymer released for 5 days, while a polymer "sandwich" released for 14 days.

New zones continued to form on each of the three agar slides for periods exceeding 100 days with ethylene–vinyl acetate copolymer "sandwiches", and for 14 and 4 days with poly(vinylalcohol) and Hydron "sandwiches" respectively. Furthermore, when the specific activity of alkaline phosphatase was tested before and after incorporation into each of the three polymers, more than 80% of the enzyme escaping was found to be active. Alkaline phosphatase activity was judged by conversion of *p*-nitrophenol phosphate to *p*-nitrophenol.

We have also used the ethylene–vinyl acetate copolymer to release insulin, heparin, and DNA (isolated from mouse SVT2 cells and labeled with H^3-thymidine). A radioimmunoassay was used to detect insulin;[20] a carbazole test was employed for heparin,[21] and TCA precipitation followed by scintillation counting was used for DNA.[22] In all cases, polymers tested contained 300 μg of the desired macromolecule. Again nearly 1 μg of each chemical was released per day for periods up to 60 days.

Release of macromolecules in vivo

Several experiments show that these polymers release macromolecules *in vivo*. The following three studies demonstrate that the pellets are releasing macromolecules in biologically active form.

In the first study, ethylene–vinyl acetate copolymer "sandwiches" containing 300 μg of tumor angiogenesis factor (TAF) were tested *in vivo* for their continuous ability to elicit a host vascular response. TAF is an extract obtained from tumors which induces new blood vessels to grow from the host. It has a molecular weight on the order of 100,000.[10] In over 100 experiments, pellets containing TAF were implanted into rabbit corneas and, in every case, a significant vascular response was observed (Figure 10). No edema or white cells were detected. After releasing TAF in corneas for more than 40 days, several of the polymers were removed and vessels disappeared over a 3-week period. When these same polymers were washed in Ringer's and transplanted to new corneas, vascular responses were again observed, although rates of vessel growth were somewhat slower than they were in the initial implants. We have also found that slow release pellets made from poly(vinylalcohol) or ethylene–vinyl acetate copolymer and containing TAF consistently produced neovascular responses when implanted on a 10 or 11 day old chick CAM. The polymers by themselves induced no such response.

A dose–response curve for different levels of TAF in ethylene–vinyl acetate copolymer "sandwiches" is shown in Figure 11. Polymers containing 3 μg of TAF either induced a minute vascular response which quickly disappeared or no vascular response at all. Polymers containing 30 μg induced a mild but sustained vascular response, while pellets containing 300 μg induced a significant response.

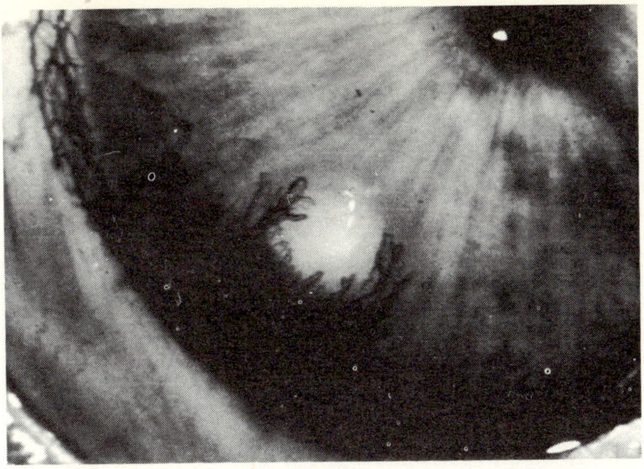

FIGURE 10 Rabbit cornea 20 days after implantation of ethylene–vinyl acetate copolymer "sandwich" containing tumor angiogenesis factor (TAF). Vessels from the corneal edge grow towards the polymer. No edema or white cells are observed.

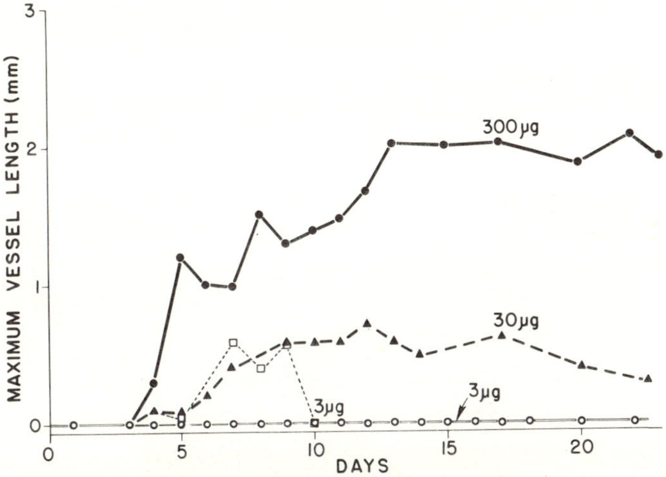

FIGURE 11 Dose response curve for tumor angiogenesis factor (TAF) in ethylene–vinyl acetate copolymer "sandwiches". Vessel length is measured from the corneal edge using a slit-lamp stereomicroscope (error ±0.1 mm).

In a second study, we tested a variety of macromolecules isolated from cartilagenous tissue for their ability to inhibit the growth of new blood vessels elicited by tumors.[16] Implants of tumor and ethylene–vinyl acetate copolymer "sandwiches" containing the different macromolecules were implanted adjacent to each other in rabbit corneas (Figure 12a). The tumor induced vessels to sprout from the corneal edge and these vessels grew toward the tumor. When no macromolecules were present in the slow release pellet or if these macromolecules had no biological activity, thousands of vessels grew over the pellet into the tumor (Figure 12b). By contrast, when the inhibitor isolated from cartilage[16] was incorporated into the polymer pellet, vessels were sparse, grew slowly, and failed to grow through a zone occupied by the polymer (Figure 12c). This inhibition of vascularization greatly restricted the growth of the tumor. The inhibitor appears to have a molecular weight near 16,000.

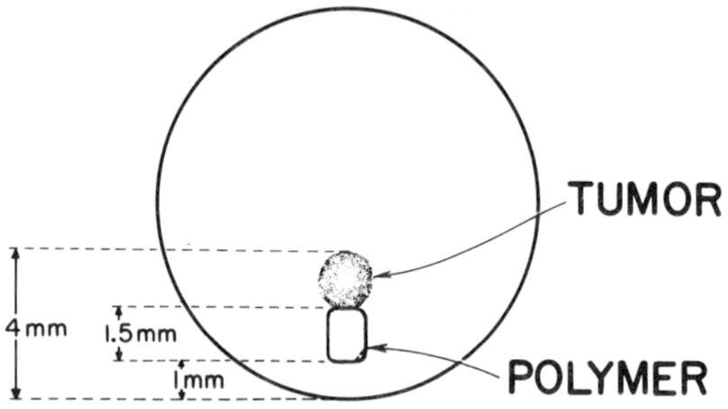

FIGURE 12 (a) Schematic diagram of rabbit cornea with tumor and polymer.

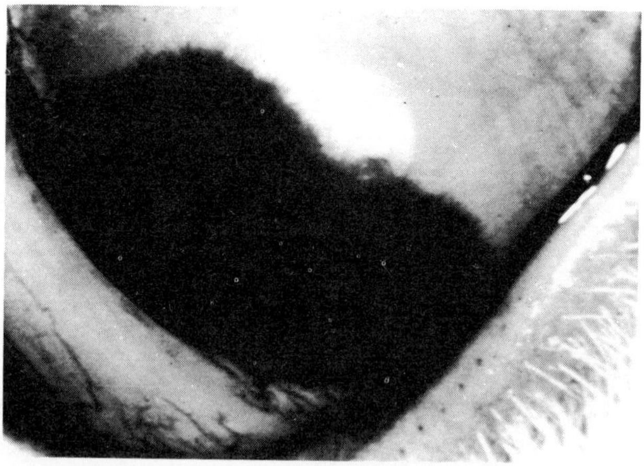

FIGURE 12 (b) Lower third of a rabbit cornea containing tumor and empty polymer, 17 days after implantation. Vessels appear as a dense carpet sweeping over the polymer. At this time, the tumor has already vascularized and is growing rapidly compared to that shown in (c).

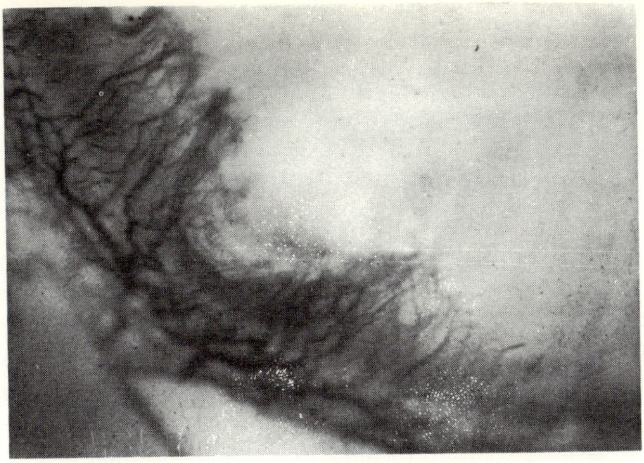

FIGURE 12 (c) Lower third of a rabbit cornea containing tumor and polymer with inhibitor. Vessels are sparse and fail to grow in a zone surrounding the polymer. The tumor has not become vascularized. This photograph was also taken 17 days after implantation.

In a third set of experiments, we implanted polymer pellets containing insulin in diabetic rats. The rats weighed approximately 100 g and were made diabetic by the injection into the tail vein of streptozotocin (70 mg/kg) dissolved in citrate buffer, pH 4.5. Ethylene–vinyl acetate copolymer sandwiches (discs, 8 mm dia., 2 mm thick) containing 100 mg of bovine insulin were implanted into the rats and blood glucose was determined using the method of Kingsley and Getchell.[23]

Diabetic rats with control pellets, or no pellets, showed blood glucose levels of nearly 400 mg %. Implantation of an insulin polymer into a diabetic rat caused a dramatic fall in blood glucose to normal or subnormal levels (50 to 120 mg %). These levels were maintained for periods between 50 and 57 days, whereupon the blood sugar levels returned to nearly 400 mg %.

DISCUSSION

The present experiments show that macromolecules with a wide range of molecular weights can be delivered in significant quantities from polymeric vehicles that are not inflammatory when implanted in animals. These polymers can release macromolecules in biochemically and biologically active form for periods in excess of 100 days as measured by direct assays.

Critical to the success of these experiments is the non-reactivity of the polymers in tissues. Ethylene–vinyl acetate copolymer and Hydron caused no detectable inflammation in the cornea, either grossly or microscopically.

An important, although unexpected, result was the finding that the rate of release from ethylene–vinyl acetate copolymer was not a close function of molecular weight. This polymer may prove useful for delivery of a wide range of macromolecules without reformulation of the vehicle.

The procedure for constructing these slow-release vehicles can be carried out simply and under sterile conditions in any laboratory. Other investigators need not be restricted by the configuration and small size of the implants we have used. Various combinations of the three polymers and extra coatings may be employed.

The capability of these polymers to deliver macromolecules may permit the development of bioassays for a variety of substances, including growth factors and chemotactic compounds. The ability of ethylene–vinyl acetate copolymer in casting solution to bind to plastic may open new possibilities for tissue culture experiments. Absence of even a mild inflammatory response in the presence of these polymers may open new avenues for *in vivo* testing and assay of macromolecules. The eventual clinical application of these polymers for delivery of macromolecules such as insulin, heparin, or enzymes may merit consideration.

Whatever the eventual applications, macromolecules can now be delivered in biologically and biochemically active form for prolonged periods *in vitro* and *in vivo* from vehicles that are not inflammatory when implanted in animals.

References

1. J. Folkman and D. M. Long, *J. Surg. Res.*, **4**, 139 (1964).
2. P. Bass, R. A. Purdon, and J. N. Wiley, *Nature*, **208**, 591 (1965).
3. K. G. Powers, *J. Parasitol.*, **51**, Section II, 53 (1965).
4. J. Dziuk and B. Cook, *Endocrinol.*, **78**, 208 (1966).
5. J. Folkman, S. Winsey, and T. Moghul, *Anesthesiol.*, **29**, 410 (1968).
6. J. Folkman and V. H. Mark, *Trans. N.Y. Acad. Sci.*, *Series II*, **30**, 1187 (1968).
7. R. W. Baker and H. K. Lonsdale, in *Controlled Release of Biologically Active Agents.* Advances in Experimental Medicine and Biology Series, pp. 15–72, A. C. Tanquary and R. E. Lacey, eds., Volume 47, Plenum Press, N.Y. (1972).
8. B. K. Davis, *Experientia*, **28**, 348 (1972).
9. B. K. Davis, *Proc. Nat. Acad. Sci. USA*, **71**, 3120 (1974).
10. J. Folkman, E. Merler, C. Abernathy, and G. Williams, *J. Exp. Med.*, **133**, 275 (1971).
11. J. Folkman, *Advances in Cancer Research*, Vol. 19, Academic Press, N.Y., pp. 331–358 (1974).
12. J. Folkman, *Cancer Res.*, **34**, 2109 (1974).
13. J. Folkman, *Ann. Int. Med.*, **82**, 96 (1975).
14. M. A. Gimbrone, Jr., R. S. Cotran, S. B. Leapman, and J. Folkman, *J. Nat. Cancer Inst.*, **42**, 314 (1974).
15. H. Brem and J. Folkman, *J. Exp. Med.*, **141**, 427 (1975).
16. R. Langer, H. Brem, K. Falterman, M. Klein, and J. Folkman, *Science*, **193**, 70 (1976).
17. C. H. Fromer and G. K. Klintworth, *Am. J. Path.*, **79**, 537 (1975).

18. R. Langer and J. Folkman, *Nature*, **263,** 797 (1976).
19. O. H. Lowry, N. J. Rosebrough, A. L. Farr, and R. J. Randall, *J. Biol. Chem.*, **193,** 265 (1951).
20. J. S. Soeldner and D. Slone, *Diabetes*, **14,** 771 (1965).
21. T. Bittner and H. Muir, *Anal. Biochem.* **4,** 330 (1962).
22. A. V. Furano, *Anal. Biochem.*, **43,** 639 (1972).
23. G. R. Kingsley and G. Getchell, *Clin. Chem.*, **6,** 466 (1960).

DISCUSSION

Dr. H. Hopfenberg (*North Carolina State Univ., Raleigh, NC*): What was the typical concentration of the polypeptide in the drug–polymer matrix?

Dr. R. Langer: In most of the studies about 300 μg of protein were put into the matrix.

Dr. H. Hopfenberg: What was the weight of the polymer?

Dr. R. Langer: The weight of the polymer was about 1.2 mg., meaning that the protein comprised about 25% of the total weight.

Dr. W. Leonard (*Dynapol Corporation, Palo Alto, CA*): What was the degree of swelling of the implant when placed in the eye?

Dr. R. Langer: It varied depending on which kind of polymer was used. We didn't measure this precisely. The hydron and the poly(vinylalcohol) swell to a considerable extent while the ethylene–vinyl acetate copolymer swells hardly at all. We always pre-treated the polymers before implanting them by placing them in an isotonic medium for about 20 min.

Dr. W. Good (*Ciba-Geigy Corporation, Ardsley, NY*): You tested several polymers for compatability before implanting. Did you purify these polymers first, or were they used as received? Also, were the polymers characterized in any manner?

Dr. R. Langer: We generally used them as we received them, other than subjecting them to the washing procedure which I discussed.

Dr. W. Good: What I find curious is the apparent incompatability of poly-(vinylpyrrolidone) which should be actually not that bad.

Dr. J. Folkman: The cornea is the tissue most sensitive to inflammation. When we say that there is no inflammation in the cornea, that means that even by electron microscopy there are no additional cells which have migrated into the area. Ten new white cells are counted as early inflammation. This is something that you couldn't even see. The cornea provides a super sensitive test. Most of these polymers would give mild or no inflammation if implanted in other tissues subcutaneously or in chick embryos. When we say there is zero inflammation in the cornea, there is absolutely no cellular infiltration, no increased permeability and no new vessels as well as no new DNA synthesis in the endothelial cells in the vessels.

Dr. R. Capozza (*Alza Corporation, Palo Alto, CA*): Could the mechanism of release from hydron be a simple leaching out by water as occurs in a poly-(acrylamide) matrix? In the case of acrylamide gels containing immobilized enzymes, the gel traps the enzyme for a while, but it is eventually leached out slowly by water.

Dr. J. Folkman: Yes, but this would not explain why the emptied films are then impermeable to the protein. The water may act to drive the proteins out of the polymer through some channels formed by the protein.

Dr. H.-G. Elias (*Midland Macromolecular Institute, Midland, MI*): In the case of the ethylene–vinyl acetate copolymer, it is important to know the vinyl acetate composition. The composition determines the glass transition temperature, which in turn may govern the release mechanism.

Dr. R. Langer: The vinyl acetate composition is 40% by weight.

Dr. H. Hopfenberg: Have you performed any studies regarding the effect of protein loading level? If so, have you found a threshold level where the effect of the protein vanishes?

Also, have you examined the temperature dependence of the process? If you are interested in discerning the mechanism controlling protein release this would be a very good first step, since you could determine the activation energy associated with the kinetic process controlling release. If the process were simply dissolution and extraction, the activation energy would be quite low; if activated diffusion of polypeptide controlled, it would be much higher.

Dr. R. Langer: If the loading is 30 μg, the protein always comes out. If the loading level is 3 μg, then the assays are difficult unless you are using sensitive assays, such as radioactively labelled protein, and we have not used this technique.

We have not studied the temperature dependence of the process. All of the work has been done at 37°C.

Dr. W. Good: I submit that what happens when you mix these protein macromolecules with the polymers is that you are forming a blend of two different types of polymers, which results in polymeric phase separation due to incompatability. This would explain the observed initial burst in release.

Dr. D. Meier (*Midland Macromolecular Institute, Midland, MI*): I feel that Bill Good's comment on phase separation is correct. It is also important to realize that the minor polymer component is not necessarily dispersed in the form of spherical-like particles but might very well be in a fibrillar form which reaches to the surface. This would allow the material to be extracted relatively quickly by dissolution along the fibrils. This morphology would indeed produce the experimentally observed initial burst in release rate.

Dr. J. Folkman: Would you expect the channels to remain after the protein is extracted, or would they swell shut?

Dr. D. Meier: They would remain. They can only grow larger during swelling, not swell shut.

Dr. H. Hopfenberg: Do you observe channels in the polymer matrix after the protein has been released?

Dr. R. Langer: We have not examined this thoroughly at this point.

Dr. H.-G. Elias: Just a minute. These channels would be on the order of 15 to 20 Å for a protein molecule with a molecular weight of about 60,000.

Dr. H. Hopfenberg: But the data for the coated devices appear to mitigate against release mechanisms based on channels since, presumably, these channels do not extend through the intact polymer coating.

Dr. R. Capozza: Could you explain again how you coated the particles, and indicate whether there is any evidence for intact, non-porous coatings?

Dr. R. Langer: There are two methods we use for coating the particles. One is just dipping them into a pot of pure polymer solution and the other is by using a mold. First we make a layer of polymer in casting solution and dry this under a mild vacuum. Then we add a center layer which is the polymer plus protein and dry that under a vacuum. Then we add a top layer which is the polymer in casting solution and again dry under vacuum.

Dr. R. Capozza: Wouldn't that process cause imperfections?

Dr. R. Langer: Possibly.

Dr. W. Larson (*Dow Corning Corporation, Midland, MI*): Can you make a film out of the loaded polymer, leach the protein out, and subsequently test the remaining film for permeation with a higher molecular weight protein? Such a test would indicate whether the channels are continuous.

Dr. R. Langer: Yes, we've done that. The point is that even the original protein will not migrate through the film once it is leached out.

Let me repeat this point. If you take the polymers which I discussed and make a film from them with a protein trapped inside and leach the protein out for 100 days or so on agar gel slides, and then put another protein of at least the same molecular weight on top of the film, it will not pass through even if it is the same protein. Thus, if there are channels, they cannot be continuous.

Dr. W. Leonard (*Dynapol, Palo Alto, CA*): Is it possible that when water swells this polymer, it forms a microphase, two phase separation where you have concentrated polymer and dilute polymer?

Dr. R. Langer: It's possible.

Dr. W. Leonard: Upon swelling, do you observe any cloudiness or increase in light scattering?

Dr. R. Langer: It's hard to say. There is so much protein in the system, about 25%, that it is a dispersion at the beginning.

Synthetic Polymeric Drugs

HELMUT RINGSDORF

Institute of Organic Chemistry, University of Mainz, 6500 Mainz, J.-J.-Becher-Weg Germany

Although the concept of using pharmacologically active macromolecular compounds as drugs is still regarded with much scepticism for both theoretical and practical reasons, interest in this field has grown in recent years because of the opportunity to take advantage of the specific properties of polymeric materials.

Using a model for pharmacologically active polymers as a guide, examples of the synthesis and study of the biological activity of various macromolecular drugs are presented in order to emphasize the importance of this cooperative effort between polymer research and therapeutic problems. Especially the importance for an effective tumor therapy is discussed.

INTRODUCTION

The investigation of biomaterials is one of the newest and most exciting areas of polymer chemistry. Especially interesting are investigations of pharmacologically active polymers which represent macromolecules which by themselves may be active as drugs or alternatively may be used as carriers for normal pharmaceutical agents.

Although polymers are already widely used in the area of biomedical engineering, e.g., to construct artificial replacements for human organs[1] or to provide biochemical functions,[2] the field of polymeric drugs is still subject to considerable scepticism. More points have been raised against the use of pharmacologically active polymers than in their favor. Nevertheless much has been accomplished in recent years as shown by the growing number of papers and review articles in this area.[3–9]

Originally one of the main ideas behind the possible use of pharmacologically active polymers was the hope that one might achieve depot effects with such drugs based on previous experience with synthetic polymeric blood substitutes. However, considering the many possibilities for specific chemical and physical interaction between synthetic or natural polymers and body membranes or involvement with various biochemical processes in the body, this depot effect is only one facet of the new horizons which are opened to chemotherapy by the switch from low to high molecular weight drug systems. In the present survey no attempt will be made to discuss the many types of drugs which have already been fixed to polymeric carriers or the biological activity of these materials.[3,4,7,8,10] Instead, on the basis of a rather simple model for the medical use of polymeric drugs,[9] a few examples of pharmacologically active macromolecules will be discussed for the purpose of demonstrating the potential importance of this new field. Special emphasis will be given to polymeric antitumor agents and the problem of cell-uptake of polymers.

The term "polymeric drug" is not yet definitely defined. In its broader sense this term comprises not only pharmacologically active polymers the low molecular weight models of which are inactive, but also substances in which a known drug is fixed to a polymeric carrier or incorporated into a polymeric chain.

PROBLEMS IN DRUG DESIGN

After the application of a drug a series of events normally takes place before the drug finally reaches the site of action. A schematic summary of such drug pathways is presented in Figure 1.

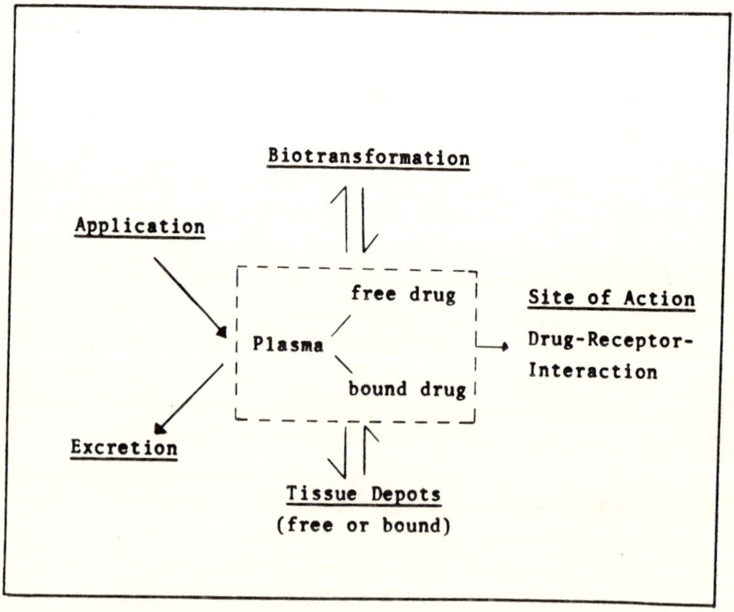

FIGURE 1 Drug pathways in biological systems.[9,103]

Between the application and excretion of a drug it will be subject to many physiological systems which will influence both its concentration and various reactions in the body. Upon entering the blood plasma the drug molecules may be free or bound to transport proteins. The free drug molecules may undergo biotransformations or may form tissue depots. Eventually the drug will enter the site of action where, for example, by a drug–receptor interaction the biological event will take place. Because of these different metabolic pathways the drug is distributed throughout the body rather than solely at the site of action, and therefore the therapeutic concentration must be greatly exceeded with the normal application forms of low molecular weight drugs. This is shown schematically in Figure 2.

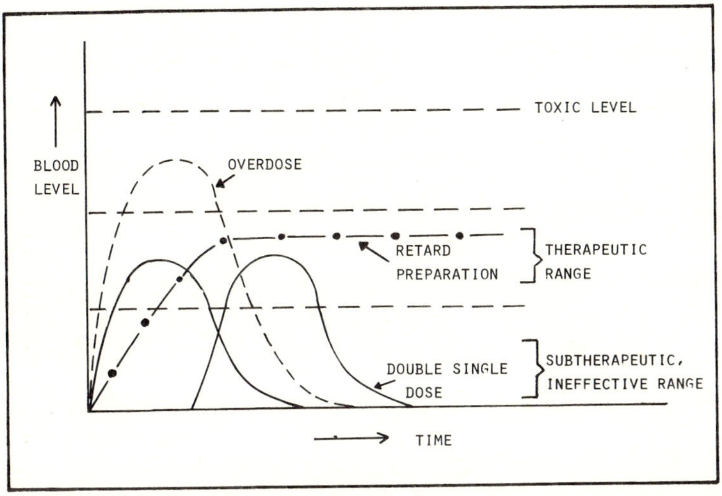

FIGURE 2 Schematic curves of blood level concentrations of drugs.

One part of this problem can be solved by dosage forms with prolonged action and this will be partially discussed in the contributions of R. Capozza and G. Banker in this volume. Preparations with prolonged action[11] are designed especially to maintain a necessary therapeutic concentration over a longer period of time in a range below the side effects and above an insufficient concentration. But even more interesting in modern drug design than to maintain a drug concentration within the therapeutic range is the goal to avoid unwanted side reaction by a cell-specific activity of a pharmacon. It would be extremely important to be able to synthesize drugs for the different diseases which would act cell-specifically and only at the desired site.

Considering the various factors to which a drug is subject it is clear that in attempting to effect their combination in order to develop an ideal drug, there is a possibility that the original biological activity of the low molecular weight compound may be lost.

A very interesting example of this difficulty was described by Connors and coworkers who studied the influence of substituents on the cell-specific uptake of sulphadiazines by malignant tissue.[12] (See Figure 3.)

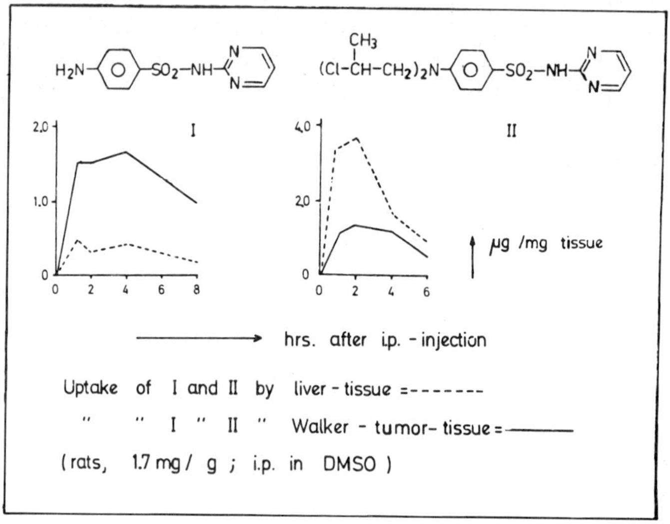

FIGURE 3 Influence of substituents on the cell-specific uptake of sulfadiazines.[12,103]

On the left side of Figure 3 it can be seen that sulphadiazine is more rapidly incorporated into Walker carcinoma tissue than into liver tissue. These findings are of considerable interest, since tumors generally take up injected materials to a lesser extent than the liver. Based on this result Connors and coworkers synthesized a sulphadiazine mustard (compound II in Figure 3) in order to combine the ability of the sulfonamide to concentrate in tumor cells with the cytotoxic effect of the lost system. Compound (II) was in fact shown to be an antitumor agent although by combination with the lost system the sulphadiazine unit lost its ability to act as a homing device for tumor tissue. As can be seen in Figure 3 uptake of the sulphadiazine mustard (II) into liver tissue is higher than its uptake into malignant tissue.

This example demonstrates a specific problem in modern drug design and shows that the alteration of a low molecular weight drug to enhance its biological activity or to incorporate a specific property may be difficult or even impossible.

MODEL FOR PHARMACOLOGICALLY ACTIVE POLYMERS

From the standpoint of polymer chemistry one can consider the possibility of devising a polymeric drug with which one can separate the different necessary requirements by using discrete areas along the polymer chain to accomplish specific effects. The principle is shown schematically in Figure 4.

In this schematic representation of a pharmacologically active polymer, the biostable or biodegradable backbone is used as carrier for at least three different units. One area of the polymer is used to make the whole macromolecule soluble and nontoxic, the second area is the region where the pharmacon is fixed and the third area incorporates a transport system the crucial property of which is represented by its ability to carry the whole polymer to the target cells. The separation of the different areas along the polymer chain may be accomplished by statistical terpolymerization or by block copolymerization.

In addition unusual and polymer-specific properties of pharmacologically active polymers may be induced by certain polymer-specific structural characteristics such as high molecular weight, coil structure, neighbouring group effects, copolymer composition, unusual binding capabilities, variable polyelectrolyte charges, and tacticity.

Considering the three phases of drug action[13] as shown in Figure 5 one can expect that all three of them will be influenced by the structure of biologically active macromolecules.

From Figure 5 one sees (a) the pharmaceutical phase, e.g., by disintegration of dosage forms; this is the typical and only field for dosage forms with prolonged action or sustained release; (b) the pharmacodynamic phase, e.g., by unusual absorption and distribution, variations in toxicity and metabolism or retarded excretion and (c) the pharmacokinetic phase, e.g., by variations in drug–receptor interactions and possible cell-specific effects.

From the model and these considerations one can predict some special properties to be expected of polymeric drugs as compared with their low molecular weight analogs. They are summarized in Table I.

A potentially significant aspect of the use of a polymeric drug which has been discussed even in the older review literature[3] is the question of a depot effect which might be brought about by either reduced absorption or reduced excretion of the drug.

More important is the possibility of varying drastically the pharmacodynamics of drug systems by the use of polymeric substrates. This is due not only to the fact that one may effect variable release of an active component, but also because the toxicity of a drug may be diminished drastically and in addition completely different metabolic pathways may result.

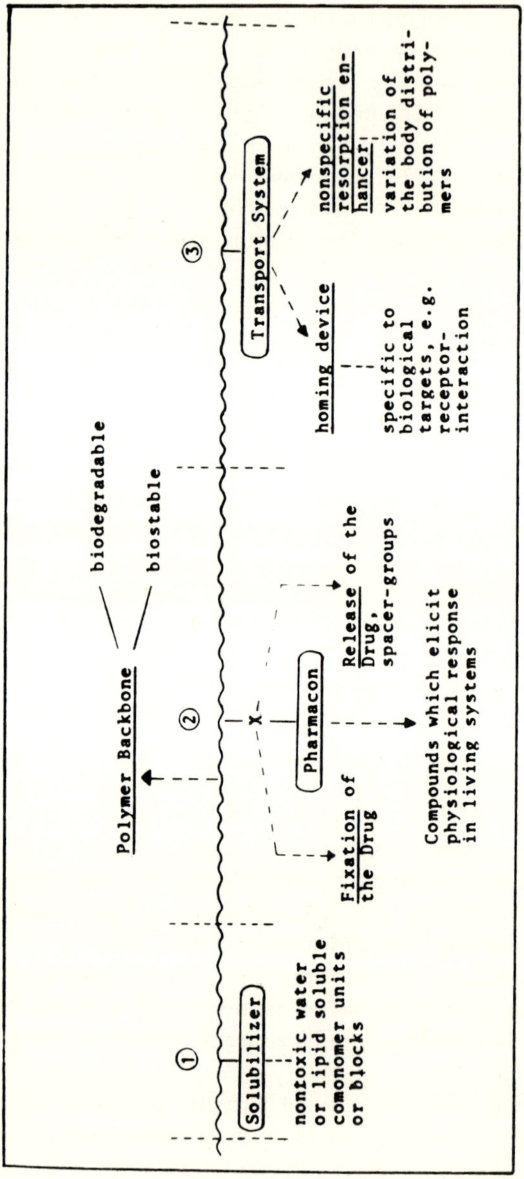

FIGURE 4 Model for pharmacologically active polymers.[9,103]

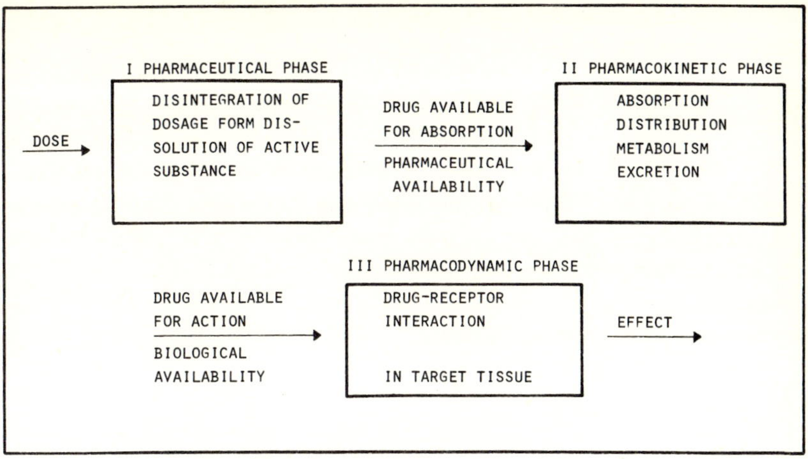

FIGURE 5 Main phases of drug action.[13]

TABLE I

Potential differences of low and high molecular weight drugs

1) Depot effects < retarded absorption / retarded excretion
2) Widely variable toxicity and solubility.
3) Pharmacokinetic variation:
   a) variable release of the active component;
   b) different metabolic pathways;
   c) influence of the structure of the polymer, of molecular weight and of comonomer units incorporated.
4) Different body distribution with regard to:
   a) protein binding;
   b) resorption;
   c) cell-specific interaction and uptake.
5) Polymer specific effects.
6) Drug combination along the polymer chain.

Another very important consideration is the possibility of effecting a different body distribution of the polymeric drug relative to its low molecular weight analog. Transforming an unspecific bioactive compound into a more selective or even cell-specific molecule is one of the most stimulating possibilities of pharmacologically active polymers. This will be discussed later in connection with the body distribution and the endocytotic cell-uptake of polymers. The attempt to develop bioactive materials acting as target seeking guided missiles is one of the topics also discussed by A. Trouet in this volume.

Last but not least one should consider the possibility of discovering drugs which show specific effects not exhibited by their low molecular weight analogs. Besides the well-known plasma expanders an interesting example comes from the work of Schlipköter and coworkers.[13–15] They found that poly(vinylpyridine-*N*-oxides) are active against silicosis although the low molecular weight models such as isopropylpyridine-*N*-oxide showed no activity. The polymer-specific effect of these *N*-oxides in protecting macrophages against silica injury can be explained by the fact that the polymer coats silica particles and thus prevents them from interacting with lysosomal membranes thereby causing their rupture.[13,16,17]

Finally it is not unreasonable to speculate that systematic studies on the synthesis and pharmacological activity of polymeric drugs might lead to new methods of drug design.

DISCUSSION OF THE MODEL AND EXAMPLES OF PHARMACOLOGICALLY ACTIVE POLYMERS

To consider further the model depicted in Figure 4 a few remarks may be made about the area of the solubilizer (area 1 in Figure 4) including the problem of molecular weight and immunogenicity of synthetic polymeric drugs.

The comonomer units introduced to render the resulting polymer water or lipid soluble must satisfy some fundamental requirements, e.g., lack of toxicity and immunogenicity. For polyvinyl derivatives such non-toxic, water-solubilizing units have generally been introduced via vinylpyrrolidone,[19] β-hydroxyethyl acrylates and acrylamides,[20] vinylpyridine-*N*-oxides[21] and sulfoxide-containing acrylates.[22] In order to make polymeric drugs lipid-soluble one can use comonomer units bearing long alkyl chains which serve to increase possible absorption at lipid phases and cell membranes.

Because of their pronounced lack of acute toxicity we have been especially interested in polymeric systems containing sulfoxide and *N*-oxide units.[23] Homopolymers as well as a number of copolymers containing these solubilizer units did not show any immunogenicity in preliminary experiments. Synthetic polymers seem to show no or only weak antigenicity.[19,24,25] The question of immunological response to synthetic macromolecules has not yet been investigated carefully enough. Systematic investigations on the antibody formation against synthetic carriers and/or fixed classical haptens were started recently.[26]

In addition to toxicity and antigenicity of specific macromolecular units the overall molecular weight of the synthetic polymers used and their normal body distribution are of significance. It may be mentioned that excretion via the highly porous glomerular membrane of the kidneys is limited for most synthetic polymers since those with molecular weights above 80,000–100,000 cannot pass the tubular epithelium. On the other hand such polymers can be slowly excreted via the liver and its biliary system into the intestine. Synthetic polymers as well as oligomers with molecular weights even lower than 1000 cannot pass the so-called "blood–brain barrier" and thus are not able to enter the brain and cerebrospinal fluid.

Clearly molecular weight and its influence on total excretion and storage in RES-system is one of the crucial problems in connection with the use of non-degradable polymeric drugs.

The second area schematically represented in the model (Figure 4) is that where the pharmacon is fixed to the polymer chain. (See Figure 6.)

Three points are of interest in this connection: (a) the question of how to fix the pharmacon to the polymer chain under conditions mild enough not to affect its biological activity, (b) the nature of the group used to fix the polymer to the chain and (c) the nature of the pharmacon itself.

For *fixation of the pharmacon* to the polymer chain, conditions must be mild enough to allow attachment without any adverse effect on the biological activity of the pharmacon. For this purpose various coupling methods which are well-known in the field of peptide synthesis can be applied.[27] To this end several polymerizable active esters and amides covering a wide spectrum of reactivity have been used in order to allow for selectivity depending on the nature of the nucleophilic center being used to fix the drug to the polymer chain.[28] Systems examined included esters of trichlorophenol, *N*-hydroxysuccinimide, *N*-hydroxybenztriazole and imidazole derivatives, such as:[28,29]

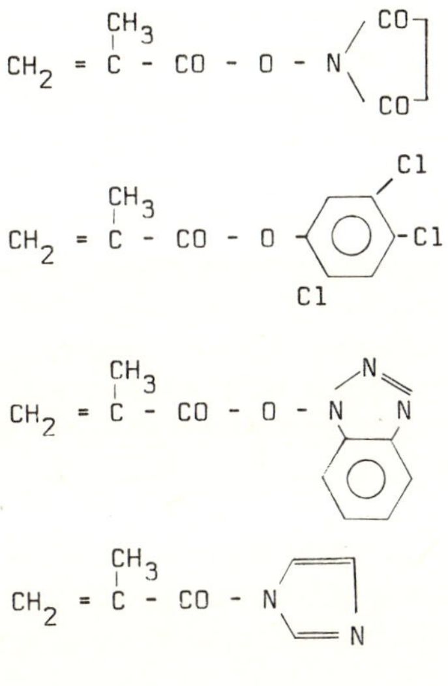

In the case of hydroxysuccinimide and hydroxybenztriazole derivatives it was easy to discriminate between hydroxyl and amino groups with only the latter reacting under mild conditions.[28] For example, the *N*-hydroxysuccinimide derivative of poly(methacrylic acid) reacts selectively with the amino group of ethanolamine at room temperature to give completely soluble non-crosslinked polymers.

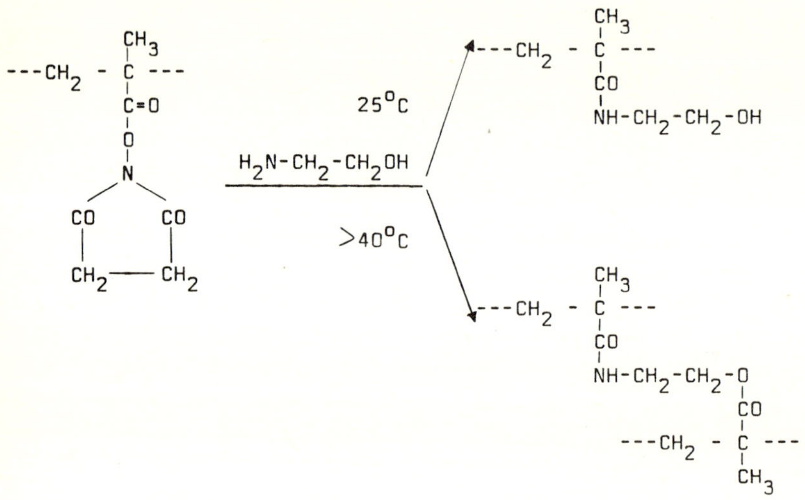

On the other hand at temperatures above 40°C reaction occurs with both the amino and hydroxyl substituents as shown by the formation of insoluble crosslinked polymers.

The question of the eventual *release of the drug* in the body is as important as the problem of its initial fixation to the polymer chain. For some purposes it might be necessary to prepare polymers in which the pharmacon is firmly attached to the polymer by means of a linkage which is completely stable under all normal body conditions. Alternatively one may wish to use a linkage from which the drug can be released rapidly in the body either by hydrolysis or via an enzymatic process. For these purposes permanent or temporary spacer groups may be used. The influence of spacer groups on the drug release is schematically summarized in Figure 7.

A permanent spacer group (distance holder) merely separates the active drug from the polymeric backbone or coil so that the latter does not interfere with the biological activity of the bound material. Such effects are common in the field of polymer-bound enzymes. In some such cases, direct fixation leads to loss of enzymatic activity, possibly because of interference with the

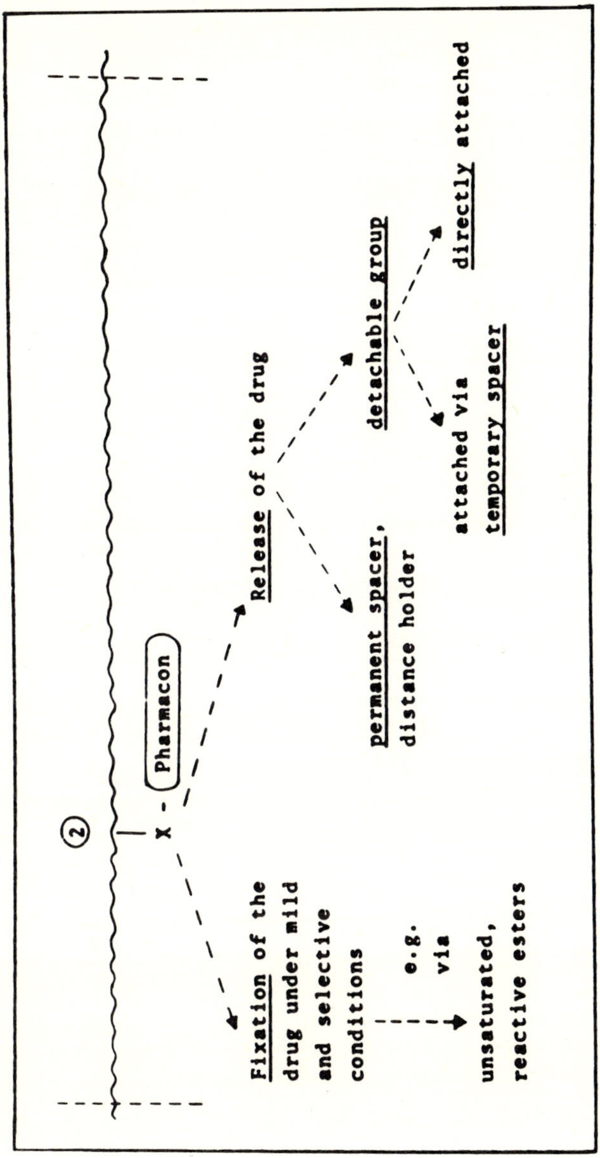

FIGURE 6 Area of drug fixation and drug release (see model in Fig. 4).[9,103]

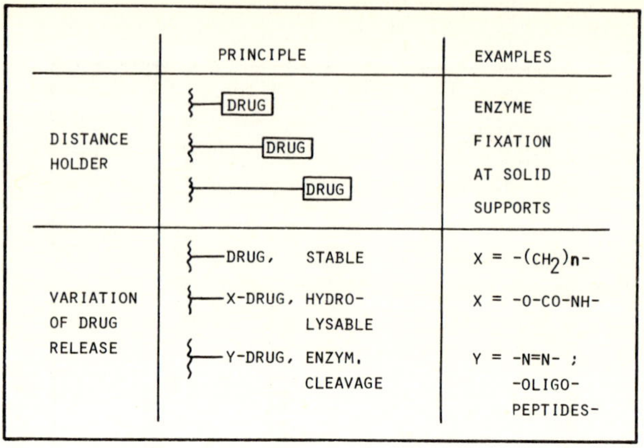

FIGURE 7 Influence of spacer groups on the release of drugs from polymeric carriers (schematic).

normal conformation of the enzyme, whereas attachment via a distance holder such as a long alkyl group restores the normal activity. An instructive example is given in Figure 8, describing the binding of β-Galactosidase by its immobilized inhibitor in dependence of the spacer group.[30]

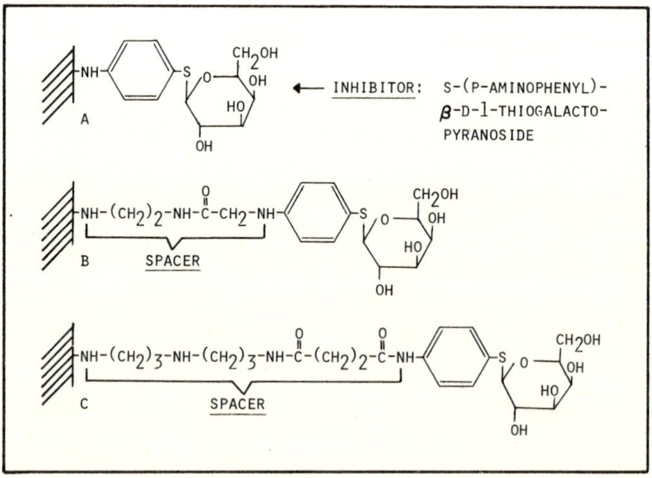

FIGURE 8 Binding of β-galactosidase by its immobilized inhibitor in dependence of the spacer groups:[30]

activity: C $\gg$ B > A

Recently Kaplan and Venter and their coworkers reported that catecholamines fixed to glass beads or polystyrene matrices by permanent spacers were able to induce muscle contractions and affect the heartbeat rates of experimental animals in a manner comparable to the effect of the low molecular weight compounds themselves.[31–33] These very interesting results seem to demonstrate a true membrane response in which the effect sensed by the drug receptor is transmitted without prior detachment of the pharmacon from the solid support. Details are discussed in the contribution of J. C. Venter *et al.*, in this volume.

A *temporary spacer group* is one from which the active material can be readily released. This is necessary because the majority of drugs are inactive in the macromolecular form and, thus, their links to the polymer must be reversible or degradable. One of the first examples of the use of such a temporary or detachable spacer group was described by Jatzkewitz as early as 1954.[34,35]

Copolymers of *N*-vinylpyrrolidone and amides formed from mescaline and acrylic acid in which the mescaline was bound directly to the main chain were found not to release the psychotic agent in the body. On the other hand when mescaline was bound, again as the amide, via the terminal carboxyl group of the dipeptide glycyl leucine which was itself bound to the polymeric acid via the terminal amino group, Jatzkewitz was able to demonstrate the continuous excretion of mescaline over a period of 17 days. Mescaline itself was completely excreted after only 20 h. Based on the discussed considerations about the drug release via temporary spacer groups Ch. Wilk[36] and J. Kopecek, J. Kalal *et al.*[37] synthesized a series of esteramide- and peptide-spacer containing polymeric drug models and studied their hydrolytic and enzymatic cleavage. In connection with the investigation of hormone containing polymers[38] the influence of the spacer group in an *in vivo* experiment was demonstrated lately.[39] In Figure 10 the androgenic effects of two polymeric testosterone derivatives are compared.

While testosterone directly fixed to the polymer chain (1) does not show any effect, the activity of the polymer with a temporary spacer group (2) is as effective as the low molecular weight model.

Regarding the *types of biologically active materials* which have been studied so far there is hardly any type of drug which has not been fixed to some type of polymer. Especially the review of L. G. Donaruma[7] summarizes sixteen areas where applications of biologically active polymers have been attempted, including macromolecules with antibacterial, antifungal, antiviral and antiparasitic properties, polymeric herbicides, polymers affecting the immune response and polymers having activity related to various circulatory, cardiac and blood problems.

One of the typical effects to be expected from polymeric drugs is the depot

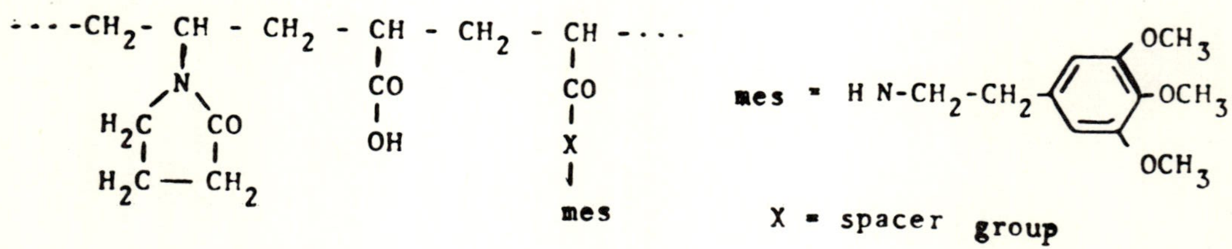

X = - : no excretion

X = -gly-leu- : continuous excretion for 17 days

free mescaline: excretion for 20 hours

FIGURE 9 Influence of a temporary spacer group on the release of polymer-bound mescaline (H. Jatzkewitz[34,35]).[103]

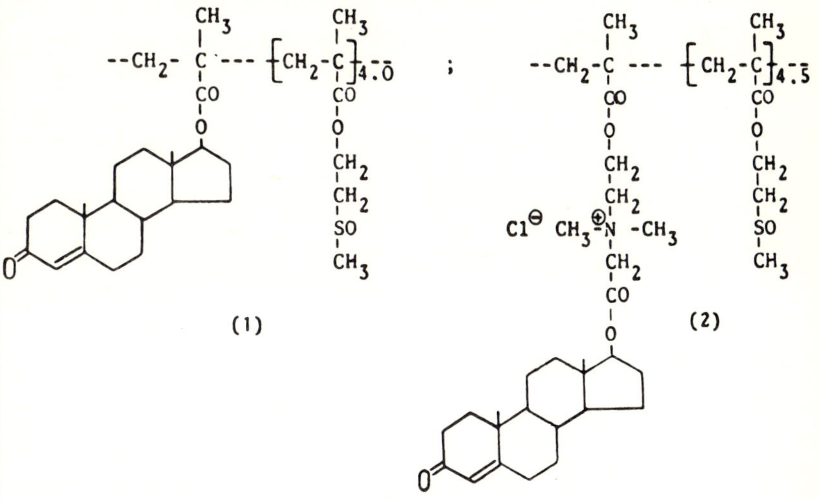

	polymer	levator ani muscle	prostate gland	sperm bladder
(1)	without spacer	-	-	-
(2)	with spacer	113 %	91 %	74 %

FIGURE 10 Androgenic effects of testosterone-containing polymers on castrated rats:[9,38,103]

S.C.—application; equimolar doses of polymer-fixed testosterone and free testosterone in control experiments. (Effect of free testosterone = 100%.)

effect. To demonstrate this one recent example from the area of polymeric anti-radiation prophylactics will be given. The great disadvantage of practically all known chemical radioprotective agents is their limited period of activity due to their low molecular weight.[40,42] They are effective only if injected 15–30 min before irradiation. Since macromolecules are only slowly excreted by the body, polymers containing potentially radioprotective functions were investigated for possible long lasting effects by C. G. Overberger and coworkers.[40–43] Recently a water-soluble terpolymer (S_3) prepared from vinylpyrrolidone, acrylic acid and acryloyl thiazolidine was tested.[9,44] In Figure 11 the effectiveness of the terpolymer is compared with its low molecular weight model (mercaptoethylamine, MEA). The Dose

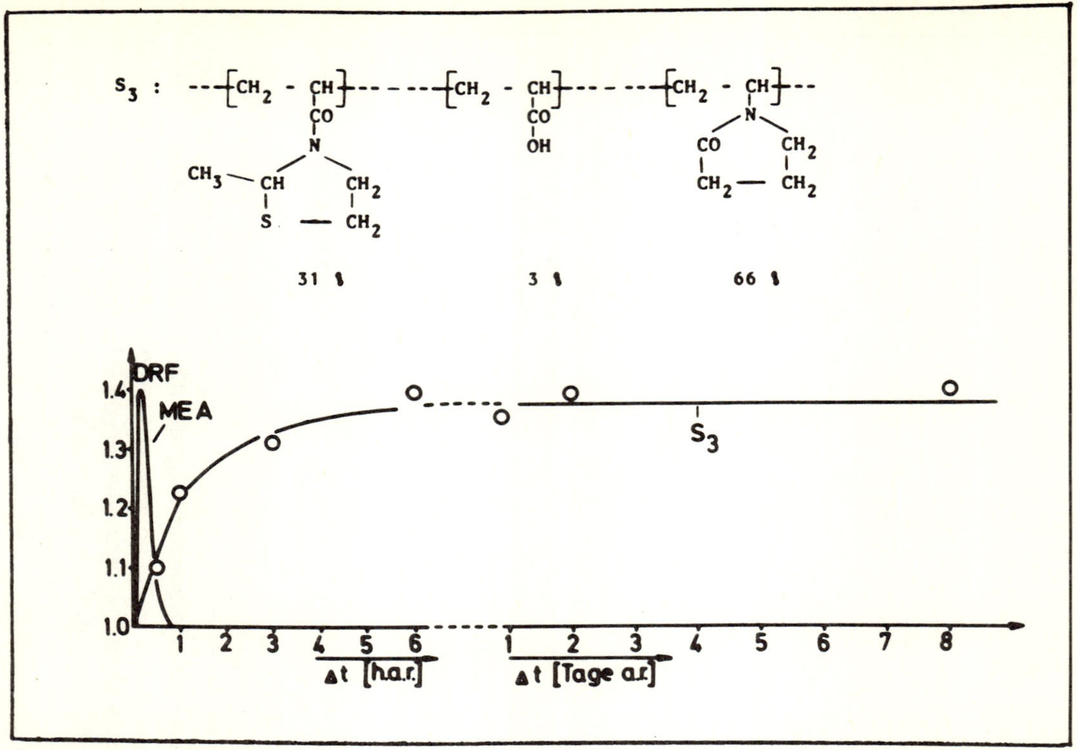

FIGURE 11 DRF values for S_3 after administration of 15 mg i.v. to rats at various time differences between injection and irradiation[43] (Δ_t, a.r.). Values for mercaptoethylamine (MEA) are given for comparison. The DRF values for the different times were taken from the corresponding dose effect curves.[103]

$$DRF = \frac{LD_{50/30}\text{ (protected animals)}}{LD_{50/30}\text{ (controlled animals)}}$$

DRF = Dose reduction factor
$LD_{50/30}$ = Lethal dose for 50% of the animals after 30 days.

Reduction Factor is plotted against the time difference (Δt) between the i.v. injection of the drugs and the onset of irradiation.

While the low molecular weight model (MEA) is only active if injected about 30 min before irradiation, the polymer (S_3) shows the same activity up to a period of 8 days.

In this special case it could be shown that the molecular weight of the polymers used did not influence their effectiveness as radio-protectants. Carefully fractionated terpolymers (S_3) with $\bar{M}_n$-values between 3000 and 80,000 were examined at two Δt values.[9,44]

To return to the schematic diagram depicted in Figure 4, the third area

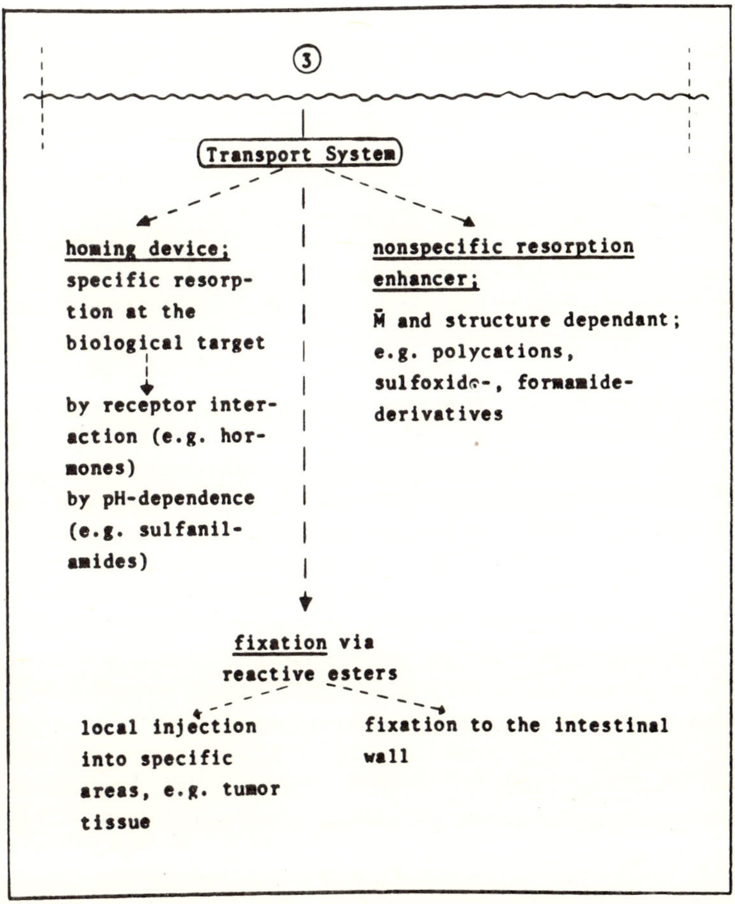

FIGURE 12 Transport system for pharmacologically active polymers (see model in Figure 4).[9,103]

of general importance for pharmacologically active polymers is that of the *transport system.*

Such transport systems may induce non-specific or specific resorption with the final goal to act as homing devices and, thus, transform unspecific bioactive materials into guided missiles seeking defined receptor sites at the target cells.

One has to keep in mind that pharmacologically active polymers, simply because of their high molecular weights, will be capable of inducing some cell-specific uptake without regard to other factors. As will be discussed later the intracellular penetration of polymers is restricted to the endocytotic route,[45,46] (see Figure 14) and one can therefore use the high endocytotic activity of certain cells to preferentially increase their polymeric drug load. This concept is the basis of recent attempts to apply polymers in cancer chemotherapy[9,38,47–52] by taking advantage of the high endocytotic activity of many tumor cells.[53–55] This is also the case for all types of diseases of the reticuloendothelial system which is in fact the main target for polymeric drugs.

Transport systems which can act, according to Figure 12, as *non-specific resorption enhancers* by inducing variations in the normal body distribution of polymers can be expected of surface-, membrane- and skin-active systems, e.g., sulfoxides and formamides. Based on studies of dimethyl sulfoxide as a carrier for drugs through the skin and various membranes[56] a series of sulfoxide-containing polymers was prepared[22,23] and their ability to enhance the uptake of certain drugs through the skin was successfully checked.[23] The body distribution of these water-soluble and very non-toxic polymers after i.v. application was established by side-chain ^{14}C-labeling using rats with Walker muscle tumors.[23,57] Some of the results are shown in Figure 13.

Although the average molecular weight ($\bar{M}_n$) of the polymeric sulfoxide was as high as 170,000, after 72 h 41% of the material was excreted in the urine (37%) and feces (4%). These results are in contrast to those obtained in comparable investigations of poly(vinylpyrrolidone) fractions of the same molecular weight which were found not to be excreted at all via the kidneys. Even more remarkable is the fact that even after 72 h the highest concentration of the polymer was found in the serum and blood rather than the liver. This may result from exceptionally high binding to transport proteins in the blood. The amount of polymer taken up by the liver, spleen and lungs was normal whereas uptake in the muscle tumor was 181% compared with 28% in normal muscle tissue. The amount found in the brain (12.1% after 48 h, 10.2% after 72 h) suggests that part of the polymer must have been hydrolyzed with consequent appearance of low molecular weight β-hydroxyethyl sulfoxides in the brain area. The fact that the polymers themselves definitely cannot pass into the brain area has been checked more recently

Organ	48 h	72 h
serum	650.2 %	434.7 %
blood	473.6 %	284.8 %
liver	107.9 %	85.4 %
spleen	106.3 %	100.5 %
lung	95.0 %	72.9 %
muscle	28.0 %	30.8 %
muscle-tumor	181.3 %	129.5 %
brain	12.1 %	10.2 %

FIGURE 13 Body distribution of

$$[CH_2{-}CH]_n \quad | \quad CO{-}O{-}CH_2{-}CH_2{-}SO{-}^{14}CH_3$$

$(\eta)=37.0$ (Ml/g); 30°C, in H_2O in rats with Walker-muscle-tumor (50 mg i.v./rat$\triangleq$0.25 mg/g tissue; 0.25 mg polymer/g tissue$\triangleq$100%.)[9,103]

by working with fractions of the same polymer labelled in the man chain rather than the side chain.[57,58]

More specific transport of soluble polymers to target cells will – as already mentioned – only be possible by the use of *homing devices*, e.g., receptor-active components like immunoglobulins, enzymes, hormones and special drugs. The targeting of drugs is an old but still fascinating and not yet solved problem. This can be demonstrated especially in cancer chemotherapy where the slow progress can be attributed to the failure of antitumor agents to act on malignant cells only. Thus, the possibility to modulate the pharmacon transport as a major factor in determining the bioavailability and, hence, the biological action of drugs by macromolecular carriers is one of the typical and stimulating aspects of pharmacologically active polymers.

These various considerations concerning the body distribution of polymers and the specific or non-specific effect of transport systems according to Figure 12 are ultimately dependent on the mechanism of uptake of polymers by mammalian cells as mentioned at the beginning of this paragraph. This problem will be discussed briefly as the last topic of this survey in connection with a short discussion of some polymeric antitumor agents to demonstrate the chances for polymeric drugs in this area.

PIGGY-BACK ENDOCYTOSIS AND POLYMERIC ANTITUMOR AGENTS

In 1968 H. J. P. Ryser considered the question of the uptake of polymers by mammalian cells an underdeveloped area of research.[46] As demonstrated recently in a review by A. Trouet, Ch. DeDuve *et al.*[59] this field is not only no longer underdeveloped but has blossomed into one of the most fascinating areas of modern pharmacology. Synthetic polymers normally cannot enter cells by diffusion through a membrane or by active processes via membrane proteins. The normal mechanism whereby such a polymer passes the cell membrane is by the process of *endocytosis*.[60–62] This represents engulfment by an infolding of the plasma membrane with formation of a cytoplasmic vacuole (phagosome).

Following uptake, fusion with the enzyme-containing lysosomes yields the digestive vacuole.

The endocytosis of polymers is initiated by their adsorption at the cell membrane. This adsorption process plays an important role in the cell uptake of large particles and is influenced by various factors such as molecular weight or charge effects.[63,64] The higher the molecular weight the higher the rate of endocytosis. Compounds which are taken up selectively into the lysosomes following formation of the phagosome are called *lysosomotropic*.[59,65] The importance of lysosomotropic agents can be explained in connection with the process of *piggy-back endocytosis*,[66] which is also depicted in Figure 14. Piggy-back endocytosis is the cell-uptake of a compound, e.g., a drug complexed with or fixed to a carrier which is itself subject to endocytosis. Such drugs are released within the cell only after digestion of the carrier or cleavage of the detachable unit (temporary spacer, Figure 6) by lysosomal enzymes. After their attachment to the carrier the drug molecules can no longer diffuse freely with the consequence that their intracellular penetration is restricted to the endocytotic route, thus, giving rise to specific uptake into those cells which display high endocytotic activity.

Recognition of the relevance of the process of endocytosis to the applicability of pharmacologically active polymers can point the way to the possible development of specifically designed polymeric drugs. By variation of the transport system (Figure 12), it is possible to enhance or inhibit the endocytotic process as well as the lysosomal fusion process.

There are a few examples in the literature showing that the concept of piggy-back endocytosis of polymers not only offers wide scope for the design of new types of drugs but is already an established reality, e.g., liposomes,[67–71] dextrans,[72] latex particles[73] and DNA[59,61] have recently been used as carriers. Especially interesting are the investigations of Trouet,

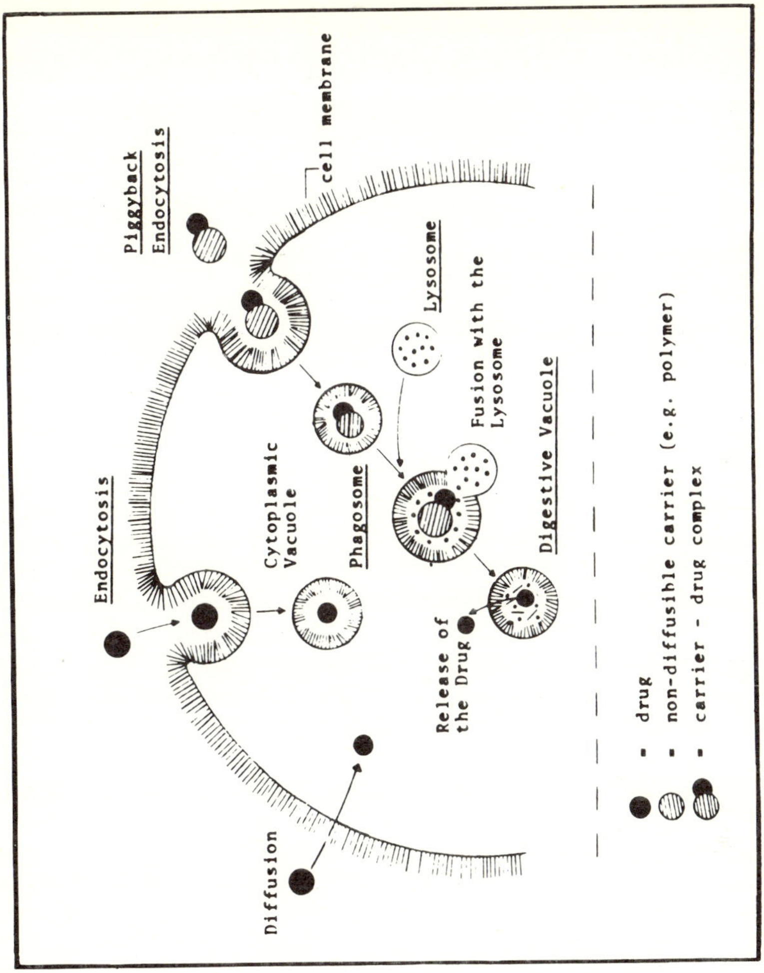

FIGURE 14 Mechanism of endocytosis and piggy-back endocytosis.[9,103]

DeDuve, Sokal *et al.* which have already reached the clinical stage.[59,74] This work involves the use of complexes between DNA and daunorubicin or adriamycin which are potent cytotoxic agents. The complexes are non-toxic inducers of endocytosis and the carrier is easily degradable by lysosomes. The results so far obtained in the treatment of leukemia, lymphosarcoma and solid tumors are encouraging and are discussed by A. Trouet in this volume.

Cancer chemotherapy[75,76] made its initial defined thrust in 1941 when Huggins and Hortages reported that the sex hormone oestrogen is useful in the treatment of prostatic cancer in men and by the discovery of the destruction of white blood cells by nitrogen mustards.[77,78] There are now about 45 anticancer drugs in medical practice with about 50 more in, or close to, clinical trials.[79]

Two of the main short comings of most tumor agents so far are the following:

1) Agents effective in killing neoplastic cells have also detrimental effects in normal cells particularly the rapidly proliferating ones, e.g., of the gastrointestinal tract and the bone marrow. Cancer chemotherapy is, thus, ultimately limited by its toxicity to these normal tissues.
2) Considering the various factors to which a drug is subject (biotransformation, receptor interaction, free or bound to transport proteins etc.) it is clear that in attempting to effect their combination in order to develop an ideal drug by alteration of a low molecular weight compound there is the possibility that the original biological activity may be lost.

Various attempts have been made to overcome these shortcomings by using polymers with carcinolytic activity.[47–52,80–84] Especially the Russian literature on this topic has been summarized already.[48] Many of the older investigations are purely empirical attempts or were based on the principle of polymer depot effect. But there is an increasing tendency to use considerations as discussed in this contribution and before[9] for a tailor-made synthesis of polymeric antitumor agents. In Table II some principal differences between normal and cancer cells are summarized.

One has to admit, however, that so far, with the exception of the lower pH in tumor cells[12,95,96] the tumor associated antigens[97–101] and the high rate of endocytosis, these differences are of more use in detecting cancer cells than in being therapeutically exploitable.

TABLE II

Differences between cancer and normal cells

Compared with normal cells, cancer cells have:
1) Lower pH (after injection of glucose)[85]
2) High rates of endocytosis[86]
3) Tumor-produced hormone peptides[87]
4) Tumor-associated antigens[88]
5) Lower calcium ion and higher potassium ion concentration[89]
6) Elevated amounts of methylated nucleosides[90]
7) Higher concentration of plasma mucoproteins and mucopolysaccharides[91]
8) Different potassium isotope ratios[92]
9) Greater need of exogenous zinc[93]
10) Some types (leukaemia cells) need exogenous L-asparagine[94]

One perfect example of the discussed model for pharmacologically active polymers[9] (Figure 4) was recently published by Rowland, O'Neill and Davies.[102] In a proceeding paper they could show [98] that successful tumor suppression with covalent conjugates of rabbit antitumor immunoglobulins and alkylating agents was dependent on using large amounts of linked material. These chemical variations of the immunoglobulin were reflected in loss of antibody activity as well as in a poor water solubility thus limiting the clinical potential of complexes in this form. They tried to overcome this by using polyglutamic acid (PGA) as the solubilizer and carrier system for the *p*-phenylenediamine mustard (PDM) and an immunoglobulin (Ig) from a rabbit antiserum against mouse lymphoma cells (EL4) as the homing device. This is summarized schematically in Figure 15.

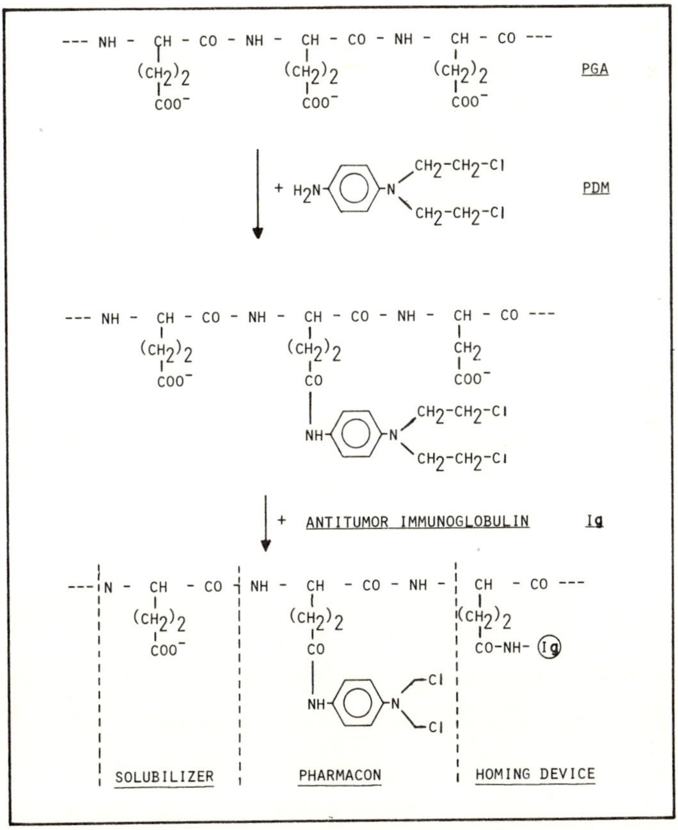

FIGURE 15 Synthesis of macromolecular tumor specific alkylating agents.[98] PGA=polyglutamic acid; PDM=*p*-phenylenediamine mustard; Ig=immunoglobulin from a rabbit antiserum against mouse lymphoma cells (EL 4).

A variety of physicochemical tests were carried out to show that both coupling steps with carbodiimide had taken place. The antibody activity of the complex was also determined. It was shown that 66% of the original anti-EL4 activity was retained in the complex without loss of specificity. To show the effectiveness of the polymer as an antitumor agent, mice were inoculated (ip.) with EL4 cells and received the first of four daily injections of the polymer or control preparation after 24 h. The results are given in Table III.

TABLE III

Effectiveness of the PGA–PDM–Ig–copolymer
Conditions: mice bearing EL 4 lymphoma; daily treatment, i.p., day 1–4

Drug	Survival time for 50% (days)	Mice alive free of tumor at day 60
NaCl–solution	13	0/5
Ig alone	19	0/5
PGA–PDM-copolymer	25	1/5
PGA–PDM-copolymer + free Ig	38	2/5
PGA–PDM–Ig-copolymer	> 100	5/5

The polymer carrying the drug and the Ig as a homing device was far more active than the components alone or their mixtures. The concept of pharmacologically active polymers has thus found an interesting and promising application.

Considering the wide variety of synthetic methods, drugs, polymeric carrier systems and homing devices including the newer developments in cellular biology, one can look forward to a quite fascinating new field of drug design, showing the importance of cooperative efforts between polymer research, therapeutic problems and cell biology.

References

1. H. Lee and Kr. Neville, *Handbook for Surgery and Artificial Organs*, Pasadena Technology Press, Calif. (1971).
2. D. J. Lyman, *Rev. in Macromolecular Chem.*, **1**, 355 (1966); *Ann. N.Y. Acad. Sci.*, **146**, 113 (1968); *Angew. Chem.*, **82**, 367 (1974).
3. K. P. Khomyakov, A. D. Virnik and Z. A. Rogovin, *Russ. Chem. Rev.*, **33**, 462 (1964).
4. H. F. Mark, *J. Pure Appl. Chem.*, **16**, 201 (1968).
5. V. Kabivanov and M. Georgieva, *Plasie und Kautschuk*, **19**, 886 (1972).
6. C. Schuerch, *Advances Polymer Sci.*, **10**, 173 (1972).
7. L. G. Donaruma, *Progress in Polymer Science*, **4**, 1 (1974).
8. I. Iliev, M. Georgieva and V. Kabaivanov, *Russian Chem. Reviews*, **43**, 69 (1974).

9. H. Ringsdorf, *J. Polymer Sci., Symp.*, **51**, 135 (1975).
10. H. G. Batz, *Advances in Polymer Sci.*, **23**, 25 (1977).
11. W. A. Ritschel, in Drug Design IV, *Medicinal Chemistry* Vol. II, 37 (1973).
12. G. Abel, T. A. Connors, W. C. Ross, Nguyen-Joang-Nam, H. Hoellinger, and L. Pichat, *Europ. J. Cancer*, **9**, 49 (1973).
13. E. J. Ariëns, in Drug Design, Vol. IV, *Medicinal Chemistry* (1973).
14. H. W. Schlipköter and A. Brockhaus, *Fortschritte d. Staublungenforschung*, 1963, 397; *Silikose Forsch.*, **6**, 213 (1965); *Klin. Wochenschr.*, **39**, 1182 (1961).
15. A. C. Allison, J. S. Harington and M. Birbeck, *J. Exp. Med.*, **124**, 141 (1966).
16. P. F. Holt and H. Lindsay, *J. Chem. Soc.* **B**, 54 (1969); **B**, 233 (1968).
17. A. C. Allison, *Arch. Intern. Med.*, **128**, 131 (1971).
18. P. Ferruti and M. A. Marchisio, *Medicina del Lavoro*, **57**, 481 (1966).
19. W. Wessel, M. Schoog and E. Winkler, *Arzneimittel Forschung*, **21**, 1468 (1971).
20. J. Kopecek and H. Bazilova, *Europ. Polym. J.*, **9**, **7** (1973); *J. Biomed. Mater. Res.*, **7**, 123 (1973).
21. M. Liefländer and F. J. Strecker, *Hoppe-Seylers Z. Physiol. Chem.*, **347**, 268 (1966).
22. H. G. Batz, V. Hofmann and H. Ringsdorf, *Makromol. Chem.*, **169**, 323 (1973).
23. V. Hofmann, H. Ringsdorf and G. Muacevic, *Makromol. Chem.*, **176**, 1929 (1975).
24. T. J. Gill and H. W. Kunz, *Proc. Nat. Acad. Sci. USA*, **61**, 490 (1968).
25. B. Andersson and H. Blomgren, *Cell. Immunol.*, **2**, 411 (1971).
26. R. Schnee, Dissertation, University of Mainz (1977).
27. Y. S. Klausner and M. Bodzansky, *Synthesis*, 453 (1972).
28. H. G. Batz, Gi. Franzmann and H. Ringsdorf, *Angew. Chem.*, **24**, 1189 (1972); *Makromol. Chem.*, **172**, 27 (1973).
29. Gi. Franzmann, Dissertation, University of Mainz (1975).
30. P. Cuatrecasas, *J. Biol. Chem.*, **245**, 3059 (1970); *Angew. Chem.*, **84** (1972). E. Steers, P. Cuatrecasas and H. P. Pollard, *J. Biol. Chem.*, **246**, 196 (1971).
31. J. C. Venter, J. E. Dixon, P. R. Navoko and N. O. Kaplan, *Proc. Nat. Acad. Sci. USA*, **69**, 1141 (1972); **70**, 1214 (1973).
32. J. C. Venter, J. Ross and N. O. Kaplan, *Proc. Nat. Acad. Sci. USA*, **72**, 824 (1975).
33. J. C. Venter, L. J. Arnold and N. O. Kaplan, *Molecular Pharmacology*, **11**, 1 (1975).
34. H. Jatzkewitz, *Hoppe-Seylers Z. Physiol. Chem.*, **297**, 149 (1954).
35. H. Jatzkewitz, *Z. Naturforsch.* **10b**, 27 (1955).
36. Ch. Wilk, Dissertation, University of Mainz (1977).
37. J. Drobnik, J. Kopecek, J. Labsky, P. Rejmanova, J. Exner, V. Sandek, and J. Kalal, *Makromol. Chem.*, **177**, 2833 (1976).
38. H. G. Batz, H. Ringsdorf and H. Ritter, *Makromol. Chem.*, **175**, 2229 (1974).
39. H. Ringsdorf, H. Ritter and J. Sandow, unpublished results.
40. Z. M. Bacq, *Chemical Protection Against Ionizing Radiation*, Springfield, Virginia, USA (1965).
41. C. G. Overberger and B. V. Bonsignore, *J. Am. Chem. Soc.*, **80**, 5431 (1958).
42. C. G. Overberger, H. Ringsdorf and B. Avchen, *J. Med. Chem.*, **8**, 862 (1965).
43. H. Ringsdorf, *Strahlentherapie*, **132**, 627 (1967).
44. H. Ringsdorf, A. G. Heisler, F. H. Müller, E. H. Graul, and W. Rüther, *Biological Aspects of Radiation Protection*, edited by T. Sugahara and O. Hug, Igaku Shoin Ltd., Tokyo, p. 138 (1971).
45. H. Holter and H. Holtzer, *Exp. Cell. Res.*, **18**, 421 (1959).
46. H. J. P. Ryser, *Science*, **159**, 390 (1968); *Proc. Fourth Int. Cong. Pharm.*, **3**, 96.
47. O. Zubova, E. Kirsh, T. Lebedeva, A. Shorokhova, A. Silaev, V. Kabanov, V. Kargin, *Dokl. Akad. Nauk. USSR*, **186**, 477 (1969).
48. J. Jliev, M. Georgieva and V. Kabaivanov, *Russ. Chemical Rev.*, **43**, 69 (1974).
49. A. Trouet, D. Deprez-De Campeneere and Ch. DeDuve, *Nature New Biology*, **239**, 110 (1972).
50. D. S. Breslow, E. J. Edwards and N. R. Newburg, *Nature*, **246**, 160 (1973).
51. P. Ferruti, F. Danusso, G. Frenchi, N. Polentarutti, and S. Garattini, *J. Med. Chem.*, **16**, 496 (1973).

52. J. Bartulin, M. Przybylski, H. Ringsdorf, and H. Ritter, *Makromol. Chem.*, **175**, 1007 (1974).
53. H. Busch, E. Fujiwara and D. C. Firszt, *Cancer Res.*, **21**, 271 (1961).
54. A. R. Poole, in *Lysosomes in Biology and Pathology*, edited by J. T. Dingle, North-Holland Publ. Co., Amsterdam, Vol. III, p. 303 (1973).
55. C. H. Sutton and N. H. Becher, *Ann. N.Y. Acad. Sci.*, **159**, 497 (1969).
56. S. W. Jacob, E. E. Rosenbaum and D. C. Wood, *Dimethylsulfoxide*, Vol. 1, M. Dekker Publishing Co., N.Y. (1971).
57. V. Hofmann, H. Ringsdorf, and E. Schaumlöffel, unpublished results.
58. V. Hofmann, Dissertation, University of Mainz (1975).
59. C. DeDuve, Th. DeVarsy, B. Poole, A. Trouet, P. Tulkens, and F. Van Hoof, *Biochemical Pharmacology*, **23**, 2495 (1974).
60. C. DeDuve and W. Wattiaux, *Ann. Rev. Physiol.*, **28**, 435 (1966).
61. P. J. Jacques, *Rev. Questions Sci.*, **27**, 99 (1966).
62. P. J. Jacques, in *Lysosomes in Biology and Pathology*, edited by J. T. Dingle, North-Holland Publ. Co., Amsterdam, Vol. II, p. 395 (1969).
63. H. J. P. Ryser, *Nature*, **215**, 934 (1967).
64. H. J. P. Ryser, *Science*, **150**, 501 (1965).
65. C. DeDuve, in *Biological Approaches to Cancer Chemotherapy*, edited by R. J. C. Harris, Academic Press, London, p. 101 (1961).
66. A. J. Sbarra, W. Shirley, and W. A. Bardawil, *Nature*, **194**, 255 (1962).
67. M. Szekerke, R. Wade, and M. E. Whisson, *Neoplasma*, **19**, 211 (1972).
68. G. Gregoriadis, C. P. Swain, A. S. Tavill, and E. J. Wills, *Lancet*, 1313 (1974).
69. G. Gregoriadis and B. E. Ryman, *Biochem. J.*, **129**, 123 (1972).
70. Y. E. Rahman, M. W. Rosenthal and E. A. Cerny, *Science*, **180**, 300 (1973).
71. W. E. Magee and O. V. Miller, *Nature*, **235**, 339 (1972).
72. S. K. Basu, *Lancet*, **1**, 1430 (1963).
73. T. M. S. Chang, *Science*, **146**, 524 (1964).
74. G. Sokal, A. Trouet, J. L. Michaux, and G. Cornn, *Eur. J. Cancer*, **9**, 391 (1973).
75. *Cancer Medicine*, edited by J. F. Holland and E. Frei, Philadelphia (1973).
76. L. N. Ferguson, *J. Chem. Ed.*, **52**, 688 (1975).
77. A. Gilman and F. S. Philips, *Science*, **103**, 409 (1946).
78. T. A. Connors, *Topics in Current Chemistry*, **52** (Medicinal Chemistry) 141 (1974).
79. Report of the Division of Cancer Treatment, National Cancer Institute, Bethesda, Md., Vols. 1 and 2 (1974).
80. C. W. Mosher, R. H. Iwamotow, E. M. Acton, and L. Goodman, *J. Med. Chem.*, **7**, 650 (1964).
81. E. M. Hodnett and J. Tien Hai Tai, *J. Med. Chem.*, **17**, 1335 (1974).
82. H. G. Batz, H. Ringsdorf, and H. Ritter, *Makromol. Chem.*, **175**, 2229 (1974).
83. G. Franzmann and H. Ringsdorf, *Makromol. Chem.*, **177**, 2547 (1976).
84. L. A. Carpino, H. Ringsdorf, and H. Ritter, *Makromol. Chem.*, **177**, 1631 (1976).
85. Z. B. Papanastassiou, R. J. Bruni, E. White, and P. L. Levins, *J. Med. Chem.*, **9**, 725 (1966); P. Weiss and B. I. H. Scott, *Proc. Nat. Acad. Sci. USA*, **50**, 330 (1963).
86. H. Busch, E. Fujiwara and D. C. Firszt, *Cancer Res.*, **21**, 371 (1961); G. C. Easty, M. M. Yarnell and R. D. Andrews, *Brit. J. Cancer*, **18**, 354 (1964); E. S. Handler, E. E. Handler and A. B. Chesebrough, *J. Reticuloendothel. Soc.*, **6**, 511 (1969); J. L. Mego and J. D. McQueen, *Cancer Res.*, **25**, 865 (1965); C. H. Sutton and N. H. Becker, *Ann. NY Acad. Sci.*, **159**, 497 (1969); J. Whang-Peng, S. Perry and T. Knutsen, *J. Nat. Cancer Inst.*, **38**, 969 (1967).
87. cf. T. H. Maugh, *Science*, **184**, 147 (1974).
88. cf. T. H. Maugh, *Science*, **184**, 147 (1974).
89. A. K. Brewer, *Amer. Laboratory*, **5**, 12 (1973).
90. C. C. Cheng, *J. Pharm. Sci.*, **61**, 645 (1972).
91. D. W. Dixon, *Cancer*, **31**, 596 (1973); V. N. Nigam and A. Cantero, *Adv. Cancer Res.*, **16**, 1 (1972).
92. A. K. Brewer, *Amer. Laboratory*, **5**, 12 (1973).

93. C. P. Li, "Anticancer Agents Recently Developed in the People's Republic of China", D. H. E. W. Publn. No. (NIH) 74, Washington, DC (1974).
94. C. Tan, *Hosp. Pract.*, **7**, 99 (1972).
95. V. Hofmann, M. Przybylski, H. Ringsdorf and H. Ritter, *Makromol. Chem.*, **177**, 179 (1976).
96. G. Abel, Th. A. Connors, V. Hofmann and H. Ringsdorf, *Makromol. Chem.*, **177**, 2669 (1976).
97. I. Flechner, *Eur. J. Cancer*, **9**, 741 (1973).
98. D. A. Davies and G. J. O'Neill, Proc. XI Int. Cancer Congress, Florence 1974.
99. R. Levy, E. Hurwitz, R. Maron, R. Avnon, M. Sela, *Cancer Res.*, **35**, 1175 (1975).
100. T. Ghose, S. T. Norwell, A. Guela, and A. S. McDonald, *Eur. J. Cancer*, **11**, 321 (1975).
101. R. D. Rubens and R. Dulbecco, *Nature*, **248**, 81 (1974).
102. G. F. Rowland, G. J. O'Neill, and D. A. L. Davies, *Nature*, **255**, 487 (1975).
103. Reprinted with the permission of John Wiley and Sons, Inc.

DISCUSSION

Mr. L. Olanoff (*Case-Western Reserve Univ., Cleveland, OH*): Have any studies been done with normal drugs attached to polymer carriers, which are highly endocytotic, to determine whether the drugs are being concentrated in specific cell systems?

Dr. H. Ringsdorf: Yes, there definitely were. We started with an investigation of radiation prophylactics as I mentioned. These compounds are taken up to a high degree by the RES like the liver, the lymph system, the small intestine and the bone marrow. This is what you get with nearly all compounds when they are in a polymeric form.

This has been very well investigated for different types of plasma expanders, and they all go into cell systems which have high rates of endocytosis.

Mr. L. Olanoff: Has there been any corresponding immunogenic response?

Dr. H. Ringsdorf: No, this is completely independent of cell uptake as far as I know. As I mentioned, the question of antibody–antigen reactions of the synthetic compounds has not been discussed thoroughly so far. We are trying to do a little.

However, in my opinion, there is no connection between the rate of uptake by endocytosis and immunological effects.

Dr. N. Kaplan (*Univ. of California/San Diego, LaJolla, CA*): How was the immunoglobulin attached to the polymeric delivery system?

Dr. H. Ringsdorf: The phenylene diamine lost as the antitumor agent as well as the immunoglobulin as the homing device were fixed to 2 poly(glutamic acid) via carbodiimides. The details are given in the paper of Rowland, O'Neill and Davies, *Nature*, **255**, 487 (1975) (Reference 102 in this contribution).

Dr. W. Leonard (*Dynapol Corp., Palo Alto, CA*): Is the cellular uptake of macromolecules, as described by H. Ryser, of sufficient magnitude to achieve the desired pharmacological effects?

Dr. H. Ringsdorf: Yes. This will be demonstrated in the contribution of A. Trouet in this meeting. There is no doubt that the rate of endocytosis is increased with molecular weight. It is known that above a certain molecular weight, say 100,000, the rate of uptake is about 100 times faster when the molecular weight is doubled.

The question of uptake is a very interesting point, because one might find an area where a polymeric drug has too high a molecular weight to diffuse in and too low a molecular weight to be taken in by endocytosis. So there may be an area which can exclude the uptake of a drug into a cell system. We are looking into this.

Dr. W. Leonard: Ryser also distinguishes between the macromolecules that are actually transported across cellular membranes and the macromolecules that serve exclusively to promote their uptake by the cell. Are these separate functions incorporated into your model?

Dr. H. Ringsdorf: No, they are not. But it is difficult to differentiate between promotor-macromolecules and transport-macromolecules. Ryser cannot distinguish between these two or a complex of both. The increase of the uptake may be due to the higher molecular weight of the complex or to an induced vescicle formation by the promotor macromolecule.

Dr. R. Capozza (*Alza Corp., Palo Alto, CA*): After a drug which is attached to a non-degradable polymer carrier is taken into the cell by endocytosis, how is the carrier eventually removed from the cell?

Dr. H. Ringsdorf: I didn't want to cover this in detail because that will be covered in Dr. Trouet's talk. I'll just point out very briefly that you have fusion of the vescicles with the lysosomes, and digestion of the carrier in the vescicles—if it is a digestable carrier. This is typical for example, for DNA.

The removal of the nondegradable carrier from the cell is a problem. This is why one could debate whether it is worthwhile to discuss synthetic nondegradable polymers as polymeric drugs. However, there is the possibility of exocytosis. The body can at least partially get rid of a synthetic polymeric compound by the endero-hepatic cycle, i.e. via liver, bile and intestine. We've found, e.g., that after half a year even high molecular weight sulfoxides are not found in the body anymore (^{14}C-labelling).

Dr. A. Trouet (*International Institute of Cellular and Molecular Pathology, Brussels, Belgium*): The rate of the exocytosis is very low, and the main result of an uptake of undigestible polymer is lysosome overloading and cellular overloading, which can of course lead to toxic effects. Exocytosis does indeed exist, although it is normally much too slow to be important. Lysosome overloading has been reported for polymers such as dextran, poly(acrylic acid), and poly(vinylpyrrolidone).

Dr. H. Ringsdorf: The toxicity of poly(vinylpyrrolidone) has been heavily debated. However, there are, as far as I know, no cases of liver diseases which have been proved to be related to the injection of poly(vinylpyrrolidone).

Dr. L. Bangs (*Dow Chemical Co., Midland, MI*): There have been some interesting studies on endocytosis of our uniform latex particles (non-degradable). In these studies, there was no exocytosis, and the cells eventually burst.

Dr. R. Capozza: How long does it take for an intravenously administered drug tethered to a polymeric carrier to be taken into cells by endocytosis?

Dr. H. Ringsdorf: This is difficult for me to discuss because I don't have all the results. As soon as you have absorption by a membrane (this was done *in vitro* with cell cultures), the rate may be very high. The uptake may occur in 0.1 s.

Dr. A. Trouet: My experience has been that 50% of DNA is taken up in 1–2 h in culture. With *in vivo* injection, it becomes fixed to the membrane in about 15 min, although 1–2 h is required for uptake into the cell.

Dr. H. Ringsdorf: The problem of endocytosis has not been studied intensively enough. There is no doubt that the absorption at the cell membrane plays an important role and may strongly influence form and rate of the uptake. Systematic studies on the structure dependence of absorption and uptake will be necessary.

Polymeric Affinity Drugs

EUGENE P. GOLDBERG

Department of Materials Science and Engineering, University of Florida, Gainesville, FL32611, U.S.A.

Polymer compatibility and thrombogenicity in the human circulatory system and the development of polymers for cardiovascular prostheses have received primary attention in biopolymer research in recent years. In general, prosthetic materials must combine tissue acceptance with desired mechanical, electrical, optical and other physical properties. Unfortunately, despite the considerable research activity devoted to blood compatibility and the somewhat lesser level of attention to tissue compatibility, polymer bioacceptance is still understood in only an empirical fashion.

Polymers designed to possess both bioacceptance and biological activity for therapeutic applications have begun to receive increasing attention. To date, polymeric drug research has been most strongly focused upon insoluble materials for drug localization, for controlled drug release and for detoxification by complex formation or enzymatic action.

We have now reached the point where the investigation of soluble polymeric therapeutic agents combining both site-specific binding properties and therapeutic biological activity represents one of the most exciting areas for biopolymer research. These macromolecular materials, based upon biospecific adsorption and affinity chromatography concepts, may be termed "affinity drugs"; drugs which incorporate complex-forming functions to promote highly specific binding to cell surface receptor sites or to particular metabolites such as enzyme substrates, antibodies, polysaccharides or lipoproteins. This paper will briefly describe polymeric affinity drug concepts directed at atherosclerosis, cancer, and urolithiasis therapy.

INTRODUCTION

The contributions of polymer science to the solution of life sciences problems have become increasingly important during the past 15–20 years. This has been due in part to the obvious need to establish more fundamental correlations between molecular structure and physiological function for biological macromolecules. Equally important have been those efforts to utilize synthetic and natural polymers for prosthesis, for immobilizing and modifying enzymes, for improving biological separations and for biopolymer synthesis. Polymer drug research is only now emerging as an outgrowth of these earlier studies. Although a recent review by Donaruma[1] cites hundreds of references to the preparation and properties of polymeric drugs, this field of research is still in its infancy.

Numerous soluble and insoluble polymeric drug compositions have been reported to demonstrate therapeutic activity for a wide range of disease conditions. However, the greatest attention and the most significant accomplishments thus far have been in the area of insoluble biopolymers for localization and controlled release of drug action.[2]

Soluble polymeric drugs have received much less attention, possibly because the opportunities for localizing action and beneficially altering drug activity have been less obvious as compared with insoluble systems. The main thrust of this paper is therefore to suggest some interesting opportunities for attacking major disease problems using water soluble polymers. These approaches to localized therapy involve attachment of biospecific complexing ligands to polymeric drugs, particularly soluble systems, to produce polymers which combine site-specific binding properties with therapeutic biological activity. These macromolecular materials, based upon concepts derived from affinity chromatography, may be appropriately termed "affinity drugs"; drugs which incorporate complex-forming functions to promote strong and highly specific binding to cell surface receptor sites or to various metabolites such as enzymes, enzyme substrates, antibodies, antigens, lipoproteins, etc. Some current studies aimed at therapy for the important clinical disease problems of atherosclerosis, urolithiasis and cancer will be very briefly noted after first discussing the general rationale for the "polymeric affinity drug."

DISCUSSION

The confluence of a number of areas of biopolymer research lead to the affinity drug concept. One important area, that of polymer biocompatibility and thrombogenicity, has received considerable attention, especially for the identification of materials suitable for cardiovascular prosthesis. Although surface phenomena underlying blood and soft or hard tissue bioacceptance are still understood in relatively empirical terms, these studies have provided some basis for the selection of polymers which may be useful in polymer drug research; polymers which are likely to have minimal tissue antagonism and antigenicity. In particular, the bioacceptability of soluble and cross-linked hydrophilic polymers based upon poly(vinylpyrrolidone) [PVP], poly(hydroxethylmethacrylate) [HEMA], poly(lactic acid) or dextran is noteworthy in suggesting appropriate polymeric matrices for drug modification.

The field of immobilized enzymes and biocatalysis pioneered by E. Katchalsky[3,4] has contributed a great deal to the identification of coupling chemistry suitable for binding proteins to polymers with retention of activity by the biological macromolecules. Furthermore, this area of research has been most instructive in suggesting the protective role of a polymer-associated microenvironment in enhancing enzyme stability, in altering pH optima, and in limiting accessability (especially from the standpoint of toxicity and antigenicity) of a covalently bound protein to a physiological environment. Soluble polymer–enzyme systems which are more directly pertinent to this discussion have unfortunately received relatively little attention. However, those few studies reported, e.g., for chymotrypsin–dextran,[5] α-amylase-dextran[6] and glucose oxidase–poly(vinyl alcohol/12% acetate),[7,8] do further the idea that a soluble polymer matrix may significantly alter the stability and accessability of an attached or associated protein.

The recent great interest in insoluble polymeric drugs prepared by encapsulation, by solid solution or admixture with linear and cross-linked polymers, or by covalent binding to polymer matrices stems from the desire to alter drug pharmacology in a selective fashion and limit or control systemic exposure to hazardous bioactive agents. Indeed the majority of papers in this Symposium on Polymeric Delivery Systems[9] are devoted to drug immobilization to (a) localize action, (b) control release kinetics and prolong action at desired concentrations, (c) enhance drug stability and (d) moderate antigenicity and toxicity.

Probably the two most important areas of research in recent years from the standpoint of providing impetus to new site-specific drug delivery concepts have been in the fields of (a) affinity chromatography and (b) enzyme deficiency diseases. Advances in affinity chromatography,[4,10] particularly the identification of highly specific cell surface receptor binding sites, provide the conceptual framework for designing therapeutic agents, "affinity drugs," with specific cell or tissue targeting properties. The increasing attention to enzymes as drugs for therapy of enzyme deficiencies or for metabolic intervention in such diseases as cancer is a collateral development of significance in that enzymatic drugs are indeed "affinity drugs" by virtue of their substrate binding specificity. That such enzymatic affinity drugs alone and unmodified are inadequate is amply indicated by clinical and pharmacological evaluations showing serious toxicity, immunogenicity and instability in such systems as (a) urokinase, streptokinase or fibrinolysin for thrombolytic activity, (b) asparaginase for lymphosarcoma chemotherapy or (c) arylsulfatase-A for metachromatic leukodystrophy.[11]

The problems with systemically hazardous drugs and with "naked" affinity drugs such as enzymes or antibodies, coupled with the potential for protecting both the patient and the drug by providing a mediating polymeric microenvironment, leads to the concept of the polymeric affinity drug (PAD).

PADs derived from insoluble matrices, e.g., dispersions, microcapsules or liposomal vesicles, by attachment of appropriate affinity binding ligands to these insoluble drug carriers will surely be interesting and important and will be considered in more detail in future affinity drug papers. However, the prime focus of this discussion is upon soluble PAD systems. A schematic representation, Figure 1, indicates the features of a generalized PAD. One or more "affinity binding ligands" and one or more "bioactive functions" may be covalently attached to a water soluble polymer. The polymer is a soluble analog of the insoluble affinity chromatography polymer matrix. The role of the affinity binding ligand (which may also be physiologically active, e.g., an enzyme) is to provide site-specific binding properties for cell surface receptors or for specific inter- or intra-cellular metabolites. Although it is believed that binding specificity will be enhanced by the combined use of several differing affinity ligands, this important point needs experimental verification. The "bioactive function" may be any moiety having the physiological activity sought at the target site ranging from hormonal action, to cytotoxicity, to immunostimulation, to metabolic inhibition, to enzymatic activity, etc.

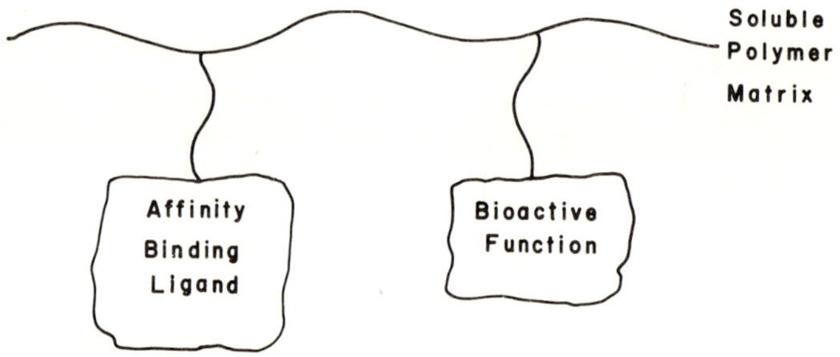

FIGURE 1 The features of a generalized polymeric affinity drug (PAD).

The PAD concept is critically dependent upon the soluble polymer matrix. The polymer must (1) stabilize and protect the attached functions from degradation, (2) protect the patient by minimizing toxicity and antigenicity, (3) provide prolonged and controlled activity through molecular weight selection (for elimination of stable matrices) and/or polymer instability (for degradation and elimination) and (4) provide a matrix for multifunctional attachment facilitating strong target binding and broad spectrum bioactivity. The soluble matrix may be derived from such relatively stable bioacceptable hydrophilic polymers as PVP, HEMA or poly(vinylpyridine-*N*-oxide) and their functionalized copolymers or from more readily degradable polymers such as soluble dextran. Molecular weight and biostability of the ligand–matrix bond as well as biostability and tissue affinity of the polymer matrix itself will contribute importantly to the pharmacological fate of the PAD. For example, the known affinity of PVP for the reticuloendothelium system (RES) will result in RES accumulation of very high molecular weight PVP. The prospect that endocytosis may be favored for higher molecular weight PAD systems or that some tissues will have an enhanced tendency for endocytosis, e.g., tumor cells, is another pharmacological factor.

The philosophy of our polymer drug research has been to work back from the clinical disease problem to seek PADs which may contribute both to the understanding of disease mechanisms as well as to the development of new types of drug therapy. Three areas of major research interest are briefly summarized in the following sections.

CANCER CHEMOTHERAPY

Although there is considerable controversy regarding tumor-specific antigenicity (TSA) and the very complex immunological character of transformed cells,[12] it seems likely that TSA concepts will be utilized to target specific cells or tissues. Chemotherapy research must increasingly be concerned with localization to limit the serious systemic toxicity associated with current non-localized chemotherapeutic practices. Some tumor-specific affinity drug examples include the study of a cytotoxic melanoma antibody[13] and the complexing of chlorambucil with tumor immunoglobulin to improved tumor cell killing as compared with the individual cytotoxic agent or antibody.[14] Most important is the work by Sela's group in which adriamycin and daunomycin have been covalently bound to B-leukemia antibody with reportedly improved cytotoxic action by virtue of drug–antibody affinity for the target cells.[15] In the context of the polymeric affinity drug concept presented in this paper, it is considered that an antibody–drug conjugate might be significantly improved by soluble polymer attachment to reduce systemic toxic effects and increase drug stability and efficacy. This type of PAD is schematically represented in Figure 2.

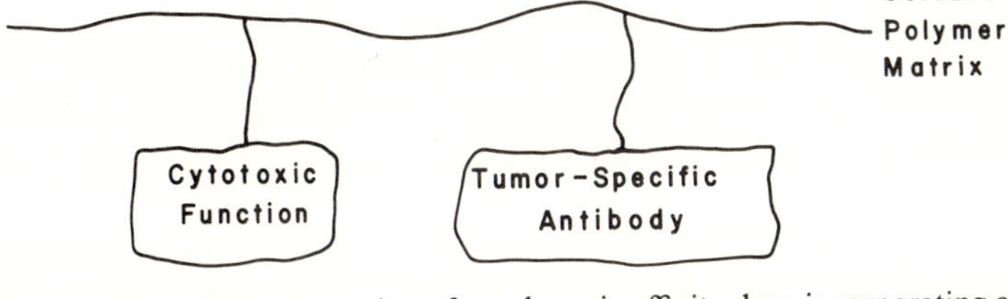

FIGURE 2 Schematic representation of a polymeric affinity drug incorporating a tumor-specific antibody and a cytotoxic function.

It seems reasonable to predict that this general class of PADs will have most important consequences for safer and more effective cancer chemotherapy in the future. One piece of evidence that this approach may indeed lead to a new class of polymeric chemotherapeutic drugs is afforded by a recent communication from the Davies-O'Neill group at Searle.[16] They reported a covalently coupled system comprised of poly(L-α-glutamic acid) as a polymer matrix, *p*-phenylenediamine mustard as a cytotoxic function and rabbit Ig against mouse lymphoma (EL4) as an affinity binding ligand to have greater chemotherapeutic activity than the individual constituents.

UROLITHIASIS

The mechanisms by which mineralization, such as kidney stone formation, occurs in animals and humans is not clearly understood. Even the microscopic morphology and chemical composition of kidney stones is somewhat unclear and rather variable.[17] For this brief and preliminary discussion it is sufficient to note that (a) calcium oxalate and phosphate stones are quite common and important clinically, (b) they are difficult to dissolve and very often require surgical attention, (c) they contain 2–5 wt. % of an organic matrix material which is generally regarded as primarily insoluble mucoprotein and (d) they show a microcrystalline structure when examined by electron microscopy.

One facet of our PAD studies dealing with this important clinical disease condition (in collaboration with B. Finlayson) is aimed at testing a hypothesis regarding calcium oxalate stone morphology; that the stone is a polycrystalline aggregate bound or "glued" together by the mucoprotein matrix. If this is the case, lysis of the matrix with an appropriate protease or a more complex enzyme system tailored specifically to the stone matrix composition should promote significant mechanical weakening or even deaggregation of kidney stones. Initial *in vitro* enzyme lysing and stone affinity binding experiments will guide the synthesis of polymer–enzyme PAD systems for urolithiasis therapy.

CARDIOVASCULAR THERAPY—PLAQUE REGRESSION

The objectives of this research (in collaboration with W. Fisher) are to devise a novel PAD therapy and to help clarify disease mechanisms for atherosclerosis. This disease is marked by progressive development of fatty plaques or lesions in the arterial intima which eventually lead to deterioration of the cardiovascular system.[18] A major element of this study is, therefore, the investigation of enzymatic drugs for plaque regression using those enzymes for which there are substrates in the atheromic tissue. Because of the complex composition of the plaques, cholesterol esterase, protease, fibrinolysin, hyaluronidase and collaginase are all candidates for investigation.

The facility with which an enormous molecular aggregate of 2–3 million molecular weight, low density lipoprotein (LDL), crosses the endothelium lining of the arteries into the arterial intima is encouraging to the idea that a polymeric drug may also be readily transported to plaque sites in the intima. For the purposes of this brief discussion which will be elaborated in greater detail in future papers, it is sufficient to simply indicate a generalized PAD structure for plaque therapy, shown in Figure 3. Site-specific delivery of this model plaque lysing PAD is directed by the affinity binding properties of the apolipoprotein (Apo-LDL) antibody as well as the enzyme substrate affinities.

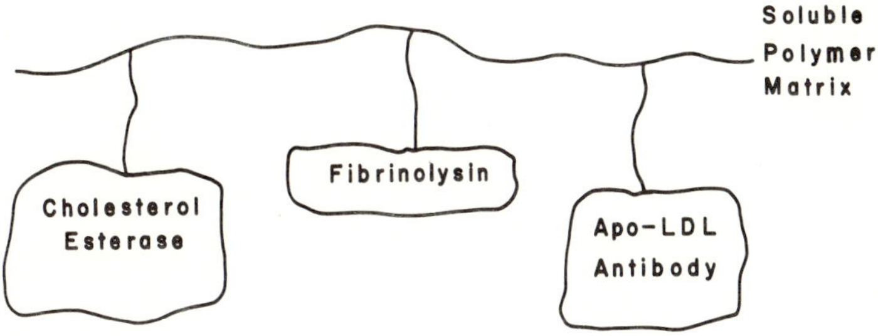

FIGURE 3 Schematic representation of a polymeric affinity drug designed for plaque therapy.

CONCLUSION

The concept of targeting the delivery of drugs by covalent coupling of a drug and a biospecific affinity binding ligand to a polymeric matrix is suggested as a potentially important new approach to improved therapy. The polymeric matrix is considered to play a vital role as both the carrier of multiple functional groups as well as providing a protective microenvironment for the bioactive functions.

References

1. L. G. Donaruma, "Synthetic Biologically Active Polymers," A. D. Jenkins, ed., *Progr. in Polymer Sci.* Volume 4, Pergamon Press, Oxford (1975).
2. Symp. on Controlled Release Polymeric Formulations, *Am. Chem. Soc. Coatings & Plastics Preprints*, **36**(1), xxiii (1976).
3. I. Silman and E. Katchalsky, *Ann. Rev. Biochem.*, **35**, 873 (1966).

4. H. F. Hixson and E. P. Goldberg, eds., *Polymer Grafts in Biochemistry*, Dekker, New York, 1976.
5. S. P. O'Neill *et al.*, *Biotechnol. Bioeng.*, **13**, 319 (1971).
6. J. R. Wykes *et al.*, *Biochim. Biophys. Acta*, **250**, 522 (1971).
7. J. J. O'Malley and R. W. Ulmer, *Biotechnol. Bioeng.*, **15**, 917 (1973).
8. H. F. Hixson, *Biotechnol. Bioeng.*, **15**, 1011 (1973).
9. Symp. on Polymeric Delivery Systems, Midland Macromolecular Institute, Midland, Michigan, Aug. 23–27, 1976.
10. W. B. Jakoby and M. Wilchek, eds., *Methods Enzymol.—Affinity Techniques*, Vol. **34**, Academic Press, New York (1974).
11. D. Bergsma, ed., *Birth Defects—Enzyme Therapy*, **9**(2), Williams and Wilkins, Baltimore, 1973.
12. J. W. Yarbro, ed., *Seminars in Oncology—Tumor Immunology*, **1**(4), Grune and Stratton (1974).
13. A. J. Bodurtha, *Cancer Res.*, **35**, 189 (1975).
14. T. Ghose *et al.*, *Brit Med. J.*, **26**, 495 (1972).
15. E. Hurwitz *et al.*, *Cancer Res.*, **35**, 1175 (1975).
16. G. F. Rowland *et al.*, *Nature*, **255**, 487 (1975).
17. B. Finlayson *et al.*, eds., *Urolithiasis*, National Acad. Sci., Washington, D.C. (1972).
18. R. Ross and J. A. Glomset, *Science*, **180**, 1332 (1973).

DISCUSSION

J. C. Venter (*State Univ. of New York at Buffalo, Buffalo, NY*): Have you plans to overcome the antigenicity of your complexes? If anything, I would think that the antigenicity should increase.

Dr. E. P. Goldberg: I think the point is conjectural. There is some evidence that simple admixture with poly(vinylpyrrolidone), for example, tends to limit antigencity. There are drug formulations with PVP which are claimed to have lower antigenic character, by virtue of that kind of protective association. Our objective is in fact to design the soluble matrix to mask the antigenic character of the carrier-drug system. That this is not unreasonable is also suggested by (1) the reported reduction in antigenicity for catalase and for bovine serum albumin when covalently coupled to polyethylene glycol [A. Abuchowski *et al.*, First Chem. Congress N.A., Mexico City, Dec. 1975—paper #57-BMPC] and (2) studies which show that antigenic competition by mixed antigens may depress antibody synthesis and alter the ratio of low to high affinity antibodies [T. K. Young *et al.*, *Immunology*, **26**, 943 (1974)].

Dr. L. Bangs (*Dow Chemical Co., Midland, MI*): Can you tell us what attempts you have made in these fine directions? What work has your group done?

Dr. E. P. Goldberg: We have initiated enzymatic lysis experiments with kidney stones and aorta tissue. This program is new and our work began only about 6 months ago. We've also started polymer coupling experiments using soluble matrices like poly(vinylpyrrolidone) and dextran, and we are considering using poly(vinylpyridine-*N*-oxide) also as a soluble bioacceptable carrier. Enzymes of choice in this initial work are proteases such as trypsin for stone matrix mucoprotein and cholesterol esterase for atheroma studies.

Dr. J. C. Venter: I'm all for polymeric drugs, but I'm not sure I see in your situation any distinct advantages of using a polymer matrix. What is your evidence that soluble enzymes injected for example with a catheter would not have the same effect?

Dr. E. P. Goldberg: I think the problems associated with using enzymes alone are well documented in the literature. In all cases, independent of the method of application whether by local injection, catheterization or intravenous injection, the problems are serious in terms of rapid degradation, toxicity, antigenicity and so on. The major goal of soluble polymer matrix binding is to solve some of these problems, otherwise it has no value whatever.

Dr. J. C. Venter: I think this has to be looked at for each particular enzyme system.

Dr. E. P. Goldberg: From the small amount of work appearing in the literature as well as related studies with antibody conjugates and natural glycoprotein–enzyme systems, it seems very likely that many soluble polymer–enzyme systems will be found to exhibit good activity and enhanced stability.

Dr. H. Ringsdorf (*Univ. of Mainz, Mainz, Germany*): Do these polymers get through the endothelium?

Dr. E. P. Goldberg: Yes, they should get through in the same way that low density lipoprotein (LDL), an enormous molecular aggregate of ~3 million molecular weight, gets through. The current models for transport emphasize two factors: (1) endothelium damage which may produce intercellular cracks and voids and (2) vesicular transport. Both of these mechanisms would be expected to move reasonable quantities of soluble polymeric drugs having very high molecular weights; perhaps even as high as 3–5 million M.W. for immunoglobulin conjugates.

Biological Activity of Isoproterenol Covalently Linked to Synthetic Polypeptides

J. CRAIG VENTER,† MICHAEL S. VERLANDER, NATHAN O. KAPLAN, MURRAY GOODMAN, JOHN ROSS, Jr., and SHIGATAKE SESAYAMA‡

Departments of Chemistry and Medicine, University of California, San Diego, La Jolla, CA92093, U.S.A.

D-Isoproterenol was covalently coupled via an azo linkage to soluble copolypeptides of molecular weight 1500, 3000 and 10,000 containing an aromatic amine. The polymeric azo-isoproterenol derivatives were purified by gel chromatography which reduced contamination by the parent isoproterenol to undetectable levels (i.e., less than 0.01%) and by 6-aminoisoproterenol (a possible decomposition product) to less than 0.4%. Both polymeric isoproterenol derivatives were found to elicit positive chronotropic responses in isolated perfused guinea pig hearts. The mean effective doses (ED_{50}) for the 1500 and 10,000 molecular weight derivatives were within 1.3 and 2.0 orders of magnitude, respectively, of the ED_{50} of *D*-isoproterenol. The responses cannot be attributed to free isoproterenol because this drug could not be detected in our preparations. Neither can the observed biological activity be attributed to 6-aminoisoproterenol, since this compound's dose-response is shifted 3 orders of magnitude to the right of *D*-isoproterenol and 1–2 orders of magnitude to the right of the polymeric derivatives. Inotropic response decay times in isolated cat papillary muscles following washouts indicate that the polymer-bound drug does not diffuse into the muscle tissues. It was found that there was an inverse relationship between the molecular weight and duration of action when the polymeric derivatives were injected into conscious dogs.

Our findings demonstrate that under controlled conditions isoproterenol can retain biological activity while covalently bound to a polymeric support, and that polymer bound drugs can exhibit a molecular weight dependent, prolonged *in vivo* activity.

One of the first uses of small molecules covalently bound to a macromolecular support can be traced to the 1920s when reactions with haptenic groups were elucidated. These studies involved *p*-amino-benzenearsenate diazotized to various proteins and their use in determining antibody specificities.[1]

---

† Present Address: Department of Pharmacology and Therapeutics, State University of New York at Buffalo, Buffalo, New York 14214.

‡ Present Address: School of Medicine, University of Kyoto, Kyoto, Japan.

Lerman and co-workers in 1951 developed the technique now referred to as "affinity chromatography" when they covalently coupled an antigen via an azo bond to a *p*-aminobenzyl cellulose matrix for the purification of an antibody.[2] In 1961 Arakawa *et al.* applied these techniques to the pharmacology of angiotensin, when they immobilized this hormone on poly(*O*-acetylserine) and concluded from biological tests that angiotensin exerted its action on the cell membrane of target cells.[3] Schimmer, Sato and co-workers in 1968 applied these techniques to adrenocorticotropic hormone (ACTH) which retained biological activity while covalently bound via an azo linkage to Sepharose beads.[4] These workers concluded that ACTH by nature of its being bound to particles larger than cells must be exerting its action on the cells by a direct interaction with membrane receptors for ACTH. Cuatrecasas in 1969 drew similar conclusions about the mechanism of action of insulin on isolated fat cells by using insulin covalently coupled to Sepharose beads.[5]

In studies using catecholamines covalently bound to glass beads, Venter *et al.*, in 1972, applied immobilized drugs directly to various cardiac preparations in an attempt to study the site and mechanism of action of these hormones. In the case of the glass-bead-immobilized catecholamines, as in the case of other immobilized drugs and hormones, it has proved extremely difficult to eliminate completely the possibility of a minute leakage of covalently or non-covalently bound biogenic amine into a micro volume created at the bead–tissue interface.[7,8] However, differences in the mode of action of the parent and immobilized catecholamines have been demonstrated, such as a lack of change in the cyclic AMP levels in cardiac muscle in response to immobilized isoproterenol,[9] which, by inference, indicate that the drug could be acting in a covalently bound form. This question of the mode of action of immobilized catecholamines has been discussed elsewhere in the present volume by Woodman *et al.*[10] (see page 251).

In an attempt to elucidate the mechanism of action of these immobilized drugs we have taken the approach of design, synthesis and biological characterization of a number of catecholamine derivatives.[11,12] An important requirement for an immobilized drug to be biologically active is that it should be bound to the matrix in a biologically active conformation. While this is an obvious point several conflicting reports have appeared concerning the biological activity of catecholamines coupled to solid supports via different functional groups on the catecholamine molecule.[10,13,14] In the case of the glass bead derivatives, the catecholamines are coupled to an aryl diazonium group which is covalently attached to the glass. It has been shown that the azo substitution on the catecholamine is predominantly at the 6-position of the catechol ring.[11] It was also demonstrated that 6-amino substituted catecholamines were biologically active.[11] To further test the hypothesis of biologically active 6-substituted catecholamines, 6[*p*-nitrophenylazo] epinephrine was synthesized and found to have biological activity.[8,11]

Our approach has been to increase the size of the substituent at the *para* position of the 6-phenylazo catecholamines—in effect to create soluble versions of the immobilized catecholamines.[8,9] To this end we have undertaken the synthesis of a series of water-soluble polypeptides to which the catecholamines are attached via an azo linkage (see Verlander *et al.*[12]). Copolypeptides of hydroxypropylglutamine with *p*-aminophenylalanine were synthesized to which *D*-isoproterenol was covalently bound. The proposed structure of the copolymer of hydroxypropylglutamine and *p*-aminophenylalanine containing isoproterenol is shown in Figure 1; since it was prepared by the random copolymerization of the amino acid *N*-carboxyanhydrides,[12] the exact structure cannot be given. In addition, the diazotized *p*-aminophenylalanine residues are not fully substituted, and we therefore assume the remainder to be hydrolyzed to tyrosyl residues during work-up of the derivatives.

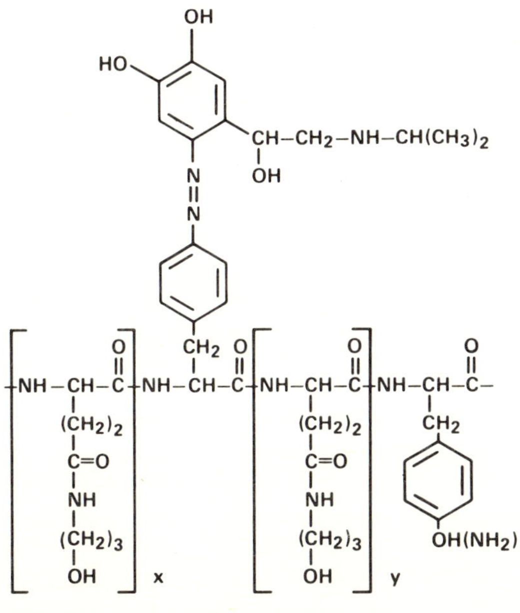

FIGURE 1 Proposed structure of polymeric azo-isoproterenol derivatives. The polymers are random copolymers of hydroxypropylglutamine and *p*-aminophenylalanine. The molar ratio of hydroxypropylglutamine is 4:1 in each case. Isoproterenol incorporation is in the range of 2.5–5% by weight. Since the diazotized aminophenylalanine residues are not fully substituted, we assume the remainder are hydrolyzed to tyrosyl residues. (Taken from Verlander *et al.*[12])

The products of the diazotization reaction were purified free of low molecular weight material by gel chromatography.[12] The column elution profiles for the lowest molecular weight polymeric azo-isoproterenol derivative are illustrated in the upper portion of Figure 2. Subsequent to separation by column chromatography the products of the diazotization reaction mixture were identified by thin-layer chromatography. The first column purification of this derivative (Figure 2-1) resulted in four peaks assigned to the polymeric derivative, an unassigned minor peak, 6-aminoisoproterenol,

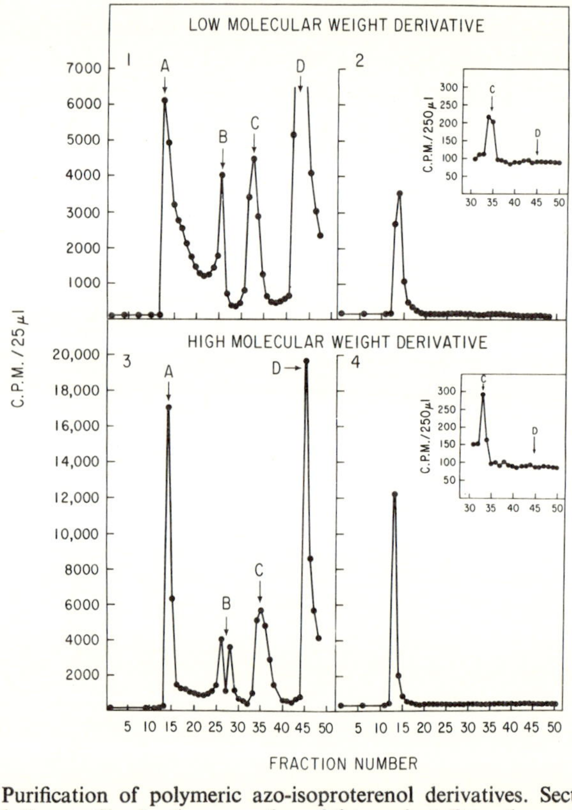

FIGURE 2 Purification of polymeric azo-isoproterenol derivatives. Sections 1 and 3 represent the elution profiles for the separation of the products from the coupling reactions of the 1500 and 10,000 mol. wt. derivative, respectively. Peak assignment by thin layer chromatography indicates: the polymeric isoproterenol derivatives (A); an unassigned lower molecular weight polymeric peak (B); 6-amino-*D*-isoproterenol (C); and *D*-isoproterenol (D). Sections 2 and 4 represent the third subsequent column purifications of the respective polymer peaks, with inserts illustrating peaks (C) and (D) at 100-fold increase in sensitivity. Gel chromatography was performed on a Bio-Gel P-2 (Bio-Rad) column (2.7 × 85 cm) at 4°, using 1 mM ammonium bicarbonate as eluent. Fractions (12.5 cm^3) were monitored by radioactivity using ^{3}H-isoproterenol; isoproterenol-containing peaks were lyophilized in the dark.[12]

and free isoproterenol. It is interesting to note that isoproterenol and its 6-amino derivative were separated by this method. Figure 2-2 represents the third subsequent purification of the polymeric isoproterenol derivative peak. The most significant point of this purification is the reduction of contamination by the parent D-isoproterenol to undetectable levels (i.e., less than 0.01%). A small but recurrent contamination by 6-aminoisoproterenol was present at each column purification, probably due to photolysis of the azo linkage of the catecholamine to the polymer during storage and handling. This contamination by the 6-amino derivative represented less than 0.4% of the polymeric catecholamine peak in each case.

The purification of the high molecular weight (10,000) polymeric catecholamine derivative was essentially identical to that of the 1500-molecular-weight derivative described above. The soluble, parent isoproterenol peak, 6-aminoisoproterenol, and polymeric isoproterenol peaks were in the identical fractions to those observed for the lower molecular weight derivative (Figure 2-3). Subsequent gel chromatography of this higher molecular weight derivative (Figure 2-4) revealed no detectable *D*-isoproterenol, indicating less than 0.01% contamination by this amine. As in the case of the lower molecular weight derivative, 6-aminoisoproterenol was detectable at each purification step, although the degree of contamination was always less than 0.05% of the polymer-bound isoproterenol. Thin-layer chromatography of both polymeric isoproterenol derivative peaks from the columns revealed no detectable free isoproterenol or its 6-amino derivative when the plates were scanned using a radiochromatograph (limits of sensitivity: 0.4%).[12]

The polymeric isoproterenol derivatives proved to be stable to storage at −15°C. Figures 2-2 and 2-4 represent chromatography after approximately 3 months storage at this temperature. Since no parent isoproterenol was detectable (i.e., less than 0.01%) in three subsequent purifications by gel chromatography and since the 6-aminocatecholamine contamination was always less than 0.4%, these values represent the maximum levels of contamination of the polymeric isoproterenol derivatives by free drug molecules.[12]

IN VITRO BIOLOGICAL CHARACTERIZATION OF POLYMERIC CATECHOLAMINE DERIVATIVES

The two polymeric isoproterenol derivatives were tested for biological activity both immediately following their final gel chromatography, and also after lyophilization.[12] The results were identical in each case. The polymeric catecholamine derivatives both produced positive chronotropic responses in a dose-related manner when injected into isolated perfused guinea pig hearts. The log dose–response relationships for the two polymeric derivatives are compared with *D*-isoproterenol, 6-aminoisoproterenol, and *D*-epinephrine in Figure 3. The responses are plotted as a function of the total number of moles of isoproterenol added, determined using radiolabeled catecholamines.

The mean effective dose (ED_{50}) for *D*-isoproterenol is 1.4×10^{-11} mol. The dose–response for the lower molecular weight (1500) polymeric isoproterenol derivative is shifted to the right of that for isoproterenol by 1.3 orders of magnitude, with an ED_{50} of 2.8×10^{-10} mol. The dose–response relationship for the higher molecular weight (10,000) derivative is shifted slightly further to the right with an ED_{50} of 1.4×10^{-9} mol. Thus, the poly-

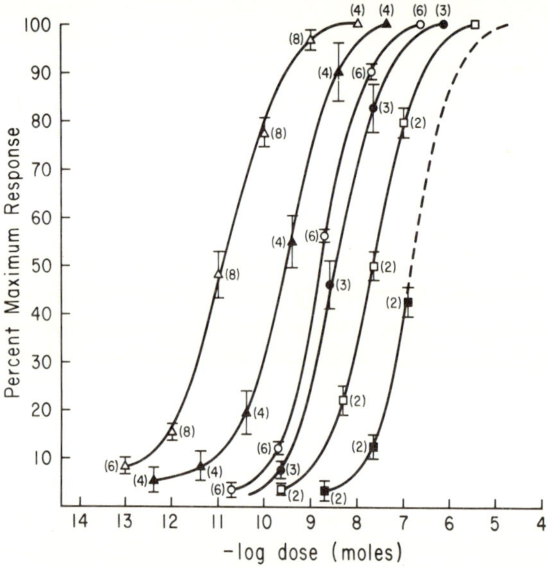

FIGURE 3 Positive chronotropic effects of *D*-isoproterenol (△); 1500 molecular weight polymeric *D*-isoproterenol derivative (▲); 10,000 mol wt. polymeric*D*-iso proterenol derivative (○); *D*-epinephrine (●); 6-amino-*D*-isoproterenol (□); and the 10,000 mol. wt. polymeric *D*-isoproterenol derivative in the presence of continuously perfused *D,L*-propranolol (5×10^{-7} M) in Krebs solution (■). The doses of the polymeric isoproterenol derivatives are plotted as the total moles of isoproterenol added. The error bars represent the standard error of the mean of each point; the number of experiments for each point is indicated in parentheses. Responses are plotted as the percentage of the maximum change in heart rate of isolated perfused guinea pig hearts which were rapidly dissected from heparinized guinea pigs (300–500 g) and suspended via an aortic cannula in a modified Langendorff apparatus. Krebs solution, maintained at 35° and pH 7.4 when bubbled continuously with 95% O_2/5% CO_2, was perfused at a constant pressure of 50 cm H_2O (4.9 kPa). Contractile rate, electrocardiogram (ECG), force contraction, and change in contraction velocity with time, d*F*/d*t*, were recorded on a Brush polygraph recorder. Hearts were equilibrated 20–30 min prior to drug testing, and solutions of the test substances were injected into the perfusing Krebs solution via a rubber diaphragm 6 mm above the orifices of the coronary arteries. All of the drugs tested gave identical maximal responses.[12]

meric isoproterenol derivatives, while less active than the parent isoproterenol, are considerably more active than both 6-aminoisoproterenol and *D*-epinephrine. The dose–response relationship for 6-aminoisoproterenol (Figure 3) indicates an ED_{50} of 2.2×10^{-8} mol which is very similar to that reported previously for this derivative.[11] The calculated ED_{50} for *D*-epinephrine is 2.0×10^{-9} mol.

The β-adrenergic antagonist *D,L*-propranolol (5×10^{-7} M) inhibits the chronotropic action of the polymeric isoproterenol (10,000 mol. wt.) derivative in an apparently competitive manner (Figure 3). We have calculated a pA_2 value of 8.3 for propranolol, from a calculated K_b value of 5.0×10^{-9} M. (pA_2 is the negative logarithm of the concentration of an antagonist required to give a dose ratio of two, and K_b is the apparent receptor–antagonist dissociation constant.) The calculated pA_2 value is similar to those reported by others for propranolol antagonism of isoproterenol on β-adrenergic receptor systems.[15]

The perfusates from the guinea pig hearts were collected within seconds of drug testing and immediately subjected to thin-layer chromatography. No 6-aminoisoproterenol or isoproterenol was detectable by this technique. The perfusate was also concentrated more than 1000-fold by lyophilization and re-subjected to thin-layer chromatography, after which no detectable contaminants were found either visually or using radioscanning.

IN VIVO ACTION OF POLYMERIC CATECHOLAMINE DERIVATIVES

D-isoproterenol and the polymeric isoproterenol derivatives produce positive inotropic and chronotropic responses when injected as a bolus into the saphenous veins of conscious dogs. The time course of the response to *D*-isoproterenol can be characterized by an immediate rise to peak in heart rate levels within 30 s of injection. This rise in heart rate is followed immediately, without a plateau phase, by a rapid return to control levels within 3–4 min. Increasing the dose of isoproterenol from 0.1 μg/kg (a moderate dose) to 1 μg/kg (a high dose) results in only a slight prolongation of the chronotropic and inotropic action. Average duration for the response to 1 μg/kg *D*-isoproterenol was 4.5 ± 0.14 min ($N=5$).

In contrast to the short time course of the parent isoproterenol's action, a bolus injection of the 1500 mol. wt. polymeric isoproterenol results in positive chronotropic and inotropic responses which are relatively well sustained, decreasing slowly to control levels usually within 30 min. The average duration of the response to the 1500 mol. wt. derivatives was 27.4 ± 4.9 min ($N=7$).

The duration of the inotropic and chronotropic responses to the polymeric isoproterenol derivatives was found to be inversely related to their molecular weight over the molecular weight range tested. With isoproterenol coupled to a polymer of molecular weight 3000 (prepared by the same procedure), the durations of inotropic and chronotropic responses were significantly shorter than those obtained with the 1500 mol. wt. derivatives. Cardiac response durations to the 3000 mol. wt. derivative averaged 12.8 ± 1.6 min ($N=5$), a value significantly different from that obtained with the 1500 mol. wt. derivative [$p < 0.01$ (Student's t test)]. By increasing the polymer molecular weight to 10,000, response durations were decreased to an average of only 5.5 ± 0.4 min ($N=5$) a value significantly different from those obtained with the 1500 and 3000 mol. wt. derivatives $p < 0.01$ (Student's test).

DIFFUSION STUDIES WITH POLYMERIC CATECHOLAMINE DERIVATIVES

In previous studies with immobilized catecholamines a differential action of catecholamine molecules has been suggested based on limits of diffusion and sites of action.[7] It was calculated that within 10 min of adding soluble isoproterenol (mol. wt. $\simeq 250$) to the bathing solution of an isolated cat papillary muscle the drug will have approached equilibrium in the inner core of the muscle by diffusion.[7] In contrast, the diffusion of a 10,000 mol. wt. polymeric derivative will be substantially reduced. Because the response onset in cardiac muscle involves catecholamine-induced propagation the drug washout times were studied, comparing the free isoproterenol to the high molecular weight polymeric isoproterenol derivative. The results were dramatic. Following a single washout of the polymeric derivative, after a 10 min incubation, the force of contraction returned to control levels, usually within 2 min. In contrast a dose of free isoproterenol which produces an essentially identical response took on average over 20 min to return to control levels following a single washout. As a control soluble isoproterenol was added together with unsubstituted (mol. wt. 10,000) polymer. The resulting washout time was essentially identical to that for free isoproterenol. These results are summarized in Table I.

Subjecting the high molecular weight polymeric derivative to simulated muscle bath conditions produced no detectable isoproterenol and less than 0.4% 6-aminoisoproterenol.[12] From these results we conclude that insufficient free drug was produced during the course of the assay to produce the observed biological activity.

The dose–response relationships illustrated in Figure 3 indicate that the responses observed for the polymeric isoproterenol derivatives cannot be

TABLE I

Polymer azo-isoproterenol (mol. wt. 10,000) vs. *D*-isoproterenol response decay times following drug washout in isolated cat papillary muscles

Drugg added (mol isoproterenol)	Incubation time, min	% Response (Δ force/ control force)[a]	Response decay time following single washout (contractions to reach control force)[a]
Polymer azo-isoproterenol (mol. wt. 10,000) (110 nmol)	10	58.9 ± 11.6 n=6	19.5 ± 2.6 n=6
D-Isoproterenol (15–60 nmol)	10	65.0 ± 8.9 n=8	239.7 ± 39.2 n=8
D-Isoproterenol and unsubstituted polymer (mol. wt. 10,000) (60 nmol)	10	40.5 ± 5.5 n=2	335 ± 25 n=2
Polymer azo-isoproterenol (mol. wt. 10,000) (220 nmol)	40	77.5 ± 12.5 n=2	91 ± 5 n=2
D-Isoproterenol (60 nmol)	40	97.0 ± 21 n=2	330 ± 12 n=2

Note: Isolated right ventricular cat papillary muscles were prepared as described.[9,13] They were subjected to the indicated compounds for 10 or 40 min. The papillary muscle bath was drained, refilled with fresh oxygenated Krebs solution (pre-warmed to 30°C), drained and refilled with fresh Krebs solution. The force of contraction was monitored until the control inotropic state was regained. The dose of soluble isoproterenol was chosen to match the inotropic response obtained with the polymer-immobilized isoproterenol. The papillary muscles were maintained at 30°C and stimulated to contract at 12 contractions per min.[12]

[a] Mean ± SEM, n = number of experiments.

due to contamination by 6-aminoisoproterenol, since 300 and 500% contaminations of this derivative would be required to explain the activity we observe for the 10,000 and 1500 molecular weight polymeric isoproterenol derivatives, respectively. Because of the shift in the dose–responses to the right of the parent *D*-isoproterenol it was critical to be able to rule out the presence of traces of this compound. A contamination of greater than 5% and 1% of *D*-isoproterenol would be required to explain the respective dose–responses of the 1500 and 10,000 mol. wt. derivatives. Gel chromatography (Figure 2) and thin-layer chromatography have indicated that the highest possible contamination by *D*-isoproterenol is less than 0.01%, which is at least 1.5–2 orders of magnitude less than the minimum amount required to explain our results.[12] The chromatography immediately prior and subsequent to biological testing indicates that no detectable free *D*-isoproterenol was present in our preparations. We feel that our controls more than adequately rule out the possibility of contamination by free isoproterenol or 6-aminoisoproterenol explaining our results.

The drug washout and response decay experiments (Table I) represent both a unique control experiment and also an example of an area where polymeric drug derivatives may find great utility. Our results show that the higher molecular weight (10,000) polymeric isoproterenol derivative can be removed from the muscle baths with a single washing, resulting in an immediate response decay to control levels. This is in sharp contrast to the results obtained with soluble isoproterenol (Table I) where, following one bath change, there is a very slow response decay to control levels. We feel that these results can be explained in terms of the different diffusion rates of the two types of drug. These observations are a further indication that we are not dealing with a "dissociated" species of isoproterenol. If the observed responses were due to an isoproterenol contaminant the washout differentials would be identical to that obtained with a mixture of soluble isoproterenol and unsubstituted polymer (Table I).

The structure–activity relationships of substituted catecholamines further support our contention that we have a biologically active, covalently-coupled polymeric isoproterenol derivative. It has previously been shown that azo substitution occurs at the 6-position on the catechol ring and that soluble azo-substituted catecholamines possess considerable activity.[10]

The positive inotropic and chronotropic responses obtained in dogs with the polymeric catecholamine derivatives suggest that the *in vivo* activity is due to the covalently coupled species of isoproterenol and not released drug. Significant differences in response duration depended only upon molecular weight of the polymer since the same total amount of isoproterenol was given in each case. In addition, the profile of the responses due to the polymeric isoproterenol were very different from those obtainable with soluble isoproterenol alone, further implicating the covalently coupled species as the biological stimulus.

At present we have no adequate explanation for the inverse relationship between the molecular weight of the polymeric support and duration of biological response. It is possible that the duration of response is related to both a function of tissue penetration (diffusion), which is molecular weight dependent, and also to the rate of metabolism of the isoproterenol. In order to explain our findings, free isoproterenol with relatively rapid diffusion properties would need to be rapidly metabolized, which it is known to be. Our results suggest that the increased duration of action of the polymeric derivatives may be related to their reduced tissue penetration, or their metabolism may be slowed; it is possible that both these or other phenomena are involved. Studies are currently under way to determine the distribution and metabolism of these polymeric drugs.

We conclude from the *in vitro* studies that the polymeric azo-isoproterenol derivatives are exerting their activity in the immobilized or conjugated form.

The activity of the polymeric isoproterenol derivatives in the conscious dogs also appears to be consistent with this mechanism of action. These soluble polymeric derivatives may be acting in an entirely different manner from the insoluble immobilized catecholamines.[7,8] However, our present findings, which indicate that drugs and hormones can retain biological activity while covalently bound to soluble polymers, suggest that insoluble, covalently bound drugs may also react directly in biological systems, although this is not proven.[10]

We feel that our results demonstrate that under controlled conditions (including rigorous purification) isoproterenol can retain biological activity while covalently bound to a polymeric support. These polymer-immobilized drugs also exhibit a molecular weight dependent, prolonged *in vivo* activity. Our studies suggest that polymeric drugs may have broad applications in pharmacology, biochemistry and medicine.

Acknowledgements

This work was supported in part by grants from the American Cancer Society (BC–60–Q) and the National Institutes of Health (USPHS CA 11683) (Nathan O. Kaplan) and a grant from Deknatel Inc. (Murray Goodman).

References

1. K. Landsteiner, *The Specificity of Serological Reactions*, Harvard University Press, Cambridge, Massachusetts (1945).
2. D. H. Campbell, E. Loescher and L. S. Lerman, *Proc. Nat. Acad. Sci. USA*, **37**, 575 (1951).
3. K. Arakawa, R. R. Smeby and M. Bumpus, *J. Amer. Chem. Soc.*, **84**, 1424 (1962).
4. B. P. Schimmer, K. Ueda and G. H. Sato, *Biochem. Biophys. Res. Commun.*, **32**, 806 (1968).
5. P. Cuatrecasas, *Proc. Nat. Acad. Sci. USA*, **63**, 450 (1969).
6. J. C. Venter, J. E. Dixon, P. R. Maroko and N. O. Kaplan, *Proc. Nat. Acad. Sci. USA*, **69**, 1141 (1972).
7. J. C. Venter, Doctoral Dissertation, University of California, San Diego (1975).
8. N. O. Kaplan and J. C. Venter, *Proceedings of the Sixth International Congress of Pharmacology*, Vol. 5, Pergamon Press, New York, (1975).
9. J. C. Venter, J. Ross, Jr. and N. O. Kaplan, *Proc. Nat. Acad. Sci. USA*, **72**, 824 (1975).
10. D. Woodman, J. C. Venter and N. O. Kaplan, this volume, p. 251.
11. J. C. Venter, L. Arnold, Jr. and N. O. Kaplan, *Mol. Pharmacol.*, **11**, 1 (1975).
12. M. S. Verlander, J. C. Venter, M. Goodman, N. O. Kaplan and B. Saks, *Proc. Nat. Acad. Sci. USA*, **73**, 1009 (1976).
13. J. C. Venter, J. Ross, Jr., J. E. Dixon, S. E. Mayer and N. O. Kaplan, *Proc. Nat. Acad. Sci. USA*, **70**, 1214 (1973).
14. B. R. Venter, J. C. Venter and N. O. Kaplan, *Proc. Nat. Acad. Sci. USA*, **73**, 2013 (1976).
15. R. F. Furchgott, *Handbuch der Experimentellen Pharmakologie*, Vol. 33, Springer-Verlag, Berlin, pp. 283–335 (1972).

DISCUSSION

Dr. D. Bennett (*Midland Hospital, Midland, MI*): What evidence do you have to distinguish between direct and indirect-acting compounds as far as these adrenergic agents are concerned? Some azo dyes are capable of releasing catecholamines. Could you have an extremely potent indirect-acting adrenergic agent in this situation?

Dr. J. C. Venter: We tried the immobilized catecholamine on papillary muscles from reserpinized cats in order to test for indirect effects due to neuronaly released catecholamines. The biological effects of the immolibized catecholamines were the same whether or not the papillary muscles were from reserpinized cats. I think we are seeing a direct receptor effect of the covalently coupled catecholamines.

Dr. H. Ringsdorf (*Univ. of Mainz, Mainz, Germany*): It would be most interesting if you would look at the action of covalently linked isoproterenol at the cellular level. Is there anything known about cell cultures which have receptors for the catecholamines? The real question is what happens when you go from the low molecular weight compound to the polymeric compound. What I wouldn't have expected was the decrease when going from the lower molecular weight polymer to the higher molecular weight polymer. One explanation you mentioned was that the compound disappears as the cell takes it in. This is not very likely in a muscle cell system, because they have a low rate of endocytosis. The other possibility is that it fits or doesn't fit at the receptor site.

Dr. J. C. Venter: It's hard to say what steric effects polymer size has on drug–receptor interactions.

We have no evidence that these materials are taken up by the cells at all. I would be surprised if they were.

In preliminary experiments, I have found that the polymeric catecholamines bind at the surface of cultured cells known to have β adrenergic receptors.

The work I described was done with a random copolymer. We are now trying to use a sequenced copolymer. If the same results occur with this carrier, then it would allow more accurate quantitation. Since we have a random polymeric system, sometimes the catecholamine is buried in the middle, and sometimes it's on the end of the polymer chain. These are important things to look at, as the drug's location on the polymer may effect receptor interactions.

Dr. M. Goodman (*Univ. of California/San Diego, LaJolla, CA*): You also have to examine the pharmacology of the polymer, since the support cannot be viewed as inert. We are synthesizing polymers and radiolabelling them. Using differently labelled catecholamines, we can then follow the fate of each component pharmacologically.

Dr. J. C. Venter: I think the polymer support will be just as important in determining the activity of these materials as the drug itself.

Dr. H. Ringsdorf: Why should the polymer be important in this regard? Isn't it just a carrier?

Dr. J. C. Venter: It can't be just a carrier because we're dealing with interactions on the cell surface presumably in this case with the β-receptor. For example, the charge on the polymer can totally affect the interaction. When we first started this work we used poly(glutamine), and we were unable to produce an active polymeric drug. So there is definitely an activity effect which is related to the polymer used as the carrier, at least in this system.

Dr. H. Ringsdorf: Your problem with the poly(glutamine) may have been the existence of the peptide linkages which may interact with proteins in the membrane. I think the case will be different if one uses synthetic polymers where one just has to deal with lipid–lipid interactions.

Dr. A. Trouet (*International Institute for Cellular and Molecular Pathology, Brussels, Belgium*): I think that the simplest explanation for the lower activity of the high molecular weight derivative is that this material does not pass through the capillary barrier between the blood stream and heart cells as easily as the low molecular weight material.

Dr. L. Miller (*A. H. Robbins Co., Richmond, VA*): If diffusion is the determining factor controlling the activity of the various molecular weight derivatives, then how do you explain the fact that the dose–response peaks are the same for the 1500, 3000, and 10,000 molecular weight derivatives?

Dr. J. C. Venter: It could be that we saturated the responses. We can't increase the heart rate on these drugs much higher. We're looking at a ceiling effect.

Dr. L. Miller: What about the onset time? If it's diffusion controlled across the capillary barrier, then the onset time would be longer for the higher molecular weight compounds.

Dr. J. C. Venter: It depends on how diffusion limited the process is. What we have is a mass of molecules moving through the heart. So you're looking at what gets in in a few seconds as it passes through the heart.

We may be looking at maximal responses, i.e., it may take only the amount of catecholamine which gets in to the tissue in a few seconds to produce a maximal response. We know this is true in the papillary muscles, but that is a much simpler system.

The lower molecular weight derivative may diffuse farther and therefore remain longer. Because propagation appears much faster than diffusion, there is no additive effect. Stimulating the surface of a papillary muscle gives the same effect as by ten minutes of diffusion of isoproterenol into the muscle. It is possible that only a very small amount of tissue need be stimulated in order to affect the whole organ.

Unidentified discussant: Have you examined the resistance of your carrier to enzymatic degradation?

Dr. J. C. Venter: No, we haven't examined this, but it is something which should be looked into.

Reproducibility of Glass Bead Immobilized Drug Preparations

DAVID C. WOODMAN, J. CRAIG VENTER,† and NATHAN O. KAPLAN

Departments of Medicine and Chemistry, University of California, San Diego, La Jolla, CA92093, U.S.A.

Catecholamines, coupled via an azo bond to a porous glass bead matrix, produce unique pharmacological responses when brought into physical contact with isolated muscle preparations. These responses are not due to leakage of threshold concentrations of free catecholamines into the whole bath. Activity of these glass bead preparations is due to either immobilized drug action or to leakage or an azo-catecholamine derivative into the microvolume created between glass bead and muscle surface analogously to a neuroeffector junction. Evidence from our laboratory supports both possibilities. Each glass bead immobilized catecholamine preparation must be standardized prior to use in pharmacological studies to eliminate the possibility of gross leakage. The finding that there are no detectable changes in cAMP levels, with glass bead immobilized catecholamines, coupled with inotropic changes in cat papillary muscle raises questions concerning current theory of drug action.

In our initial studies we coupled catecholamines via a reaction with a diazonium salt of *p*-butyramide *N*-aniline which was covalently attached to the surface of control pore glass beads.[1] Initial biological tests indicated that when a few of these glass beads containing catecholamines (glass bead–immobilized catecholamines) were placed in contact with various cardiac muscle preparations in isolated tissue baths or in tissue culture, increases in the frequency of cardiac contraction or in the contractile force were observed. This finding has stimulated some controversy concerning the exact mechanism and henceforth the significance of this result. Subsequent studies by other workers[2] suggested that glass beads were merely releasing threshold levels of free catecholamine, into the bathing solutions of the various muscle preparations and, therefore, causing biological responses as would any free drug in solution. However, our ongoing studies suggested that the most likely explanations for the glass bead catecholamine action would be either the

† Present address: Department of Pharmacology and Therapeutics, State University of New York at Buffalo, Buffalo, New York 14214.

catecholamines, while still covalently attached to the bead surface, were interacting with receptors on the muscle surface or the beads could be acting by releasing minute amounts of drug into a microvolume created between the muscle and the glass bead catecholamines functioning analogously to a neuroeffector synapse. A combination of these mechanisms might also exist as discussed previously.[3]

DEMONSTRATION OF REQUIREMENT FOR CONTACT BETWEEN GLASS BEAD-IMMOBILIZED CATECHOLAMINE AND TISSUE SURFACES

Extended studies in our laboratory have eliminated the possibility of gross leakage of threshold concentrations of catecholamines into the bathing solution of the physiological preparations as the mechanism of action of the immobilized catecholamines. We have reported numerous control experiments establishing that physical contact between glass beads in the hearts, papillary muscles, or heart cells was absolutely necessary to elicit a biological response.[1] These included the experiments in which large amounts of catecholamine glass (125 mg or more) were added to small volumes (2 cm^3 or less) of medium surrounding the biological preparations. The beads had no effect on the tissue when present in the bath without contact with the tissue. Further studies demonstrated that only one isoproterenol glass bead was necessary for stimulation of contractile responses in cat papillary muscles. One isoproterenol glass bead gave the same response in the cat papillary muscle as did 0.1 μM solution of isoproterenol. However, this bead preparation released labelled isoproterenol at a rate of 0.008% per h.[3] This would yield only 4×10^{-16} mol of catecholamine in a 40 cm^3 bath bubbled continuously with a mixture of 95% oxygen and 5% carbon dioxide for 1 h. The amount of catecholamine released in one hour or in fact the amount of catecholamine contained on one bead if released entirely into the solvent would still be below the threshold concentration for biological activity for this agent.[3] Routine studies have demonstrated that glass bead catecholamines can be washed off the muscle surface into the bottom of the muscle baths allowing the contractile responses to return immediately to control levels. The catecholamine glass beads can remain in the bath solution without effecting the response of the preparation further. Subsequent experiments in a double muscle bath allowed two muscles to be studied simultaneously with only a few millimeters distance separating the two papillary muscles in the bath. Glass bead–immobilized catecholamines placed on one muscle produced contractile changes only in that muscle without affecting the adjacent muscle, further demonstrating that contact is necessary for biological activity.

LACK OF DETECTABLE cAMP CHANGES

Stimulating interest in glass bead–immobilized catecholamines as a tool in studying pharmacological mechanisms, were the recent findings that glass bead–immobilized catecholamine can stimulate a contractile force in cat papillary muscle without a concomitant rise in the cAMP levels as measured with a radioimmunoassay.[4] This result is contrary to accepted theory concerning inotropic actions of catecholamines. It has been thought that catecholamines bind to increase the level of cAMP in the whole tissue. The increased cAMP levels, through pathways as yet unelucidated, presumably result in the observed increase in contractile force in cardiac preparation. If we are to assume that cAMP is involved in glass–bead immobilized catecholamine action, then we must also assume only a small percentage of cells of the tissue respond directly to drug with increased cAMP levels, and these changes in cAMP levels for the whole tissue are undetectable with a radioimmunoassay. Furthermore, these few cells responding directly to drug must communicate the inotropic stimulus to the rest of the tissue via a mechanism that does not produce detectable changes in cAMP levels.[5]

PHARMACOLOGICAL ANALYSIS OF TWO TYPES OF CATECHOLAMINE DERIVATIVES

The finding of no detectable change in cAMP levels with inotropic responses prompted further experiments to elucidate mechanism of action. Studies were undertaken to determine the type of release products of catecholamines from the glass beads in order to determine whether the released product could be responsible for biological activity in the microvolume created at the bead-tissue interface. It has been previously shown that 0.1 N HCl acid washing removes free or oxidized catecholamine from the glass beads.[3,6] This finding suggests that any subsequent drug released following acid washing is a cleavage product. The possibilities for the cleavage include the azo bond, the amide bond, the carbon chain, and the glass matrix itself. Depending on the cleavage mechanism involved, two types of products could be released. An azo bond containing derivative of the catecholamine and an amino derivative. Our studies have shown that with reduction of the azo bond a six amino derivative of the catecholamine is produced.[3] This derivative along with a synthesized azo substituted catecholamine, has been examined for pharmacological activity. The amino derivative was found to be an order of magnitude less active than the parent catecholamine, but has an increased period of action. The low molecular weight 6-(*p*-nitrophenylazo)-catecholamine has activity identical to the parent catecholamine.[3] This data suggests that the glass bead–immobilized catecholamine stimulated response is not due to leakage of the amino catecholamine because of its lower activity and prolonged duration of action. The azo derivative still maintains the active conformation relative to receptor activation. This suggests that if the geometry between glass surface and muscle surface allows interaction of the immobilized catecholamines with a sufficient number of receptors, then stimulation of the muscle would occur. Other work has shown that cells bind tightly to glass bead immobilized catecholamines only when these cells have the catecholamine receptors, indicating an interaction of receptors on the cell surface with immobilized catecholamine.[7]

CONFIRMATION OF cAMP FINDINGS

Other workers[8] have confirmed the cAMP findings on both cat papillary muscles and guinea pig papillary muscle with our original bead preparation[3] [550 Å (55 nm) pore diameter that bound 1.65 μmoles per g of glass and had a leakage rate of 0.008% per h]. However, when using a different glass bead immobilized catecholamine preparation synthesized in our lab, they found small changes in cAMP levels. Since the procedure followed in synthesizing the glass bead immobilized catecholamines was identical, this suggested variability in the pore glass beads themselves.

ANALYSIS OF THE BEAD MATRIX IN GLASS BEAD IMMOBILIZED CATECHOLAMINE PREPARATION

This has led us to examine several different bead preparations to further elucidate the factors involved in producing glass bead immobilized catecholamine preparations which are active pharmacologically in a defined manner; increased contractile force without changes in cAMP levels. This work is now in press.[9] The four bead preparations used were all obtained as gifts from Corning Glass Works. They were a 3000 Å (300 nm) pore diameter, a 2300 Å (230 nm) diameter and two 550 Å (55 nm) pore diameter bead preparations which were labeled Y 117 and N 693. All four preparations differed in total amount of isoproterenol bound determined by scintillation counting after hydrofluoric acid digestion.[3] The amount of isoproterenol bound should depend on two factors; the surface density of silanol groups, and the surface area per unit weight.[1]- The surface density of silanol groups is dependent on the acid leaching process used in making the beads prous, and the surface area is dependent on the pore diameter; as pore diameter decreases, surface area increases. As expected the 3000 Å (300 nm) beads bound less (1.3 μmoles per g) than the 2300 Å (230 nm) beads (2.89 μmoles per g), and the latter less than either or the 550 Å (55 nm) beads (Y 117 bound 3.48 μmoles per g and the N 693 bound 4.63 μmoles per g). The N 693 beads, having the same pore diameter, must have a higher silanol density on their surface than the Y 117 beads.

The beads were washed with 0.1 N HCl to remove adsorbed catecholamine or oxidized product from the glass surface. This process is not dependent on the amount of acid wash, as previously reported, but on the duration of the acid wash.[9] A slow wash release of label of 0.002% per h is reached only after 100 h of acid washing. Experiments of Yong and Richardson[11] which use a rapid vacuum wash to remove adsorbed drug may be ineffective, and this makes interpretation of their work difficult.

These bead preparations were examined for leakage by placing one gram of ^{3}H-isoproterenol-labelled glass bead–immobilized catecholamines in 10 cm^{3} of distilled water and taking aliquots at time intervals thereafter.[3] The bead preparations all differed in the absolute amounts of label released per hour, though the percent release of total bound was similar to three of the bead preparations.[9] The 2300 Å (230 nm) bead preparation had a percent release rate only half that of the others. This may indicate a more thorough silanization process,[10] and a more protective hydrophobic film on the surface of the glass preventing solvent attack on the glass matrix.

From these findings we conclude that any bead preparation must be defined by the two parameters of amount of hormone bound and their leakage rates prior to their use as pharmacological tools. Control experiments must be run to eliminate the possibility of gross leakage into the bath. All of these bead preparations were active in generating contractile force in a localized manner.

SEQUENTIAL RELEASE STUDIES AND LOSS OF BIOLOGICAL ACTIVITY

When we performed a number of sequential release studies on these bead preparations, we found a change in the form of the release profiles.[9] In the initial studies, the release profile appears to consist of two phases, a fast one, and a slow one.[3,9] In later release studies the fast phase disappears leaving only the slow phase. Glass bead immobilized catecholamines that show this latter profile in release studies are inactive on cat papillary muscle, although they still have 80–90% of the originally bound isoproterenol. This would suggest that leakage of an azo derivative into the microvolume between bead and muscle surface accounts for a major fraction of the biological activity. An alternative explanation is that the fast phase of the release profile represents surface bound isoproterenol and when this is stripped off no activity remains because the isoproterenol bound to the interior of the beads is inaccessible to receptor and is more tightly bound. Further work is now in progress to confirm these findings.

IDENTITY OF RELEASE MATERIAL

We have begun analyzing the release derivative, and initial findings[9] suggest a silicon monomer. This monomer condenses to form hydrophobic polymers, sensitive to hydrofluoric acid, as would be expected from an alkyl trihydroxy silane.[12] Furthermore, the polymer has been examined by x-ray elemental analysis, and we have found that the polymer contains a large amount of silicon. Further work is now in progress examining the chemistry and pharmacology of this monomer, and how it is released from the glass matrix.

We cannot yet distinguish between the two possible mechanisms of glass bead–immobilized catecholamine action; either direct action of bead bound azo-catecholamine or action or released azo-catecholamine into the micro-volume between bead and muscle surface analogous to a neuroeffector synapse. Our results suggest that biological activity may be a result of a combination of the two. However, if glass bead–bound azo-catecholamine is acting directly, the nature of this bound azo-catecholamine is a silicon derivative which is ionically attached to the glass matrix and is more easily released than most of the bound catecholamine. These studies support our previous contention that glass bead–immobilized catecholamines are acting as a local delivery system. Furthermore, glass bead–immobilized catecholamines may prove to be a useful tool in analyzing pharmacological mechanism where localized drug action is involved.

Acknowledgements

This work was supported in part by grants from the American Cancer Society (BC-60-R) and the National Institutes of Health (USPHS CA 11683).

References

1. J. C. Venter, J. E. Dixon, P. R. Maroko, and N. O. Kaplan, *Proc. Nat. Acad. Sci. USA*, **69**, 1141 (1971).
2. M. S. Yong, *Science*, **182**, 157 (1973).
3. J. C. Venter, L. J. Arnold, Jr., and N. O. Kaplan, *Mol. Pharm.*, **11**, 1 (1975).
4. J. C. Venter, J. Ross, Jr., and N. O. Kaplan, *Proc. Nat. Acad. Sci. USA*, **72**, 824 (1975).
5. J. C. Venter, Doctoral Dissertation, University of California, San Diego (1975).
6. G. Vauquelin, M.-L. Lacombe, J. Hanoune, and A. D. Strosberg, *Biochem. Biophys. Res. Commun.*, **64**, 1076 (1975).

7. B. R. Venter, J. C. Venter, and N. O. Kaplan, *Proc. Nat. Acad. Sci. USA*, **73**, 2013 (1976).
8. W. R. Ingebretsen, Jr., E. Becker, W. F. Friedman, and S. E. Mayer, *Circ. Res.*, submitted (1976).
9. D. C. Woodman, J. C. Venter, and N. O. Kaplan, *Proc. Nat. Acad. Sci. USA*, in press (1976).
10. H. H. Weetall and A. M. Filbert, *Methods in Enzymology* V. XXXIV, Chap. 4, p. 59 (1974).
11. M. S. Yong and J. B. Richardson, *Can. J. Phys. Pharm.*, **53**, 616 (1975).
12. W. Noll, *Chemistry and Technology of Silicons*, Academic Press, New York (1968).

DISCUSSION

Dr. E. Goldberg (*Univ. of Florida, Gainesville, FL*): The comparison that you made for control purposes was either with the free catecholamine or with the glass bound material. Have you made other catecholamine derivatives (without silicon) and examined their pharmacological behavior?

Dr. N. Kaplan: We've looked at several catechol compounds (I don't know how pure they are) but at least for the serotonin derivative, they show completely different properties from the one that has the silicon in it. In other words, these control compounds behave more like the free serotonin although it has a little less activity.

Dr. H. Ringsdorf (*Univ. of Mainz, Mainz, Germany*): I initially believed that the bound (to glass) catecholamine was the effective species. Now you seem to be saying that the amine must be released from the glass matrix in order to be effective. Is this correct?

Dr. N. Kaplan: Not necessarily. We get activity at concentrations of drug which are much below the active concentration of free drug. If the drug is localized in the microvolume then we have a high local concentration.

Dr. H. Ringsdorf: What keeps the drug localized in the proper area so that it can produce its effect?

If the old explanation (immobilized amine) holds, then one would expect a definite spacer effect on activity. Have you examined very large spacer groups, and are these derivatives more active?

Dr. N. Kaplan: There could be ionic bonds which still maintain the drug on the beads.

Dr. I. Parikh (*Burroughs Wellcome Co., Research Triangle Park, NC*): There are at least two ways in which the catecholamine could be released from the glass beads: either by cleavage of the silane bond or by cleavage of the diazolinkages, which are very susceptible to the reducing agents present in biological media. If you had selected a substrate other than glass, at least one of these two alternatives could be excluded. Have you examined any other substrates in beaded or any other form?

Dr. N. Kaplan: I really don't recall why we started with glass beads. It would be interesting to look at other matrices. This is something which should be tried.

Dr. J. C. Venter (*State Univ. of New York/Buffalo, Buffalo, NY*): With regard to the problem of immobilized vs. local release of catecholamine, I would like to mention some of our recent data on binding cells and pulling specific cells out of populations through the use of sepharose- and glass bead immobilized catecholamines as an affinity matrix. [B. R. Venter, J. C. Venter, and N. O. Kaplan, Affinity isolation of cultured tumor cells by means of drugs and hormones covalently bound to glass and sepharose beads, *Proc. Nat. Acad. Sci. USA*, **73**, 2013 (1976)]. In these studies we were able to selectively bind glial (C6) cells out of a mixed population of cells to isoproterenol immobilized on sepharose or glass beads. The glial cells have a high catecholamine responsiveness whereas the other cells present have little or no catecholamine responses. The glial cell binding to the immobilized isoproterenol could be inhibited by catecholamine receptor antagonists such as propranolol. These results suggest that the immobilized form of isoproterenol is capable of receptor interactions. The cell binding also suggests that the concentration of released catecholamines is not sufficient to block the immobilized isoproterenol-cell receptor interactions.

Dr. J. Cardinal (*Univ. of Utah, Salt Lake City, UT*): Would you comment on the compatability of your proposed mechanism for the action of the glass–isoproterenol system with the known pharmacology of this drug?

Dr. N. Kaplan: Our mechanism is compatable with the pharmacology of isoproterenol. Those cells which have receptors respond while those without receptors do not respond.

Dr. J. Robinson (*Ciba-Geigy Corp., Summit, NJ*): You said that the beads lose activity when they still retain about 90% of the original isoproterenol. Have you been able to determine whether the 10% loss has occurred from the physically-accessible surface area, or does it occur uniformly throughout the bead? Would this data help to distinguish between your proposed mechanisms B and C?

Dr. N. Kaplan: No, we haven't been able to determine this, I don't know how this could be done. Perhaps this could be done by x-ray diffraction.

Dr. J. Robinson: Wouldn't the fact that activity is lost although 90% of the catecholamine is left mitigate against your mechanism B?

Dr. H. Ringsdorf: When you remove the glass beads, you have a rapid decrease in the beating rate. This is evidence against a non-bound drug since the nonbound drug could stay at the receptor for a while and give the same curve as for the low molecular weight compound.

Dr. N. Kaplan: We would have to suggest that if mechanism B is working, the non-bound compound is maintained physically at the receptor site and is liberated to the whole solution when the bead is removed.

Dr. A. Trouet (*International Institute for Cellular and Molecular Pathology, Brussels, Belgium*): You could also explain your results if you assumed that the peptide-like bond of the spacer group is split by enzymes which are localized on the surface of the cell. I am not predicting that this happens, but it certainly is a possibility to be considered.

Dr. E. Goldberg: The conclusion of this discussion, therefore, appears to be: there is no evidence which is contrary to the notion that you have a classical affinity-chromatography type ligand attached to an insoluble matrix which is active by a receptor interaction. Do you agree?

Dr. N. Kaplan: Yes.

An Implantable, Self-Powered Pump for Drug Delivery use in Laboratory Animals

RICHARD C. CAPOZZA

ALZA Corporation, 950 Page Mill Road, Palo Alto, CA94304, U.S.A.

A miniature implantable pump for use in small experimental animals is described.
The minipump delivers solutions continuously for a period of up to one week in animals as small as mice, without the need for external connections or frequent handling of the animals. This implantable pump has been successfully used in experimental studies of the effects of continuous administration of cancer chemotherapeutic agents, addictive drugs, hormones, and antigens. The minipump is a one-time use device. During its functional lifetime, the minipump acts as a constant source of drug within an animal's body.

The minipump is an implantable, self-powered device that continuously delivers solutions of experimental agents into laboratory animals at closely controlled rates for periods up to one week. Once inserted into an animal's body, the device will deliver liquids systemically, without external connections, and will continue to function automatically for the time specified.

The minipump is a cylinder that consists of a collapsible reservoir of flexible, impermeable material, surrounded by a sealed layer containing an osmotic agent, all of which is contained by a semi-permeable membrane (Figures 1 and 2A).

When a filled minipump is exposed to an aqueous environment, the osmotic agent imbibes water at a rate that is controlled by the semipermeable membrane. The imbibed water generates hydrostatic pressure on the flexible lining of the reservoir, gradually compressing it, and producing a constant flow of its contents, through the delivery portal. Only the material placed into the reservoir by the investigator leaves the pump. The osmotic agent cannot pass into the reservoir or out through the semipermeable membrane. The pump delivery rate is therefore fixed, and predetermined by the selection of the appropriate membrane.

Following implantation of the minipump, there is a brief start-up period of several hours before delivery reaches constant flow. Thereafter, the pumping rate is essentially constant until the reservoir collapses and the minipump is spent.

The reservoir of the minipump is filled by the user, by means of a syringe attached to the filling tube (Figure 2B). The flow moderator (Figure 2C) is

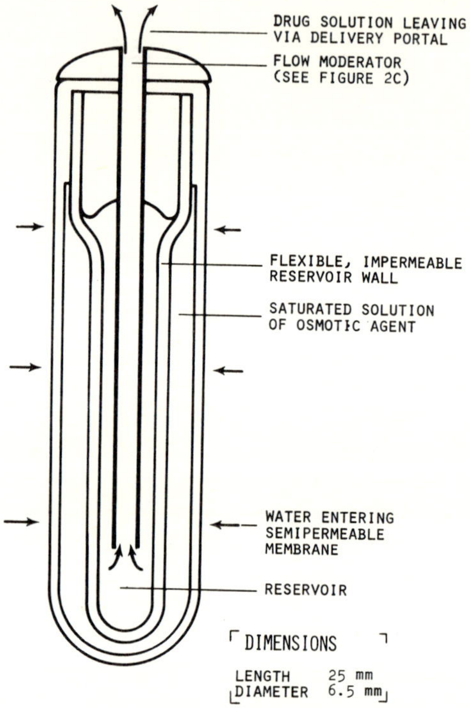

FIGURE 1 Cross section of a functioning minipump.

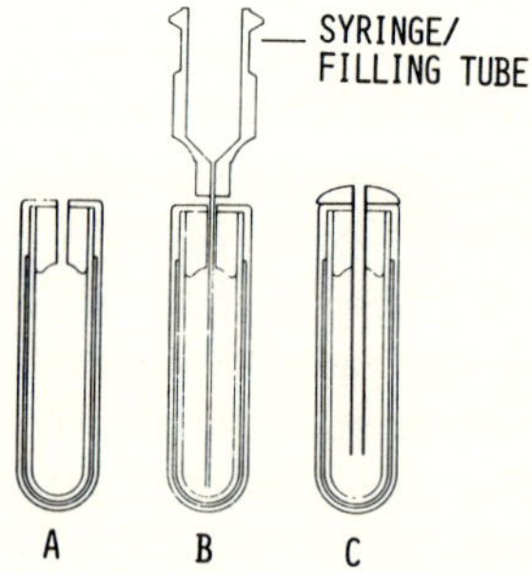

FIGURE 2 Filling the minipump: (A) cross section, (B) cross section showing syringe and filling tube attached and (C) cross section showing flow moderator in place.

then inserted. This component minimizes losses due to diffusion, as well as minimizing air bubbles that might otherwise interrupt flow. The filled device can be placed into the peritoneal cavity or subcutaneous sites in mice or larger experimental animals with only a minor surgical procedure. While specific test substances may elicit local tissue reactions, the surface of the minipump is biocompatible, and does not produce inflammation either during its use, or when left *in situ* after it is spent.

The minipump is the only self-contained, implantable infusion system small enough to be placed subcutaneously or intraperitoneally in mice or rats. It produces constant blood and tissue levels of experimental agents over extended periods. In effect, the Minipump can be viewed as an artificial drug- or hormone-secreting source within an unrestrained animal's body.

The minipump's controlled, continuous mode of action prevents the extreme blood level fluctuations that are inescapable when agents are given as tablets, capsules, injections, or implanted pellets of solid drug. With the device, the investigator need not be concerned that side effects resulting from an overdose–underdose pattern will obscure or distort the true dose-dependency of the effects of the agent being tested.

The minipump is particularly useful for testing highly potent drugs that are effective in minute doses. When such agents are introduced by conventional methods of drug administration, they are often metabolized too rapidly to permit them to be effective. In contrast, the Minipump operates at a low, continuous, and constant infusion rate that allows the biological activity of the infused substances to be observed in the steady state, over extended periods. It has been used to study the effects of continuous delivery of cancer chemotherapeutic agents, addictive drugs,[1] hormones, and antigens. The minipump can be used to deliver tracers to achieve steady-state labelling of metabolic pools.

The minipump reduces the time spent in giving experimental agents by eliminating the need for the frequent handling of the animals necessary with conventional dosing methods.

The data in Figure 3 show the performance of the minipump, as well as its similar delivery characteristics *in vivo*, in rats and mice. Each point represents the data from five minipumps. The cumulative amount pumped was computed by subtracting the remaining volume from the initial fill volume. The vertical lines show the standard errors of the means. At any one time point, the standard error of the mean of only one group is shown to avoid overlap. The solid line has a slope of 1.0 μl/h. Exhaustion is designated by the abrupt change in slope of the solid line from 1.0 μl/h to zero. The animal data indicate a pumping rate of approximately 0.85 μl/h, perhaps reflecting temperatures less than 37°C in subcutaneous tissues.

Figure 4 illustrates an example of material with biological activity, delivered

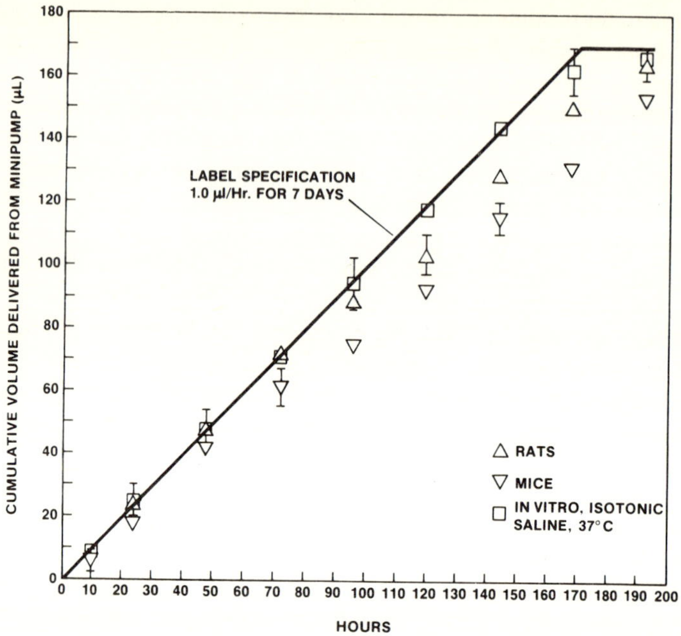

FIGURE 3 *In vivo* and *in vitro* performance of the minipump.

subcutaneously by the minipump, producing steady biological effects for the duration of the pumping. In this experiment, antidiuretic hormone (lysine vasopressin) was delivered by a minipump, placed subcutaneously in rats with congenital diabetes insipidus. The steadily maintained reduction in urine volume reflects the steady delivery of the hormone. A clear relation is also apparent between the three different rates of ADH delivery and the anti-diuretic effect. In the figure, 24-h urine volumes are expressed as a percentage of the day 4 value. Each value is the mean of the data from 6 rats; vertical bars indicate standard errors. During days 1 through 7, only the standard errors of the group with the largest standard errors are shown, to avoid overlap. Minipumps with a nominal pumping rate of 1.0 μl/h were used in all studies; the three different delivery rates were achieved by filling the minipump reservoirs with correspondingly different concentrations of lysine vasopressin. The average day 4 urine volume in the three groups of rats was 184.1 ± 44.9 ml/day ($n=18$).

Minipumps are designed to deliver homogeneous fluids with viscosities up to 100,000 centipoise. (The viscosity of water is approximately one centipoise.) Within this range, the effects of viscosity on delivery are negligible. (Suspensions may or may not be delivered reliably from the minipump, depending upon particle size, shape, and density.) The effects of temperature (T) and osmolality (π) of the environment on the pumping rate are linear in the range of $\pi=0$ to 26 atm; $T=2$ to 50°C. The change in flow rate is 2.8% per °C and 3.3% per atm.

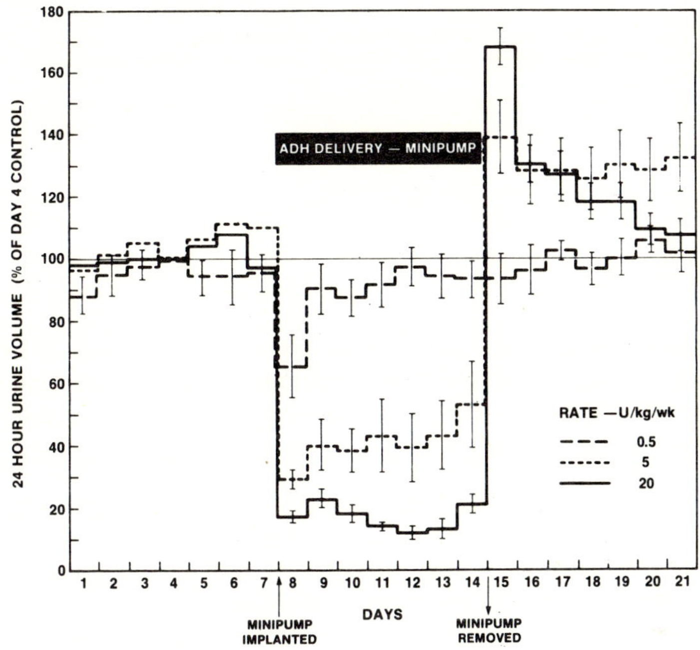

FIGURE 4 Delivery of antidiuretic hormone (ADH) from a minipump.

In general, the minipump is compatible with aqueous solutions, dilute acids and bases, alcohol–water mixtures (up to 5% alcohol by weight), and various ionic materials. The reservoir membrane is not compatible with aliphatic or aromatic hydrocarbons such as heptane, toluene, and natural oils. Contact with such substances will cause the system to malfunction due to reactions between these solvents and the membrane. Table I lists solvents found to be compatible with the minipump.

In conclusion, the osmotically powered minipump can be used for prolonged and controlled delivery of biologically active agents to experimental animals. These pumps can be implanted easily for infusion of agents continuously without external connections or frequent handling of the animals.

TABLE I

Examples of compatible solvents

1) Distilled water
2) Isotonic saline (with or without 0.9% (w/w) benzyl alcohol)
3) Ringer's solution (with or without lactate)
4) Dextrose injectable (5% w/w, in water or 0.9% NaCl)
5) Water–ethyl alcohol (5% w/w)
6) Polyethylene glycol 300 (neat or diluted with water)
7) Propylene glycol (neat or diluted with water)

Reference

1. E. Wei and H. Lok, *Science*, **193**, 1262 (1976).

DISCUSSION

Dr. R. Nevin (*Eli Lilly and Co., Indianapolis, IN*): You gave a list of compounds that are compatible with the minipump system. Are there compounds which are not compatible, and if so, what criteria do you use to exclude these materials?

Dr. R. Capozza: The reservoir material is not compatible with aliphatic or aromatic hydrocarbons such as heptane, toluene, and natural oils. Contact with such substances will cause the system to malfunction due to reactions between these solvents and the reservoir material.

Dr. R. Nevin: I didn't mean solvents, but rather proteins and substances of this nature.

Dr. R. Capozza: In general, we have found to date that if the test material is soluble in the vehicles I listed then no compatibility problems arise. We recommend, however, before beginning an experiment that the user of the minipump determine the chemical compatibility of their test material.

Dr. H. Hopfenberg (*North Carolina State University, Raleigh, NC*): What about dispersions?

Dr. R. Capozza: Suspensions or dispersions may or may not be delivered reliably from the minipump, depending on particle size, shape and density. However, solutions with a viscosity of up to 100,000 centipoise may be delivered reliably.

Unidentified discussant: How many millimeters of mercury will the device pump against?

Dr. R. Capozza: The minipump has been shown to pump against a pressure head of up to 200 mm of mercury.

Macromolecular Control of Food Additives

WILLIAM J. LEONARD, JR.

DYNAPOL, Palo Alto, CA94304, U.S.A.

The concern over the possible long-term biological effects of food additives in man has led to research on the synthesis and testing of a new class of macromolecular additives. Conventional low molecular weight additives such as the FD&C azo dyes and the food antioxidants are significantly absorbed by the gastrointestinal system. In contrast, the new additives, by virtue of their macromolecular size and their stability to relevant biological and chemical degradative processes are wholly eliminated in the feces. Thus, since they cannot come into contact with the internal organs and tissues, they have a greatly reduced likelihood of adverse effects. The macromolecular control over absorption has been demonstrated *in vivo* through metabolic fate studies of radiolabeled polymers.

Polymeric dyes and antioxidants have been developed which satisfy the non-absorption principle. The new materials show excellent functionality when added to a variety of systems designed to simulate a range of food applications. The results show that the polymeric additives are at least equivalent to their low molecular weight analogs. Often, by virtue of their macromolecular nature, they possess important functional advantages over the currently used food additives.

INTRODUCTION

One of the most controversial questions about food safety today centers on the use of food additives. A 1974 Harris poll in the United States found that 23% of the surveyed group felt that artificial additives pose a definite health danger, while 65% believed too many additives were being used in food.[1] The concern by the consumer is at least partly understandable in view of the recent history of FDA actions to remove certain additives from food based on new findings regarding their safety. Since 1960 nine food dyes have been delisted by the FDA either because they were proven unsafe or because of inadequate evidence establishing their safety under all use conditions.

In 1958 Congress expressed its concern for the safety of food additives by passage of the Food Additive Amendment to the Food, Drug and Cosmetic Act of 1938. The legislation required that the safety of all food additives in use at that time be evaluated by qualified panels for safety. As evidence, the expert panels could use whatever scientific data was available to them plus *prima facie* evidence based on the safety history of the additives in food products. Additives that were deemed acceptable by these criteria were considered to be "generally recognized as safe," thus establishing the GRAS list of approved materials which remains in use today. The amendment further specified that those additives which failed to win approval by the experts and all new additives proposed for future use in foods must be scientifically proven safe before they could be used legally in foods. This placed a demand on the science of safety testing which continues to generate a great deal of controversy.

Simplistically, the question posed is what effect, if any, will the additive have on the health of man after 60–70 or more years of daily consumption. The toxicologist is expected to answer using animal models having life spans in the range of 3–7 years. Clearly, the extrapolation of evidence between species, especially ones with such widely different life spans, is difficult to justify with the degree of certainty appropriate to the seriousness of the question posed.

In the most general sense, a food additive is any ingredient added to food for the purpose of achieving a technical effect. Also included in this definition, however, are those substances acquired by food indirectly from packaging, etc., or accidentally as contaminants, e.g., pesticide residues. Table I arranges the food additives which are intentionally added to food for technical purposes into four classifications based on their function.[2,3]

Viewed from this perspective it is clear that only the nutritional supplements have a functional purpose after the food is ingested. The others contribute either to the aesthetic quality or to ease of processing and distribution. As such they are nutritionally irrelevant once the food is past the

TABLE I

Classification of intentional food additives by function

Function	Technical objectives	Examples	Number in category
Aesthetic	Enhance palatability	Colors Flavors Non-nutritive sweeteners Surface finishing agents	1580
Economic	Prevent spoilage and facilitate distribution	Antioxidants Preservatives Fumigants	110
Technological	Facilitate processing Assist formulation of prepared foods	Emulsifiers Humectants Lubricants Propellants	770
Nutritional	Supplement nutrition	Vitamins Amino acids	85

palate and has entered the digestive system. Thus, if they could be chemically constructed so as not to be absorbed from the gastrointestinal tract during the digestive process, concern regarding their long-term safety in man could be substantially reduced. The reasoning is that the major toxic risk is realized only after the substance is absorbed. Only then can it or its metabolic products move to internal organs. On the other hand, if it is wholly retained within the intestinal lumen, it will be passed out with the feces, with a much reduced likelihood of adverse effects.

An important key to absorption control is molecular size. It is well known that polymers of both synthetic and natural origin can pass out of the gastrointestinal tract without systemic absorption, provided they are large and resistant to breakdown during the digestive process. At Dynapol, we are applying this concept of macromolecular control to the development of new non-absorbable food additives. The objective is to design macromolecules which are (1) at a minimum, functionally equivalent to their low molecular weight counterparts, and (2) not absorbed by man. Currently, our programs are focused on research in the areas of color additives, antioxidants and non-nutritive sweeteners. Development work is complete on a polymeric antioxidant which satisfies the above criteria. We are now producing the large amounts of material required to complete the approximately three years of chronic toxicity testing. Similar progress is being achieved in the development of color additives. One polymeric dye has been process-developed, with at least two more in the early stages of process development. Work on non-nutritive sweeteners is less well advanced because of the extremely difficult requirement to achieve balanced organoleptic properties. Consequently, the discussion in this paper will be restricted to results from our work in the antioxidant and colorant areas.

BIOLOGICAL CONTROL

One of the earliest systematic studies showing the important role molecular weight plays in controlling the transport of molecules from the gastrointestinal tract was made by Höber *et al.*[4] Transport rates of polyhydric alcohols ranging from ethylene glycol to mannitol and of amides ranging from acetamide to succinimide were measured using isolated rat intestinal loops *in situ*. The results clearly demonstrated that increasing the molecular weight decreases the absorption rate. However, because all of the compounds were of relatively low molecular weight, only the trend with molecular weight was obtained, rather than a lower limit for measurable absorption.

Subsequent studies have shown the marked reduction in transport that can be achieved with increasing molecular size into the polymer range.[5,6]

Polyethylene glycol is probably the most extensively studied polymer in this regard since it is a commonly used marker for monitoring fluid movement in the intestine. Its reliability as a marker requires that its recovery in the feces be quantitatively established. Biological fate studies have been performed in rats[5] and in man.[7] The results show that 4000 molecular weight polyethylene glycol is over 98% excreted in the feces by both organisms.

Unfortunately, no systematic studies are published showing the trend of gastrointestinal absorption of polymers as a function of degree of polymerization. It is possible to provide some insight on size control, however, by bringing together published results on several antioxidants of low and intermediate molecular weight.[8,9] Biological fate data are available for butylated hydroxyanisole (BHA) and butylated hydroxytoluene (BHT), two commonly used food antioxidants, and for the three antioxidants used as polymer stabilizers included in the set of structures given below.

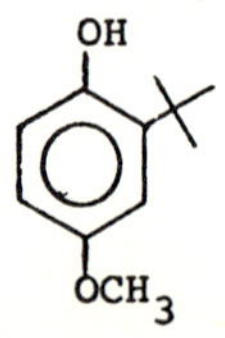

BHA

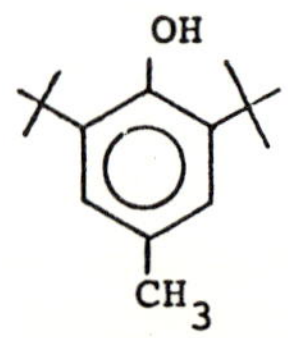

BHT

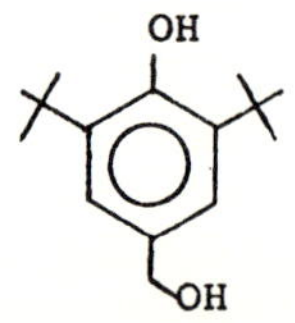

"Ethyl" 754

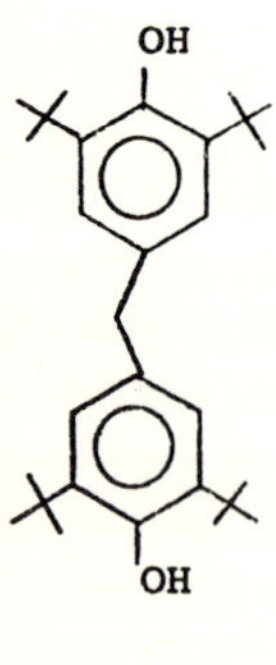

"Ethyl" 752

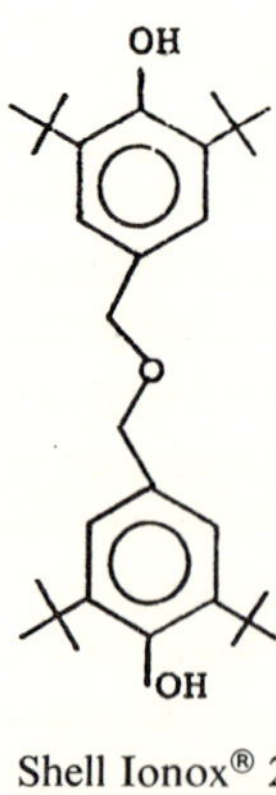

Shell Ionox® 201

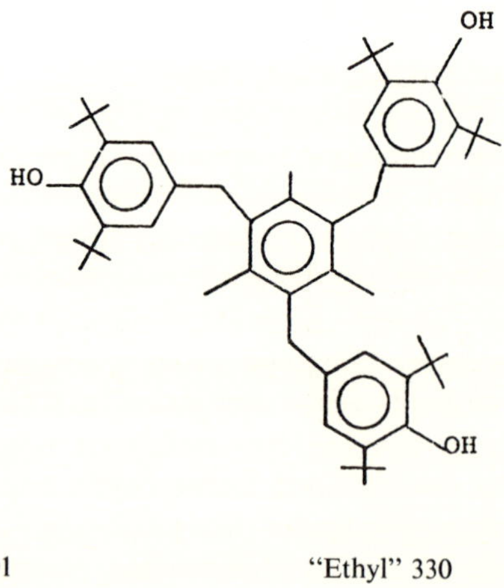

"Ethyl" 330

Figure 1 presents the GI-absorption data obtained for these antioxidants in rats. It is apparent that the materials with molecular weights below about 300 are almost completely absorbed, while those with molecular weights above approximately 800 are essentially not transported across the gastrointestinal membrane barrier.

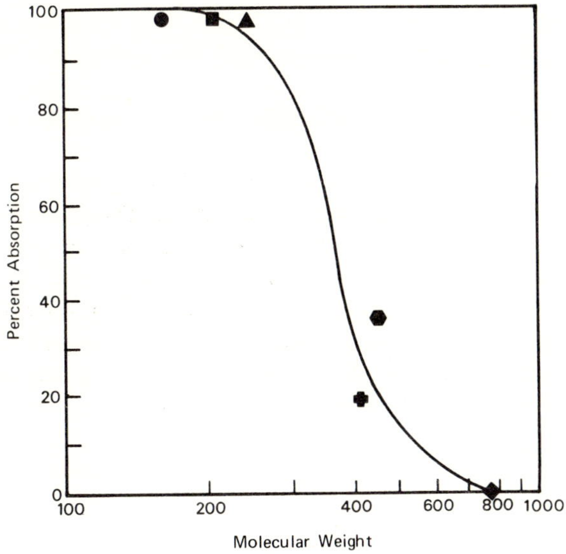

FIGURE 1 Gastrointestinal absorption of hindered phenolic antioxidants as a function of molecular weight. ● BHA, ■ BHT, ▲ Ethyl 754, ● Ionox 201, + Ethyl 752, ♦ Ethyl 330.

The molecular size dependence of GI-absorption will be affected by the polarity and electrostatic charge of the transported compound.[11] Also, the amount of a given compound absorbed can differ from one species to another. For example, the absorption of BHA in man is 80–90%, while its absorption in rabbits and dogs is 76 and 32%, respectively.[8]

At Dynapol a number of biological transport studies have been performed on rats which clearly demonstrate that chain length controls the intrinsic potential for synthetic polymers to transport across the intestinal barrier. One example, showing essentially no transport,[10] has been obtained for a polymeric dye model compound having the following composition.

The polymer used in the above metabolic fate study was determined by gel permeation chromatography (GPC) to be free of oligomers hydrodynamically equivalent in size to polystyrenes below DP=9. Since this

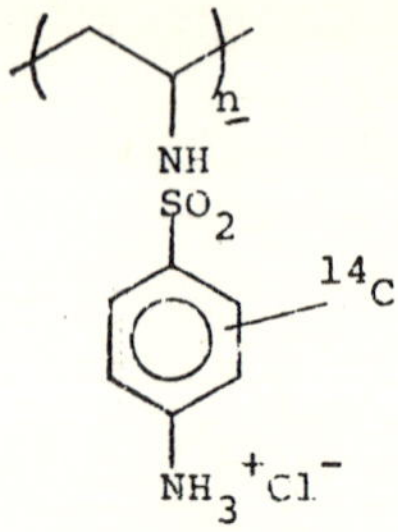

corresponds to a lower molecular weight limit of about 900, the results are consistent with the trend indicated by the data in Figure 1.

An example showing the effectiveness of macromolecular control on biological activity has recently been obtained at Dynapol using the Ames mutagenicity test.[12] This test utilizes mutant strains of *Salmonella typhimurium* which have lost their ability to synthesize the amino acid histidine. Since histidine is essential to their growth, the bacteria are dormant in a "histidine-free" culture medium. However, when exposed to mutagens, they undergo mutations that can have the effect of reversing the original genetic defect. Substances giving positive results are, by implication, potentially carcinogenic.

The following pair of dyes was included in our mutagenicity screening program.[13]

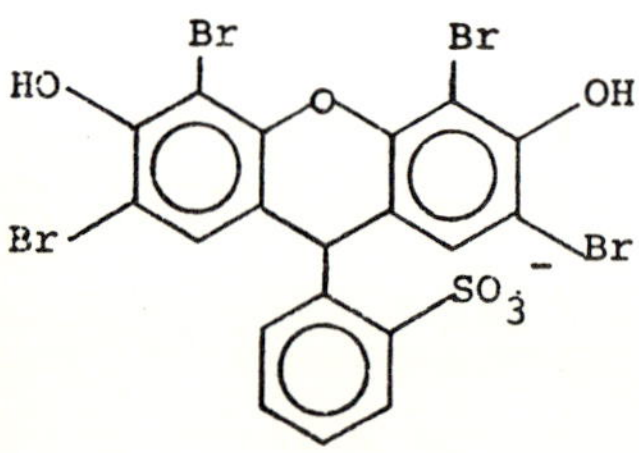

[I]

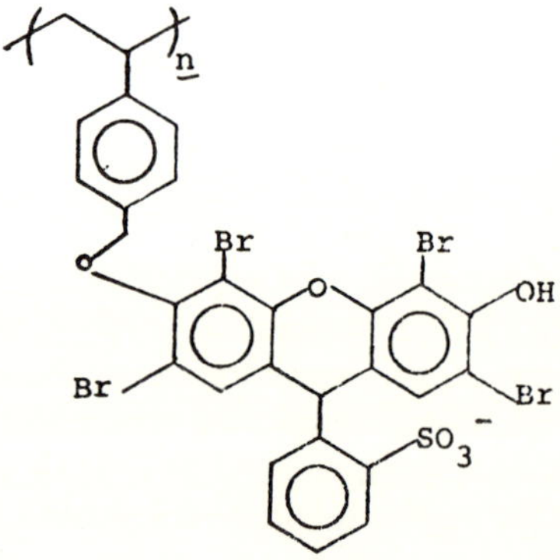

[II]

The polymer [II] is a macromolecular analog of [I] and has essentially the same color properties. Two strains of *Salmonella typhimurium* (TA 100 and TA 1537) were employed in the mutagenicity test. Both strains gave negative results for the polymer. The free chromophore, however, was strongly mutagenic to both strains, indicating a potential for effecting DNA base pair substitutions (TA 100) and frame shifts (TA 1537).

It seems most likely that the elimination of mutagenic properties in the chromophore [I] in the polymer [II] arises directly from the inability of the polymer to pass through the bacterial cell membrane. Alternatively, in the less likely event that the polymer is transported, the juxtaposition of the chromophores along the polymer backbone could prevent them from effectively interacting with the DNA bases, a requirement for the reverse mutations needed to repair the original defect.

While the control of GI-transport is a major consideration in the design of non-absorbable food additives, skin transport is also important since food dyes, food antioxidants, etc., are commonly employed in cosmetic and drug preparations intended for topical use. Consequently, we have tested the ability of Sunset Yellow (FD&C Yellow No. 6) [III] and its polymeric analog [IV][14] to penetrate human skin.

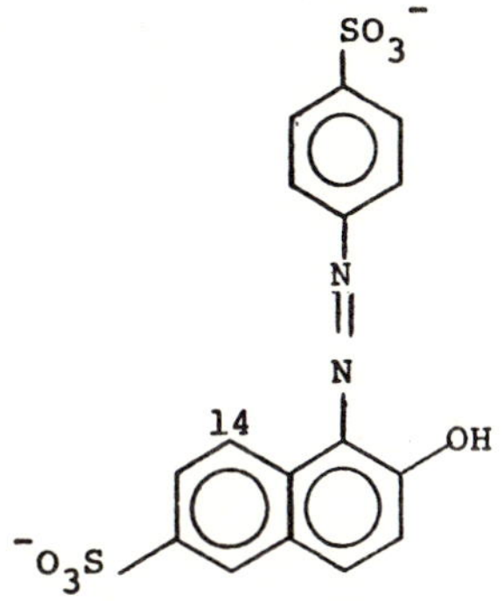

[III]

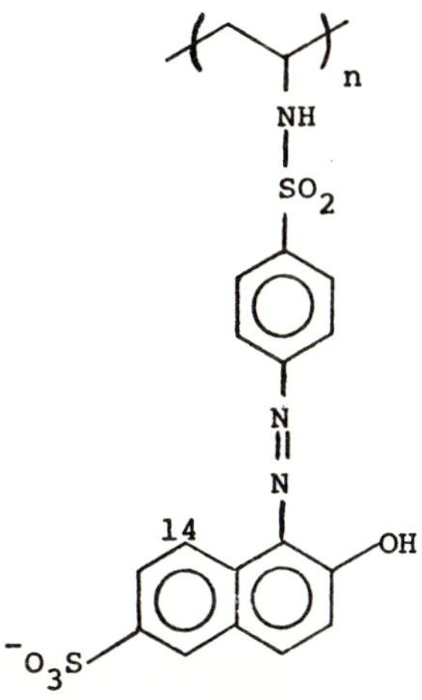

[IV]

For ease of detection both [III] and [IV] were synthesized with ^{14}C-labels in the C-8 position of the naphthalene ring. The experiments were conducted by mounting cadaver skin in permeation cells designed to allow the test solution to be applied to the stratum corneum side while maintaining the other side in continuous contact with Ringer's solution. The dyes were dissolved at 50 ppm in vehicles representative of those frequently used on skin, i.e., water, 0.1 M sodium lauryl sulfate, Jergen's lotion, and 10% glycerin soap. The results of this experiment show that neither Sunset Yellow nor its polymeric analogue [IV] permeates intact human skin, *in vitro*. However, when the epidermis is removed from the skin, Sunset Yellow freely penetrates the dermis and passes into the Ringer's solution. In contrast, the polymer [IV], apparently by virtue of its size, is essentially completely retained. The results of the latter experiment are shown in Figure 2.

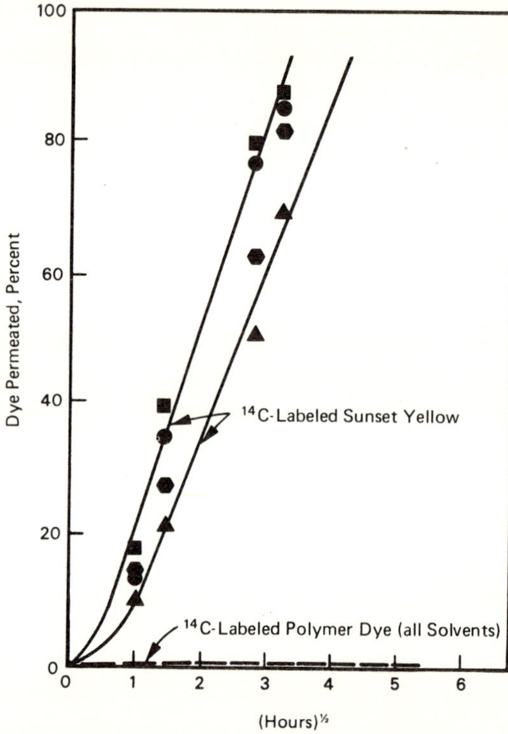

FIGURE 2 Transport rate of ^{14}C-labeled Sunset Yellow [III] (———) and ^{14}C-labeled polymeric dye [IV] (– – – –) through human dermis *in vitro* at 30°C. Carrier solvents employed: ▲ 0.1 M sodium laurel sulfate, ⬢ water, ● Jergen's lotion, ■ 10% glycerin soap in water. Data for the polydye in all solvents is given by the single dashed line as drawn.

The study on human skin transport is significant in view of the fact that the epidermis is readily removed by mild abrasion. Undoubtedly, some of the skin conditions leading to the use of topically applied drugs and cosmetics are ones that involve damage to the protective epidermal layer.

In designing a non-absorbable polymeric food additive, the distribution of chain sizes must be tailored such that the low end does not overlap the molecular weight region where absorption takes place. If, indeed, the cut-off molecular weight for intestinal absorption is about 800, as suggested by the data in Figure 1, this criterion is not a difficult one to satisfy. To illustrate this, Table II presents oligomer contents calculated for an ideal 1:1 condensation polymerization carried out to various degrees of conversion.[15,16] For a polymer with a repeat unit weight of 115, the calculations show that the %<800 (7 mer) is less than 1 wt. % where the conversion exceeds 98%. It is possible that in some cases conversions higher than 98% may be either difficult to achieve or undesirable for other reasons. A separation step would then be needed to remove the unwanted low molecular weight component.

TABLE II

Weight percent of oligomer content calculated for ideal 1:1 condensation copolymer

Degree of conversion	DP_n	DP_w	wt. % < 5 mer	wt. % < 7 mer
0.99	100	199	0.15	0.27
0.98	50	99	0.57	1.03
0.95	20	39	3.28	5.72
0.90	10	19	11.43	18.69

Other polymerization mechanisms can, of course, yield chain size distributions other than the "most probable" one discussed above. In the case of free radical polymerizations where chain transfer and termination reactions are important, distributions broader than the most probable one are generally obtained. In this event, the encroachment of the low molecular weight end towards the region of transport can be a critical problem in the design of a non-absorbable food additive. Conversely, polymerization conditions leading to "living polymer" reactions can give a product with a much narrower distribution, i.e., a Poisson distribution, thus allowing excellent control for avoiding overlap with the intestinal transport region.

In the preceding discussion the thesis has been presented that macromolecularization is requisite for controlling the biological transport of a synthetic food additive. Indeed, this is a necessary criterion but not the only one needed to guarantee non-absorption. The maintenance of structural integrity throughout the additive's use history is also required. After manufacture, the food additive is confronted with a number of chemical and biological challenges, any one of which can result in its breakdown to smaller units, thus defeating the original non-absorption objective. The probable sequence of events in the use of a food additive after manufacture is: storage prior to use→processing in a food product→storage of the food product→additional processing by the consumer→consumption and digestion→excretion. If the macromolecule contains bonds that are weak to relevant chemical and biological attacks at any stage of its use, then it is possible that structural breakdown will occur.

Clearly, then, the chemical structure of the macromolecule must be selected to withstand at least the consequences of normal use. Preferably, it should go beyond that, maintaining its structural integrity when subjected to abusive conditions.

Probably the most exacting measure of structural stability is the macromolecules' resistance to degradation in the gut during the digestive process. The enzymes produced by intestinal microflora have a versatility that is seemingly unlimited in their ability to attack chemical bonds. The enzymes are facile in the following types of reactions:[17]

1) Hydrolytic reactions. Examples: hydrolysis of esters, amides, etc.,

2) Removal of substituent groups. Examples: dehydroxylation, decarboxylation, dehydrogenation, *O*-dealkylation.

3) Modification of substituent groups. Examples: reduction of azo bonds, reduction of double bonds.

4) Modification of basic skeletal structure. Examples: aromatization of ring structures, fission of heterocyclic rings.

To be sure, the presence on the polymer of a chemical group falling within one of the above classes does not necessarily mean that the corresponding reaction will take place. To a considerable extent, neighboring chemical groups mediate the metabolic reactions that can be brought about by the intestinal micro-organisms.

FUNCTIONAL CONTROL

While achieving control over intestinal transport is a fundamental objective in Dynapol's food additive research, it is equally important that the products from this research have the needed functional activity in food. For example, the polymeric antioxidants must stabilize fats and oils against oxidative breakdown and the polymeric dyes must have colors that contribute to the aesthetic quality of food. Further, these basic technical properties should survive the processing and storage conditions encountered by foods.

In the following sections, some of the properties of polymeric antioxidants and colorants are discussed in terms related to potential food applications.

To the extent possible, comparisons are made to their low molecular weight counterparts which are currently listed by the FDA for use in foods. The functional polymers used in the examples are in many cases materials which were synthesized and tested in the course of arriving at the best materials for development, and may not be the final products selected for development. Where necessary, proprietary interests are protected by omission of structural detail.

Polymeric antioxidants

Much of our work in the development of a polymeric antioxidant has centered on materials having the generalized structure given in [V]. Polymeric antioxidants containing specified comonomers are referenced in the following text as V-a and V-b.

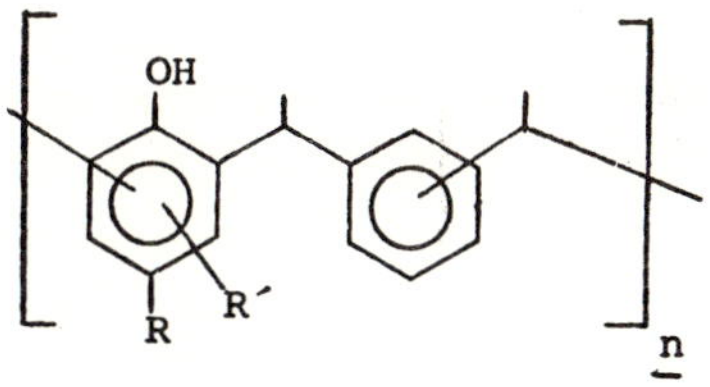

[V]

The polymers are prepared by the aluminium catalyzed polycondensation of divinylbenzene with various phenols.[14,18,20] Typical products contain substituents in which R is either a phenolic, a methoxy, or a short chain alkyl group and R′ is either H or a short chain alkyl group.

A variant of the active oxygen method (AOM) was employed to test antioxidant activities in vegetable oils.[21] In this method the oil samples are held at 80°C in an oven with a steady stream of dry air passing over them. Aliquots of the oils are periodically measured for peroxide content. The time required for the peroxide content to reach 100 meq/kg is taken as the end-point for the test.

Figure 3 shows the results for BHA, BHT and a polymeric antioxidant in a cottonseed–soybean oil containing 200 ppm citric acid.[21] The polymeric antioxidant tested is one in which the phenolic components are *p*-cresol and mono-tertiary butylhydroquinone [V-a]. The results show that the polymer is a better inhibitor to peroxide formation than the presently available food antioxidants BHA and BHT.

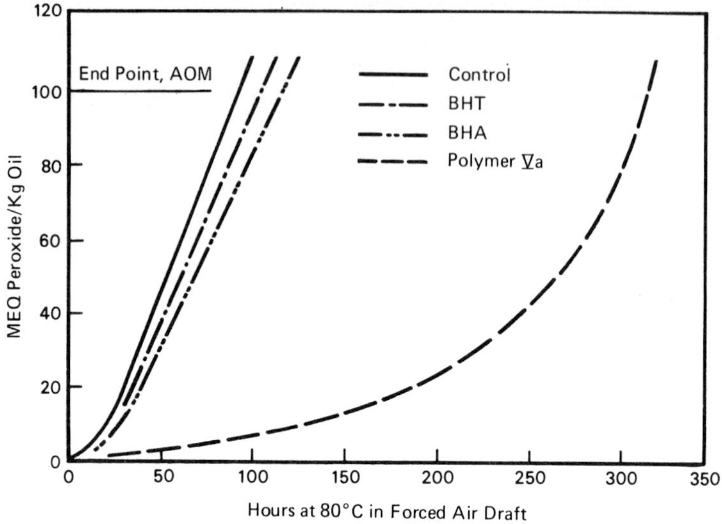

FIGURE 3 Modified AOM test results comparing the activity of food grade antioxidants with the polymeric antioxidant V-a. All measurements were made at 200 ppm antioxidant in cottonseed–soybean oil with 200 ppm citric acid added.

The standard AOM test[22] was employed in a study comparing the activity of the polymeric antioxidant to mono-tertiary butylhydroquinone (TBHQ), a food antioxidant. (The procedure differs from the one described above only in that the air is percolated through the sample and the test temperature is 97°C.) In Figure 4 the activity comparison is made on an equal weight basis while Figure 5 compares their activities on the basis of equivalent phenolic group concentration. Here it is seen that the performance of the polymeric antioxidant is less compared to TBHQ at equal weight, but clearly superior at equal functional group contents.

Tests comparing the effectiveness of polymeric and low molecular weight antioxidants have also been made using the oxygen bomb test (OBT).[23] The method operates on the principle that fats and oils take up oxygen as they deteriorate. The process is accelerated by subjecting the sample to pure oxygen under pressure, generally at 100°C. Antioxidant effectiveness is gauged by the time required to exhaust its activity, a point signaled by the onset of a pressure decrease as oxygen begins to react with the oil. Data comparing a polymeric antioxidant with BHT and TBHQ for the stabilization of lemon essence oil is presented in Figure 6. The results show the polymer to be much more effective for stabilizing the oil than either of the low molecular weight antioxidants at all concentrations tested.

The polymeric antioxidant employed on the above OBT test has a different

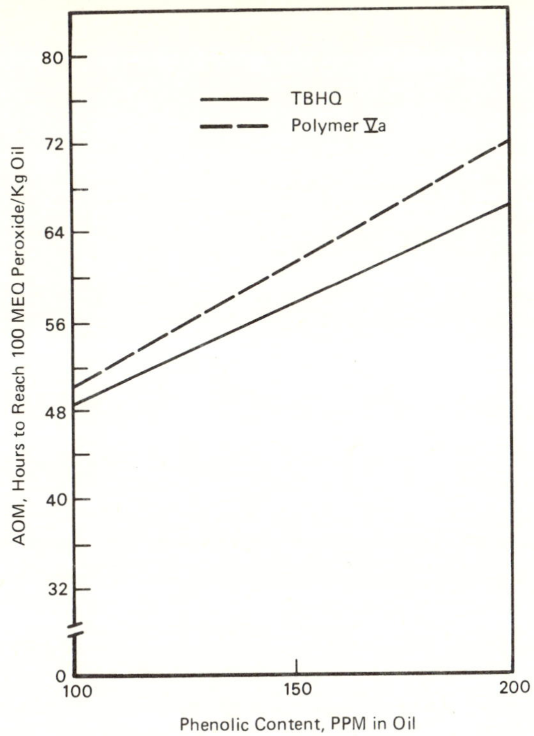

FIGURE 4 AOM test results comparing the activity of the polymeric antioxidant V-a with TBHQ at equal weights in cottonseed–soybean oil with 200 ppm citric acid added.

chemical structure [V-b] in which the phenolic comonomer is a multi-component blend containing, principally, *p*-tertiary butylphenol, *p*-hydroxy-anisol, and mono-tertiary butylhydroquinone. Similar OBT tests have been performed using a polymeric antioxidant with the V-a structure which contains only *p*-cresol and mono-tertiary butylhydroquinone comonomers. The activity of the latter polymer was considerably less than obtained in the case of the V-b polymer. The difference reflects the variable potential of phenolic comonomers for stabilizing, in this case, lemon essence oil. The observation, however, is far more general and parallels what is known about low molecular weight phenolic antioxidants. Food processors frequently select from the available list of approved antioxidants a combination of types that are most active in the food system of interest. Similarly, we have found that the selection of the various hindered phenolic comonomers and the relative ratios with which they are incorporated in [V] is critical for achieving the desired level and breadth of functional control.

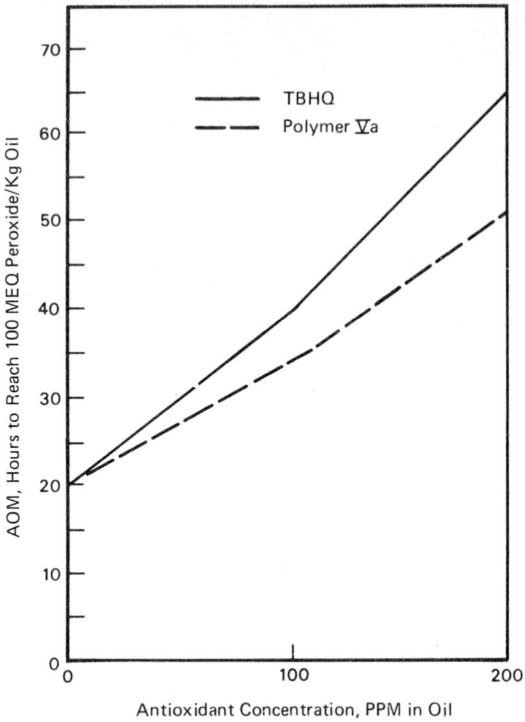

FIGURE 5 AOM test results comparing the activity of the polymeric antioxidant V-a with TBHQ at equivalent hindered phenolic contents in cottonseed–soybean oil with 200 ppm citric acid added.

Since the structure/activity profile determines optimum functionality, a unique advantage can be achieved from a polymeric antioxidant conforming to the general structure [V]. Our work has shown that a range of phenolic types can be incorporated in virtually any combination into the polymer. Thus, an optimum ratio of phenolic comonomers can be selected to best fit the food applications of major importance. This flexibility in structure design can give a product with applications versatility impossible to achieve in a single non-polymeric material.

One of the most striking properties of a polymeric antioxidant is its thermal stability. A thermogravimetric study of the polymer having the V-b composition is presented in Figure 7. For comparative purposes data on BHA, BHT and TBHQ are also included. The tests were performed in air at a heating rate of 5°C/min. The results show that the low molecular weight antioxidants completely volatilize and/or decompose at temperatures of 150–170°C. In contrast, the polymer is stable at temperatures at least up to 300°C. Further, analysis by GPC of a sample of the polymeric antioxidant which had been held at 190°C for 90 min gave no evidence of breakdown of the primary structure into smaller units. The increased thermal stability of the polymeric antioxidant over its food grade counterparts is important

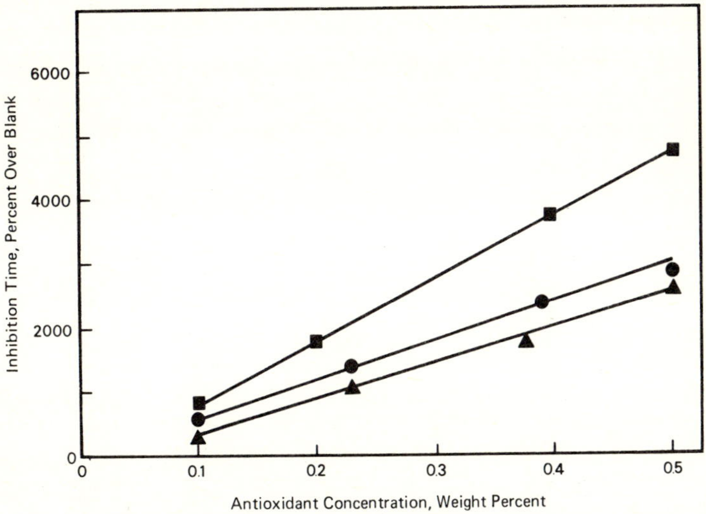

FIGURE 6 Oxygen bomb test results comparing the activity of the polymeric antioxidant V-b with BHT and TBHQ in lemon essence oil at 100°C and 65 PSIG O_2. ■ Polymer V-b, ● TBHQ, ▲ BHT.

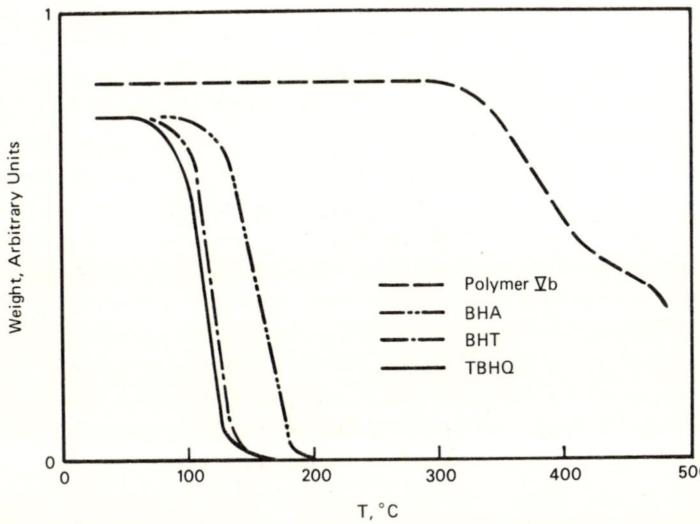

FIGURE 7 Thermal gravimetric analysis of polymeric antioxidant V-a and BHA, BHT and TBHQ. Heating rate 5°C/min.

technically, since many applications, e.g., frying and baking, employ high temperatures where direct volatility or steam volatility of the antioxidant can lead to substantial losses.

Polymeric colors

The initial objective in the "polydye" program was to produce polymeric colors with color properties similar to selected food dyes in current use. Two examples are the azo dyes, FD&C Yellow No. 6 [**VI**] and FD&C Yellow No. 5 [**VII**].

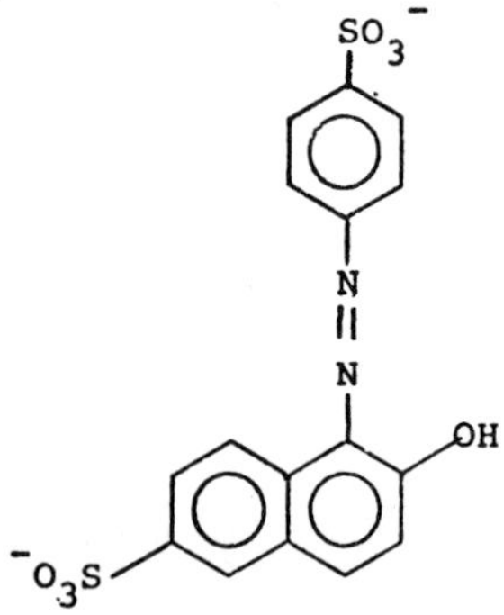

[VI]

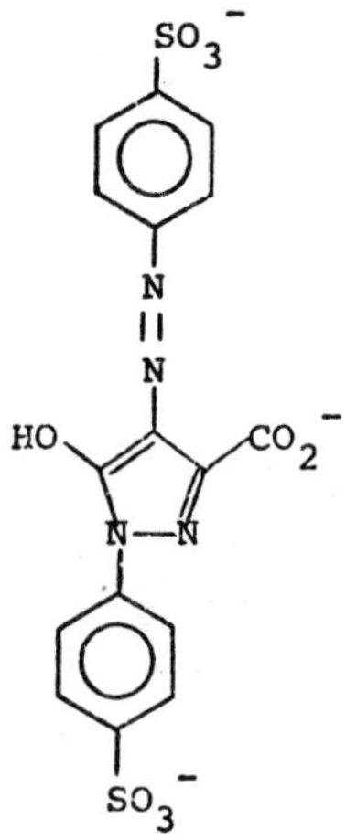

[VII]

Although our program subsequently departed from the approach employing azo chromophores, the findings serve quite well to define the type of functional controls that can be achieved for potential food applications.

Polymeric dyes were prepared by a series of reactions which affixed the chromophore to a polyaminoethylene backbone via a sulfonamide bond.[14] A series of polydyes were synthesized by this method, including the polymeric analogs of the above FD&C colors. The structures of the latter [**VIII**] and [**IX**] are given on the following page. Generally, the degrees of substitution attained were in excess of 90% of the available amine sites of the backbone.

Figures 8 and 9 present the visible spectra of the polydyes and the low molecular weight chromophores. For purposes of comparison, the concentrations of each pair were adjusted so that the chromophore concentrations, rather than weight, were equivalent. In both graphs, it is seen that the

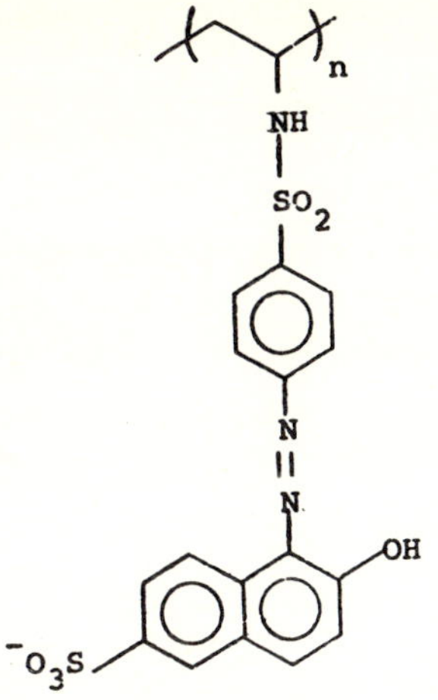

[VIII]

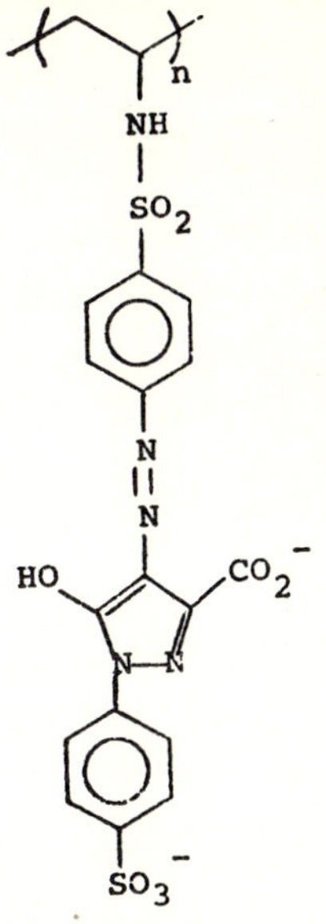

[IX]

optical densities at λ_{max} are lower for the polymer and that their λ_{max} shifts to lower wavelength. In addition, the spectral absorption bands of the polymers are slightly broader than for the free chromophore.

We attribute the difference in absorption coefficients to structural effects characteristic of the sulfonamide derivative, on the basis of a spectral analysis of compound [X]. This compound, which is a better chromophoric analog of the polymeric derivative than the FD&C dye, has a molar absorption coefficient of 18,600 which is in close agreement with the value of 19,000

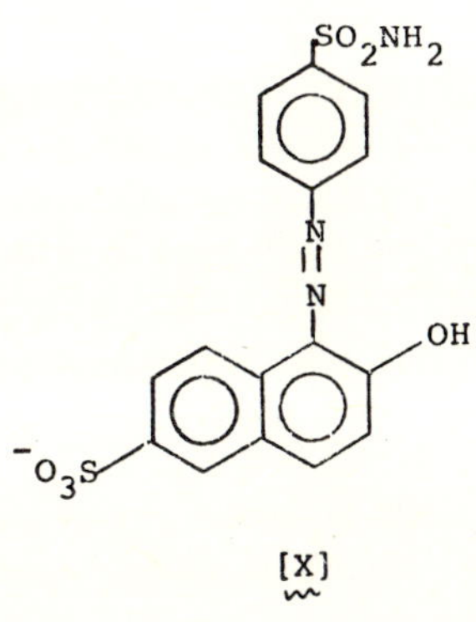

[X]

obtained for the polymer (based on equivalents of attached chromophore). In contrast, FD&C Yellow No. 6 has a molar absorption coefficient of 27,600. The λ_{max} for the model compound [X] and FD&C Yellow No. 6 were 476 nm and 475 nm, respectively.

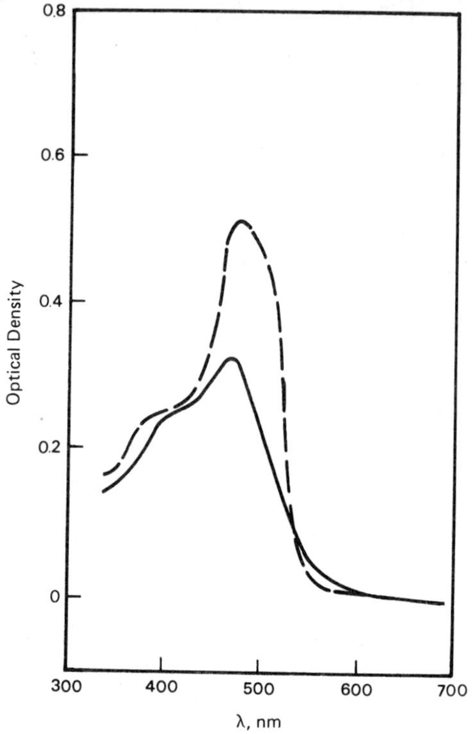

FIGURE 8 Visible spectra comparing the polymeric dye [VIII] (———) with its color and structural analog FD&C Yellow No. 6 [VI] (– – – –).

Intuitively, it might be expected that the depressed absorption coefficients of the polydyes would result in a lower perceived color intensity or "tinctorial strength." In fact, polymer **[VIII]** is virtually indistinguishable from FD&C Yellow No. 6 when compared at 50 ppm in water. Furthermore, at higher concentrations, the color of the polymer appears much more intense. A solution of polymer **[VII]** at 100 ppm requires over 150 ppm of FD&C Yellow No. 6 to develop a visual match. Similar trends have been obtained for polymer **[IX]** with FD&C Yellow No. 5.

The origin of the tinctorial enhancement by the polymers is not clearly understood, though it is probably associated with the spectral broadening

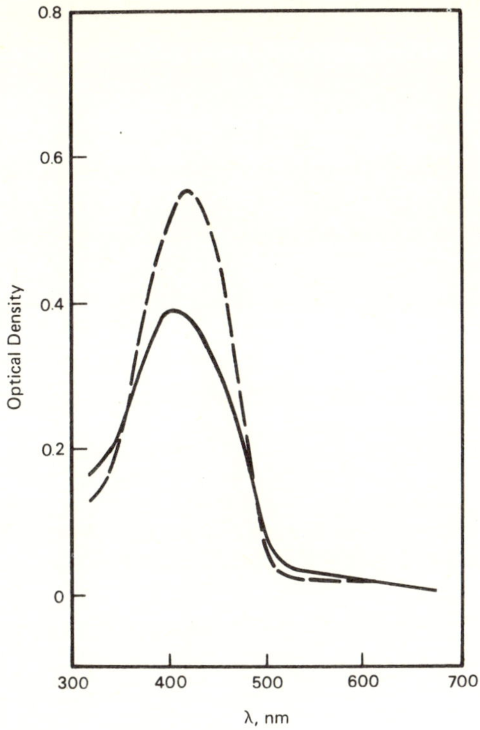

FIGURE 9 Visible spectra comparing the polymeric dye [IX] (———) with its color and structural analog FD&C Yellow No. 5 [VII] (– – – –).

and peak shift obtained for the polymers. It is known that the derivative *per se*, i.e., the sulfonamide bond, is not responsible for the observation since the model compound [**X**] did not show similar color enhancement.

The question of color enhancement as a general property of polymeric dyes is not easily answered on the basis of our research. Most of our subsequent observations on non-azo polydyes were made without comparison to the appropriate low molecular weight compound. Consequently, even though it has enormous technical importance, the color enhancement issue has not been further resolved.

One important applications property in which the azo polydyes clearly excel over their FD&C counterparts is their low metal corrosivity in canned beverage systems. The FD&C azo dyes are known to have a corrosive effect at concentrations at which they are normally used.[24] As a result, on storage, the pin hole defects in the interior resin liner lead to corrosion of the metal container. In order to avoid problems over the lifetime of the product, the soft drink manufacturers currently limit the concentration of the azo dye used and the time the product remains on the shelf.

A comparative corrosion study was made using a beverage formulation similar to a commercial soft drink. A polydye with the structure **[VIII]** was tested. The polymer also contained acrylic acid comonomer units in a ratio approximately equal to the chromophore units. The control was FD&C Yellow No. 6 **[VI]**. Corrosion rates were measured for each solution using a polarographic technique[25] which employs a tin–steel electrode representing the material used in constructing cans. The data were extrapolated to give the estimated perforation-free shelf life (EPFSL). Extrapolations were not taken beyond 12 months, since this is considered to be the upper limit of the test. The extrapolated results are given in Table III. The significantly reduced corrosivity of the polymeric dye suggests that the shelf life of a canned beverage can be increased by at least a factor of two if polymeric dyes were to be used in place of the FD&C azo dyes.

TABLE III

Estimated perforation free shelf life (EPFSL) of polymeric dye [VI] and FD&C Yellow No. 6

Dye	Concentration ppm	EPFSL (months)
FD&C Yellow No. 6	30	9
	60	6
[VI]	30	⩾12
	60	⩾12

Another polydye property that contrasts markedly with the FD&C dyes is the diffusibility of the colors in viscous media. Early in our work, it was found that gelatins containing different polymeric colors could be sequentially layered and held indefinitely without observable migration of dye from one layer into another.[26,27] In contrast, the low molecular weight dyes inter-diffused readily. The property is an important one that should lead to the design of new products and the improvement of ones currently in use.

Our evaluation of the important functional properties of polymers **[VIII]** and **[IX]** and others bearing chromophores of the azo class, shows them to perform at least equivalently and often superior to their single chromophore analogues. Tests performed in foods confirm their compatibility when processed with solids, emulsions and homogeneous solutions.

One especially important property of the polydyes is their ability to form alumina lakes that are much more resistant to color bleed than lakes made from the FD&C colors. The lakes are thus prepared more efficiently since wash-off after contacting the polydye with the alumina is essentially nil. It also expands their potential use range into product areas that are currently unsuitable for the FD&C lakes. The theoretical basis of this difference in stability follows from the general statistical considerations of polymers and comparable low molecular weight species at the solid–liquid interface, where the absorption of flexible molecules has been shown to be exponentially proportional to the polymer chain length.[28] Further, the amount of polymer adsorbed increases markedly with favorable surface interactions. The polydye lakes benefit from the effects of both the chain size and the ionic interactions between its negative charges and the positively charged alumina.

The technical success of the azo-polydyes, however, was overshadowed by the ease of azo bond reduction. Our studies have shown that both the low molecular weight azo FD&C colors and their polymeric derivatives are reduced by the action of the intestinal enzymes to give the amino cleavage products. For the polydyes [VIII] and [IX] this means that the distal parts of the chromophores are split off to give the corresponding naphthalene and pyrazolone derivatives. Further, similar reactions leading to the reduction of the azo bond take place in food systems containing reducing agents, e.g., Vitamin C.

Since it is known that the cleavage products which obtain on reduction of the azo bonds of the FD&C colors are absorbed systemically from the GI-tract,[29] we changed our course to pursue a different class of polydyes.[30] After extensive research, we found that the anthraquinones, when attached to a polymer backbone, produce a product exceedingly stable to both biological and non-biological processes.

The anthraquinone polymer products possess the functional advantages, including the ease of lake formation and low diffusibility, which were found for the azo-based polydyes. With this basic approach, we are proceeding to develop a color palette that will achieve our basic goal of a food color system having the desired functional and biological controls discussed above.

Acknowledgements

The author wishes to acknowledge the Dynapol research staff who provided much of the information presented here. In particular, special recognition is given to Drs. J. B. Brown, T. M. Parkinson, P. L. Dubin, A. R. Cooper, N. M. Weinshenker, and R. C. Phillips; and to T. E. Furia and M. Kronstadt for their permission to present data in advance of publication and for their many suggestions.

References

1. M. Edison (Lou Harris and Associates, Inc.), Oral Presentation at Food Update XIII. Scottsdale, Arizona, April (1974).

2. R. Walter, *J. Biosoc. Sci.*, **8**, 211 (1976).
3. Naturally occurring toxicants, food additives, and cosmetics, in *Chemicals and Health*, Chapter 8. (Report of the Panel on Chemicals and Health of the President's Science and Advisory Committee, Science and Technology Policy Office, National Science Foundation, Washington, D.C., (1973).)
4. R. Höber and J. Höber, *J. Cell and Comp. Physiol.*, **10**, 401 (1937).
5. D. L. Miller and H. P. Schedl, *Gastroenterology*, **58**, 40 (1970).
6. K. H. Soergel and W. J. Hogan, *Gastroenterology*, **52**, 1056 (1967).
7. R. Shields, J. Harris, and M. W. Davis, *Gastroenterology*, **54** (1968).
8. J. C. Dacre, in *Metabolic Aspects of Food Safety* (F. J. C. Roe, Editor), Chapter 5, Academic Press, New York (1969).
9. T. H. Wilson and T. N. Vincent, *J. Biol. Chem.*, **216**, 851 (1955).
10. T. M. Parkinson, private communication.
11. D. F. Magee, *Gastro-Intestinal Physiology*, Charles C. Thomas, Springfield, Illinois (1962).
12. B. Ames, J. McCann, and E. Yamasaki, *Mutat. Res.*, **31**, 347 (1975).
13. J. P. Brown, private communication.
14. D. J. Dawson, R. D. Gless, and R. E. Wingard, *J. Am. Chem. Soc.*, in press.
15. W. H. Stockmayer, *J. Polymer Sci.*, **9**, 69 (1952).
16. M. Kronstadt, J. A. Tyburczy, and P. L. Dubin, *Polymer Preprints*, **17**(2), 446 (1976).
17. D. E. Hathway, Metabolic fate in animals of hindered phenolic antioxidants in relation to the safety evaluation and antioxidant function, *Adv. Food Res.*, **15**, 1 (1966).
18. A. J. Kolka, J. P. Napolitano, A. H. Filbey, and G. G. Ecke, *J. Org. Chem.*, **22**, 642 (1957).
19. U.S. Patent No. 3,546,173.
20. J. A. Dale, S. Ng, and P. C. Wang, to be published.
21. T. E. Furia and N. Bellanca, *JAOCS*, **53**, 132 (1976).
22. *Official and Tentative Methods of the American Oil Chemists' Society*, Vols. I and II, 3rd Edition, AOCS, Champaign, Illinois, 1973, Method Cd 12-57.
23. B. N. Stuckey, E. R. Sherwin, and F. D. Hannah, Jr., *JAOCS*, **35**, 581 (1958).
24. J. Noonan, Color additives in food, in *Handbook of Food Additives*, 2nd Edition (T. E. Furia, Editor), CRC Press, Cleveland, Ohio (1972).
25. G. E. Dean, *Am. Soft Drink J.*, Continental Can Company, Inc., Chicago, Illinois, February, 1966.
26. U.S. Patent Number 3,937,851.
27. C. H. Havighorst, *Food Engineering*, **46**, 46 (1974).
28. C. V. Hoeve, E. A. DiMarzio, and P. Peyser, *J. Chem. Phys.*, **42**, 2558 (1965).
29. J. J. Roxon, A. J. Ryan, P. G. Welling, and S. E. Wright, *Fd. Cosmet. Toxicol.*, **5**, 447 (1967).
30. Dynapol's Colors Bless Ties That Bind, in *Chemical Week*, 21–23, August 4, 1976.

DISCUSSION

Dr. W. Babcock (*Bend Research, Inc., Bend, OR*): Is the optical density a linear function of the concentration of dye attached to the polymer or of the monomer by itself?

Dr. W. Leonard: Beer's Law is obeyed by both the uncoupled azo chromophore and its polymeric analogue up to a few hundred ppm, which is the usual instrumental upper limit for measuring a strongly absorbing dye.

Dr. W. Babcock: The reason I asked is because dyes are known to aggregate at low concentrations, say 10^{-5} molar, and the polymer might be keeping them apart.

Dr. W. Leonard: If these chromophores do aggregate in aqueous solution, their spectra apparently are not affected.

In any event, chromophore aggregation in the polymer would probably be less favorable due to bond constraints between neighboring groups. Interactions between more distant chromophores would also involve other chains and possibly lead to gelation. All of the polymeric dyes discussed here form homogeneous solutions in aqueous media.

Dr. J. Cardinal (*Univ. of Utah, Salt Lake City, UT*): I would expect a shift in the absorption spectrum upon aggregation.

Dr. W. Leonard: The chromophores shift spectrally when coupled to the polymer.

Dr. J. Cardinal: It's not very evident from your figure. Something like methylene blue shows dramatic shifts.

Dr. W. Leonard: We have no evidence that the spectral shifts we observe are due to aggregation of the uncoupled chromophores. In fact, the spectral differences could be entirely due to other electronic interactions between adjacent chromophores in the polymer.

Dr. D. Beyer (*Dow Chemical Co., Midland, MI*): One of the advantages of your polymeric dyes seems to be their lack of mobility. You should have the same lack of mobility in the case of your polymeric antioxidants. Have you seen any problems with antioxidant capacity in systems where the antioxidant is diffusion limited?

Dr. W. Leonard: No, though we have not looked specifically at this since our polymeric antioxidant is molecularly dispersed as a homogeneous solution in the oil systems we have studied. Further, our tests show that its activity for stabilizing oils and oil-containing food systems against rancidity is generally equivalent, and in many cases superior to that obtained by the low molecular weight food antioxidants.

The question is an interesting one, though, since there must be an upper limit on molecular weight where the antioxidant, on a weight basis, becomes increasingly less efficient.

Dr. A. Trouet (*International Institute for Cellular and Molecular Pathology, Brussels, Belgium*): I would like to question the validity of the Ames test with regard to the mutagenicity of polymeric drugs in mammalian cells.

It is easy to understand that small molecules can diffuse through the bacterial membrane as well as plasma membranes of the cells. The situation becomes completely different when you consider polymers which will be prevented from getting inside of bacteria, but can enter mammalian cells by endocytosis. So this is a general question of the Ames test with regard to polymeric drugs.

Dr. W. Leonard: Yes, I feel that the Ames test is probably completely useless in the case of polymeric drugs or polymeric additives, because you must have transport across the bacterial membrane. Our major use of the Ames test at this point is to screen our polymers for low molecular weight, mutagenic impurities or mutagenic degradation products arising through reasonable stress conditions. Such mutagenic agents could seriously reduce the safety margin of our polymers if they could not be removed or prevented. As no one test can assure the genetic safety of any agent, polymer or non-polymer, our product candidates will be evaluated in a battery of tests which can detect point mutations and chromosomal effects in both microbial and mammalian cells.

Dr. H. Ringsdorf (*Univ. of Mainz, Mainz, Germany*): During your attempts to avoid uptake of polymer by the small intestine, did you run into problems of cellular uptake by the small intestine?

Dr. W. Leonard: No, we have no evidence for an inverse molecular weight dependence on transport. All of our GI-transport studies thus far have given results consistent with the general trend of Figure 1.

Status of Delivery Systems for Drugs and Food Additives as Viewed by FDA

ARMAND R. CASOLA

Bureau of Drugs, Food and Drug Administration, Rockville, MD20857, U.S.A.

Federal regulatory authority over foods and drugs first came into being with the Pure Food and Drug Act of 1906 and was subsequently broadened and increased with the Food, Drug and Cosmetic Act of 1938 and its Amendments concerning Antibiotic Certification, Factory Inspection, Food Additives, Drug Efficacy and Medical Devices. A new drug may be marketed only after a new drug application (NDA) has been submitted to and approved by the F.D.A. The F.D.A. must be satisfied that the NDA contains the data, developed by means of investigational new drug (IND) studies, which adequately demonstrate the safety, usefulness, integrity and stability of the drug. Polymeric drug delivery systems employ unique methods of administering drugs so that both the rate and site of release of the drugs may be controlled in order to elicit certain predetermined and expected responses in the body.

The F.D.A. requires assurance, before approving such products, that they are manufactured and controlled so that the patient does not receive an unprecedented overdose of a powerful drug or on the other hand too little of the drug to bring about the desired pharmacological effect he expects and needs.

It certainly is an honor and a privilege for me to be invited to address you today on a subject both timely and consistent with the tenor of this symposium.

As you know there are several routes of administration of drugs to the human system in order to elicit a desired or expected response. One route is the oral route by such means as tablets, capsules, elixirs, etc. A second route, which is a special oral route, is the sublingual or buccal route in which tablets are used. A third route is the rectal route in which suppositories and retention enemas are used. A fourth route is by parenteral administration which includes the intravenous, intramuscular and subcutaneous routes. A fifth route is by means of inhalation for the delivery of gaseous or volatile substances into the system. A final route is the topical one, mainly to the skin but also to the eyes, nose, throat, ears, vagina, etc.

Polymers have been used in the preparation of dosage forms, imparting to them properties or characteristics of a highly desirable nature. Hydroxyethyl methylcellulose, hydroxypropyl methylcellulose and other substituted celluloses are used to film-coat tablets. Polyvinylpyrrolidone is used as a tablet binder and as a coating agent. Polyethylene glycols are not only useful as coating agents but also as bases for ointments and suppositories, etc.

Polymers have also played a unique and important role in bringing drugs to specific areas of the body by several routes just mentioned. By the use of various techniques these polymers have been used in the preparation of controlled release drug preparations.

Some examples of these special preparations are:

a) Coated granules or pellets incorporated into compressed tablets or hard gelatin capsules;

b) microcapsules as free flowing powders suitable for formulation as compressed tablets, hard gelatin capsules, suspensions and other dosage forms;

c) drugs enclosed in a polymeric membrane through which they diffuse at a controlled rate, such as Ocusert applied topically to the eye, and Progestasert applied intrauterinely.

Before any drug utilizing such a unique delivery system or any new drug, for that matter, can be marketed, it must be the subject of an approved new drug application.

At this point I think it would be both interesting and informative to present in a general way some of the highlights in the history of the drug provisions of the Federal Food, Drug, and Cosmetic Act and regulations.

A perusal of that history strongly implies that each of the major components of the Federal drug controls was enacted by Congress after its need was widely recognized by consumers, the regulated industry and the scientific community.

It is interesting to note that these congressional enactments, which resulted in substantial increases in federal regulatory authority, were brought about by certain sensational events or episodes which provoked a public outcry heard in Congress.

The first act regulating foods and drugs was the Pure Food and Drug Act of 1906. The need for this law was dramatized by Dr. Harvey W. Wiley's "poison squad" and Upton Sinclair's book, *The Jungle*. Under this act, drugs were required to conform to their declared standards of strength, quality and purity. Also, quite importantly, drugs listed in the United States Pharmacopeia and the National Formulary were required to meet the standards prescribed therein.

In 1937, the tragic sulfonilamide incident occurred in which more than one hundred people died because a new drug had been dissolved in a toxic solvent, the so-called Elixir of Sulfonilamide in diethylene glycol, and marketed without proper clinical testing. The public clamor that arose resulted in the enactment of the Federal Food, Drug, and Cosmetic Act of 1938 which required the manufacturer of a new drug to test it for safety before marketing.

There have been five amendments to the 1938 act which have greatly influenced the regulatory provisions dealing with drugs.

In 1945, 1947 and 1949 the act was amended to provide for batch certification of penicillin and certain subsequently discovered antibiotics and their derivatives because of the problems involved in assuring their identity, strength, purity and quality as lifesaving drugs.

In 1953 the act was amended to provide factory inspection authority since the authority implied in the original act of 1938 was found to be unenforceable by the Supreme Court in a decision in 1952.

In 1958 the act was amended to provide for the regulation of food additives. I shall spend a few minutes on this subject since it is part of the title of this presentation and also because it has some bearing on the review aspects of drug applications. The term food additive may be defined broadly to include any substance the intended use of which results or may result directly or indirectly in its becoming a component of or otherwise affecting the characteristics of any food.

The FD and C Act defines a food as adulterated if it is, or it bears or contains any food additive which is unsafe within the meaning of section 409 of the act. This definition is applicable to foods containing or bearing a food additive unless it is intended solely for investigational use by qualified experts or there is a regulation (part 121 of the regulations) in effect prescribing the conditions under which the additive may be safely used.

The regulation of plastics for use in food containers or in other materials which may come in contact with food is under this section of the act (section 409). Plastics used for food packaging and processing receive a great deal of attention by FDA scientists. This is understandable when we consider that the ingestion of a chemical as part of food may continue over the entire life span of the individual. Such food may be eaten by all age groups; or it may be consumed by individuals suffering from various pathological conditions or existing in borderline states of nutrition.

As you may very well know, the problem common to all plastic or polymeric materials intended for use in a manner that involves food contact, whether rigid molded utensils, films or piping, usually concerns the use of plasticizers, stabilizers, resins, antioxidants, colorants and adhesives or release agents. These must not release a harmful substance to foods wrapped, packaged, conveyed or stored in plastic containers, since the contamination of foods, animal and human, with harmful substances is a violation of the act whether such substance results from indirect additives during processing, from contact with the container or wrapper, or from deliberate incorporation in the food.

When safety has not been previously determined for authorization for use of plastic food containers, this can be secured under the provisions of section 409 where the facts demonstrated safety. Where there is no reasonable expectation of migration and the characteristics of the food in the container are not otherwise affected, there would be no need to petition for approval from FDA. However, when migration does occur, or may be reasonably expected to occur, then the identity and safety of the migrant should be established pursuant to the provisions of section 409.

The Food Additive Amendment requires that a new additive be proven safe before it may be used under any conditions. It also provides a system for the establishment of safe tolerance for substances that may be harmful in large quantities. Appropriate tests are required to establish the safety of a food additive and to determine tolerance levels that will be safe. Such tests include acute toxicity studies, allergic responses of the product, subacute and chronic toxicity studies, and absorption and metabolic fate and excretion of the substance, and pathology of the animals in which the tests are made.

Many food additives have a long history of apparently safe use. However, with the advent of more sophisticated, sensitive methods of detection and analysis, and more adequate animal models, some of the substances formerly accepted as safe for their intended uses now are being reevaluated in order to obtain reassurance of their safe use.

Packaging materials which have been found to be acceptable for food packaging are not always suitable for drug packaging. Some may be found suitable with additional testing. The food additive regulations are not applicable to drugs. The scientific information supporting these regulations, however, may be useful as part of the evidence required to show suitability for use in drug packaging. As we shall see, studies of the stability of each drug product in the market package are among the additional requirements necessary to establish the suitability of the container material in question.

In 1960 the act was amended to provide for the regulation of color additives. In effect all colors were brought within the jurisdiction of the law, requiring reevaluation of their safety by newer scientific methods, and permitting FDA to set limits and amount of each color used by means of regulations.

In 1962 the Kefauver–Harris Drug Amendments to the act of 1938 were unanimously passed by Congress, greatly increasing federal control of drugs. Again, these amendments were influenced by a tragic occurrence, the so-called "thalidomide incident." Here, too, the public demanded stricter controls. The definition of a "new drug" was enlarged to include drugs not generally recognized as effective for the purposes claimed in the labelling. The amendments strengthened the basis for the investigational new drug regulations. They brought all antibiotics intended for administration to man under the certification provisions of the act. Federal control of prescription drug advertisement was strengthened to assure that such ads are truthful and informative.

Recently, the act of 1938 was again amended by Congress with the Medical Device Amendments of 1976 which now clearly bring devices under a more stringent regulatory arm of the FDA. Thus devices, like drugs, must be shown to be safe and effective before they can be marketed, and preclearance authority has been granted to the FDA.

So much for the historical background of the laws and regulations administered by the FDA. Let us now consider the FD and C Act and the regulations pertaining to new drugs which are the areas of primary interest for us today.

First, the act specifically prohibits the shipment in interstate commerce of any new drug unless a new drug application is approved for it or it is exempted for investigational use in accord with the new drug regulations. The intent of the FDA in the control of new drugs is to insure that adequate safety and effectiveness testing has been accomplished before marketing. This is achieved by prohibiting interstate commerce of new drugs until they are approved as safe and effective in use under the conditions prescribed in the labelling.

Although the law prohibits interstate distribution of a new drug without an approved application, it does allow an exemption for shipping it solely for investigational use to experts qualified by scientific training and experience to investigate the safety and effectiveness of drugs, and authorizes the Secretary to issue regulations in this connection. The regulation promulgated for this purpose requires that the label of an investigational drug bear the statement that it is limited to investigational use. Before a sponsor of an investigation may distribute a drug for testing in man he must submit to the FDA certain specified information as part of a "Notice of claimed investigational exemption for a new drug" (IND) (Form 1571) and then allow FDA 30 days to review such information. If after the 30 day waiting period the sponsor has not received a communication from FDA requesting he withhold his studies, he may then proceed with the distribution of the drug. The information required by the IND includes the following:

1) The complete identity and composition of the drug together with its source, and sufficient manufacturing data to insure that appropriate standards are followed so as to assure safety and to give significance to the clinical trials.

2) All the information available to the sponsor derived from preclinical studies made on laboratory animals, and any clinical studies and experience with the drug from which the sponsor has concluded that it is reasonably safe to initiate clinical trials. Any consideration for the use of a polymer substance that is absorbed into the circulatory system must provide assurance that the polymer is metabolized and/or eliminated from the system within a reasonable time such as with the Dextrans which are used as blood expanders. For example, a small part of Dextran 75, corresponding to the low molecular weight molecules, is excreted during the first 1 or 2 days whereas the remainder is taken up by the reticuloendothelial system and is later metabolized requiring approximately 10 days. On the other hand, PVP or polyvinylpyrrolidone or povidone is taken up by the reticuloendothelial system and stored in macrophages in various organs without being metabolized. Low molecular weight molecules present are rapidly excreted.

Initially, FDA is primarily concerned with safety; however, the preclinical investigations shall not be considered adequate to justify clinical testing unless they give proper attention to the conditions of the proposed clinical testing.

3) An outline of the planned clinical investigations. Usually, it is developed in three phases which may overlap and may be contingent upon or modify one another:

Phase I, clinical pharmacology, is intended to develop data with regard to human toxicity, metabolism, absorption, elimination, preferred route of administration and safe dosage range. This represents the initial introduction of a drug into man. The levels of toxicity are usually determined in "normal" volunteers. Early dose-ranging studies for safety and effectiveness are also carried out in this phase. For some new drugs, ethical or scientific considerations dictate that selected patients be used as initial recipients. In phase I the usual number of people receiving the drug is between 20–80.

Phase II, clinical investigation, covers the initial controlled clinical trial on a limited number of patients for specific disease control or prophylaxis. Usually 100–200 patients are involved in this phase.

Phase III, clinical trials, in the form of controlled and uncontrolled trials, provide the assessment of a drug's safety and effectiveness and optimum dosage schedules in the diagnosis, treatment or prophylaxis of a group of subjects involving a given disease or condition.

4) The training and experience of the investigators deemed necessary to

carry out the investigations. The investigators are required to submit to the sponsor completed and signed Form FD 1572 for clinical pharmacology studies or Form FD 1573 for clinical trial.

5) All data generated by the investigators—to be submitted to the FDA by the sponsors. Emergency reports of severe, unexpected, untoward, reactions to them, must be promptly transmitted to the FDA, and their progress reports made at regular intervals, not to exceed one year.

When the sponsor decides he has developed the data and information which he feels demonstrates the safety, usefulness, integrity and stability of the new drug, he assembles it in the form of an application and submits it to the FDA. The act requires the new drug application to contain:

1) full reports of investigations which have been made to show whether or not the drug is safe and effective;
2) a full list of the components of the drug;
3) the quantitative composition of the drug;
4) a full description of the methods used in, and the facilities and controls used for, the manufacture, processing, and packaging of the drug;
5) required samples, and
6) proposed labelling.

The new drug application, form FD-356H, outlines the kind of information required to be submitted in accordance with section 314.1 of the New Drug Regulations.

Both INDs and NDAs are reviewed by various professional disciplines, namely, medical officers, pharmacologists, chemists; and when needed, by microbiologists and statisticians.

So much for the general overview of the history of the FD and C Act and requirements for new drug clearance. We shall now discuss in more specific terms some of the important requirements in the manufacturing and controls aspects of INDs and NDAs.

In an IND, Parts 1 through 5 detail the manufacturing and controls information that is needed. The reviewing chemists are responsible for evaluating this information as well as the technical portions of labelling covered in Part 7. In an NDA, Parts 5 through 9 pertain to manufacturing and controls while Part 4 pertains to labelling.

As the studies of the investigational drug progress, and as the sponsor continues to gain experience, both in the clinic and the laboratory, the IND must be amended, so that at the time of the NDA submission, the new drug is completely controlled. The FDA does not give a formal approval to the submissions in the IND, but it does monitor and review them. If deficiencies exist, the sponsor will be notified. If the deficiencies are of such a nature as to present a serious safety problem, the IND will be terminated by the FDA. However, in many cases the sponsor himself will promptly discontinue the IND. On the other hand, when adequate data are not submitted in an NDA to show safety and effectiveness of the drug for the intended use, a *not approvable* letter is sent to the applicant.

A. PREPARATION OF THE NEW DRUG SUBSTANCE

Phase I clinical investigations are one of the most important stages in drug development because an "unproven" drug entity is introduced into humans for the first time. Therefore maximum scientific information on all aspects of preclinical drug development is necessary.

Original INDs should include a detailed description of the method(s) of synthesis used for the preparation of the new drug substance. This information should include the preparation of the new drug substance used in animal studies, as well as that intended for human trials. The method(s) should be presented in a conventional form acceptable by scientific journals. It should include a flow-chart description of the synthetic paths, as well as a step-by-step write-up of each reaction. Adequate information should be included on all chemicals used in the manufacture, as well as physical–chemical data to support and confirm the structure of the intermediates formed in the synthesis. Pertinent literature references should be included. If the new drug substance is obtained by extraction from animal or plant sources, the same basic information should be submitted as described for synthetically produced drugs. If the new drug substance is a *complex* mixture, such as a natural product prepared by extraction or fermentation, whose characteristics are not adequately defined by properties readily determined in laboratory test procedures, then the method of preparation must be described as completely as possible.

Any change in a method must be submitted to the FDA with proper documentation. Polymorphism, particle size and salt/acid/base form are important characteristics and must be considered. By Phase III, the actual production procedure that will be followed should be submitted in detail.

B RAW MATERIAL CONTROLS

1 New drug substance

Since the activity of the intended product lies in the new drug substance, its identity, quality, strength and purity are of vital concern. Its control from the very first introduction into animals is important in order to have full and adequate data for making sound judgments as to its suitability for its intended use.

In the initial submission of an IND, full and adequate information should be submitted that characterizes the structure of the new drug substance. In addition, the purity and identification and quantitation of foreign substances should be included. Proposed provisional specifications should be based on laboratory experience with representative batches of the substance and detailed studies of the tests conducted and proposed.

In the first stages of development, a provisional (or interim) reference standard should be prepared. It should possess the highest purity attainable by reasonable effort. Full information should be submitted on the tests used and the results obtained from these tests in order to completely characterize the standard. An independent method of determining purity must be included.

By phase III, the specifications and tests for the new drug substance should be set and backed by good scientific data and information. They should assure absolute identity and highest quality and purity of the new drug substance. This information should then be sufficient in satisfying the requirements for Part 8d of the NDA.

2 Inactive ingredients

The tests and specifications for the inactive ingredients should be adequate to assure their identity, strength, quality and purity. All the tests should be described in detail. If the substance is recognized in an official compendium, it should be so stated.

Part 8(d) of the NDA should include full information on the specifications and tests for all raw materials. Each ingredient shall be tested for conformity with all appropriate written specifications for identity, purity, strength and quality. In lieu of such testing by the manufacturer, a complete report of analysis may be accepted from the supplier of a component, provided that at least one specific identity test is conducted on such component by the manufacturer, and provided that the manufacturer established the reliability of the supplier's analysis through appropriate validation of the supplier's test results. At this point it is important to note that if any one of the drug components is of a polymeric nature – used either for film-coating, microencapsulations or as a platform for the release of an active ingredient – then full information must be given on the method of manufacture, the components involved, etc., including the presence of any residual monomer in the polymer. If some of this information is not available to the applicant from the supplier because of reasons of confidentiality, then it may be submitted either directly by the supplier to the FDA for incorporation into the subject application or be submitted by the supplier in the form of a Drug Master File.

C DOSAGE FORMS

1 Preformulation

In order to maximize and/or control desorption of the new drug substance, an extensive study should be undertaken prior to the preparation of a dosage form. The formulation cannot provide the investigator with an active ingredient in some arbitrary dosage form. It is well established that such properties as particle size, pK_a, excipients and crystalline form are capable of affecting the rate, the extent and nature of the biological response. Application of certain physical–chemical principles in the design and development of a dosage form is of considerable importance. Using this approach, one should be able to optimize formulation requirements, predict stability, foresee drug degradation (its mechanism and products), and assure to a greater degree, a quality, stable drug formulation. Some of these physical–chemical properties in addition to the ones mentioned are:

a) solubilities in various solvents
b) dissolution profile
c) stability to light, temperature and humidity
d) compatability with excipients, lubricants, stabilizers, preservatives, suspending agents, dispersing agents, and colors.

2 Preliminary formulation screen

The rationale for the selection of a dosage form is based on a number of scientific and practical parameters. Preformulation studies together with product development studies designed to test formulation, related problems, equipment, etc., may influence a dosage form decision. Each formulation and batch produced should have identifying numbers which should be correlated with the studies in which they are used.

In phase II a reevaluation of the formulation may be necessary because of information obtained during phase I. Physical–chemical changes in the new drug substance, due to changes in preparation, coupled with added preformulation data may dictate changes in the formulation.

3 Dosage form analysis

Chemical analysis of a dosage form is one of the parameters by which quality of formulation can be accurately measured. Therefore, the development of an efficient and sensitive method is most important. Again, as in the case of a good formulation, the physical–chemical properties of a new drug substance play an important part in the development of an acceptable assay. The assay should provide:

a) unequivocal identification of the active ingredient in the dosage form.

b) specific quantitation of the active ingredient with, or without, the presence of degradation products.

c) proof that excipients do not interfere in the assay of both the freshly prepared dosage forms and those stored at stability stations.

In phase II, changes in the assay may be dictated by changes in the formulation. In any event, any proposed changes should be directed at making the procedure scientifically more precise and technically better.

By phase III, the assay should in general be finalized. It must be a regulatory method which does not call for the use of a formulation blank. The applicant may choose other quality control methods provided he can demonstrate the equivalency of these methods to the regulatory method.

4 Dosage form specifications

When a dosage form is proposed for use in humans, there must be adequate control over the various physical–chemical characteristics of the drug. The lack of adequate product specifications constitutes an important obstacle in properly assessing clinical efficacy and safety. Some examples of these important factors are

a) Weight variation for tablets, capsules, and other solid dosage forms; also disintegration times and dissolution tests.

b) Content uniformity for tablets, capsules, sterile solids and suspensions.

c) Clarity, presence of particulate matter, preservative assay, isotonicity, and pH determination of liquids.

d) Sterility of ophthalmic solutions and ointments.

e) Sterility, non-pyrogenicity and container fill of injectables.

f) Leakage test for ampules and aerosols; also metering tests and container pressure for aerosols, etc.

Product specifications reflect the inherent quality of the intended product and the degree of accuracy with which it can be repeatedly made. Rigid specifications thus imply the confidence in the manufacturing process and associated testing methods. The product specifications should be developed, as experience is gained, so that it would approach NDA standards by Phase III.

At this point I would like to address specifically polymeric drug delivery systems. Two preparations that I have mentioned earlier belong to this group, one an implant and another a membrane which are designed to release a controlled amount of drug over a prolonged period of time.

Now the rationale for the development and use of controlled drug release dosage forms may be described as follows:

1) Decrease of toxicity and occurrence of adverse reactions by control of the level of the drug and/or drug metabolites in the blood and at depot sites.

2) Better drug utilization by enabling a smaller drug dose in a controlled drug release form to produce the same clinical effect as the conventional dosage form.

3) Controlling the rate and site of release of a drug so that the drug is released where the activity is needed rather than at other sites where it can cause adverse reactions.

4) Provide a more uniform blood concentration and/or provide a more predictable drug delivery system.

5) Provide greater patient convenience and/or better patient compliance by significantly prolonging the interval between administration.

In the development testing, and setting of specifications for these products, the FDA is concerned about the following factors:

1) The rationale for the dosage form and the elucidation of the mechanism upon which the mode of release is based.

2) The capability of the applicant to produce the drug in a uniform manner, batch after batch, repeating the identical chemical and physical characteristics so that its performance will continuously occur within specified limits.

3) An adequate *in vitro* test should be defined that is capable of routinely checking the characteristics or specifications of the product.

4) There should be an adequate correlation of the *in vitro* test results with those obtained *in vivo* so that the desired effects of future batches can always be predicted.

5) The methods utilized in determining the *in vitro* characteristics should be amenable to duplication in the FDA laboratories as required for any of the tests for any new drug.

The points that have been enunciated are important because the FDA is necessarily concerned that such products always present potential dangers in that either the patient receives an unprecedented overdose of a powerful drug, or on the other hand, receives much too little of the drug to bring about the desired pharmacological effect he expects and needs.

D MANUFACTURING

In phase I, the proposed manufacturing process should be precise and reliable so as to guarantee the physical and chemical integrity, and sterility, when applicable, of the drug. The information should present a detailed description of each step, including necessary equipment and appropriate in-process controls. Quality assurance procedures should be specified. In phases II and III, additional information should be submitted adequately describing the methods. The information must be coherent and well organized in order to permit a meaningful review and verification. Of course this also applies to the information submitted in the NDA.

E CONTAINERS

The most carefully produced formulation can be no better than the quality of the container in which it is held. Whether or not a drug is safe, contains the correct dose and is therapeutically effective depends to a significant extent upon the suitability of its packaging materials and the adequacy of the packaging. Of course, the closure is regarded as part of the container.

Some of the many characteristics that must be considered in selecting a drug package are:

1) Nontoxic nature of the container materials.
2) Reproducibility of the container material.
3) Extractables from container material and their toxicity profile.
4) Ability of the container material to withstand environmental conditions of temperature, humidity and light.
5) Compatability of the container material with the drug.
6) Ability of the container to protect the drug during extended and exaggerated storage conditions.
7) Container resistance to impact.
8) Container tamperproofness and transparency.
9) Ease of opening and resealing.
10) Protection of sterility and precautions against resterilization, as aoplicable.

Provisional container specifications should include the following, when applicable:

1) Type of materials—whether glass, aluminum, plastic, etc.
2) Form of the container—whether it is a jar, bottle, barrel, tube, etc.
3) Dimensions of the container.

4) Weight and/or fill variation limits.
5) Permeability to gas, moisture, or light.
6) Limits for extractables, to impact resistance and to transparency.
7) Tamperproofness.

Changes in containers may be necessary and occur during phase II. However, they should be accompanied by appropriate justification and data. By the time phase III is reached the specifications and tests for containers should be of the type suitable for an NDA.

Packaging with plastic material offers several advantages over that with glass: namely, reduced breakage, economy of storage and transport. However, as you know, plastics are usually complex mixtures, containing additives necessary for proper manufacturing and modification of physical and chemical properties. These properties can be varied over wide ranges, giving rise to varying characteristics with regard to moisture and gas penetration, leaching of ingredients, adsorption, absorption and chemical reactivity with drugs, any of which could change with aging.

Because of the complex nature of these containers and of the confidential nature of the processing procedures and formulations, the information required in INDs and NDAs may be submitted in the form of Drug Master Files (DMFs). A large number of bulk resin producers, fabricators and processors have taken advantage of this method of preserving confidentiality, and at the same time this centralization of information has proven very helpful to the FDA reviewer. However, the DMF holder is required to keep the DMF current and must not implement any changes unless and until the holders of INDs and NDAs, to which the DMF refers, have been informed and have given approval to the proposed changes.

One more point concerning containers—several tests for containers have been incorporated in the United States Pharmacopeia and the National Formulary and should prove helpful in determining the adequacy of containers for their intended use:

1) light transmission;
2) biological tests for plastic containers;
3) physico-chemical tests for plastic containers;
4) chemical resistance for glass containers;
5) moisture permeability of plastic containers.

F STABILITY

To assure its safety and effectiveness, a pharmaceutical product must retain its identity, strength, quality and purity for the entire shelf life of the product.

Assurance that the drug dosage form in its container will be adequately stable for the anticipated shelf life must come from an accumulation of data on the packaged drug. These data involve chemical, physical, microbiological, therapeutic and toxicological considerations. Information from initial and subsequent tests performed on drug substances, the various dosage forms, and containers provide the basic data upon which the stability profile is constructed.

A stability profile must first be obtained for the new drug substance at a very early time in order to support standards of identity, strength, quality and purity proposed for it. The analytical methods used to obtain these data should be capable of accurately measuring the characteristic physical and chemical properties associated with the molecule.

The sponsor of an IND should submit preliminary data that would support the stability of the dosage form during the proposed phase I studies. For parenterals, it is essential that initial stability data be incorporated in the original submission. As data is accumulated, it should be submitted throughout the phase II and III studies so that by the time the NDA is ready for submission a complete stability profile is available.

Thus the NDA submission should contain full and complete information on the stability of both the new drug substance and the finished dosage form. The stability studies should be submitted in a manner that will include all details and make them conducive to easy review. Data should be cumulative with all conditions, code numbers, lot numbers, dates, methods clearly defined and displayed.

Stability studies should be monitored by test methods which are capable of distinguishing the intact drug from its degradation products. The experimental evidence to support this capability, as well as the accuracy and precision of the methods, must be included in the submissions. Degradation products should be qualitatively and quantitatively determined, and the chemical mechanisms by which they are formed should be elucidated, if known. In addition, stability studies should include adequate testing, with regard to controlled release dosage forms in order to determine that aging does not adversely alter the release characteristics of the drug.

As is well known, the scale and method of manufacture, the change in the source of excipients of the dosage form can have an effect on the stability of the product. Drugs that are susceptible to heat, light and air should be very carefully monitored. Changes in manufacturing procedures may expose the product to these factors for longer periods of time and in a different manner. The presence of trace quantities of metals may affect stability. All factors should be considered at an early stage so that proper specifications and methods may be developed. Changes in the sterilization procedure may have a profound effect on the product. Thus, changes for example, from heat sterilization to aseptic membrane filtration or to ethylene oxide gas sterilization must be evaluated very carefully from all aspects. Doing so will provide substantial insurance against the possibility of batch failures and recalls. It should be noted that we are talking not only about chemical degradation, that is, failure to maintain approved specifications, but also physical factors, such as emulsion separation, microbiological factors such as mold growth, therapeutic factors such as a decrease in dissolution time rate, and toxicological factors such as an increase in degradation products that are toxic.

When a final container has been selected for marketing, it should be used during the stability studies in the exact same manner as that proposed for the marketing of the drug. For example, in the case of plastic containers, the label should be attached or imprinted, as the case may be. Studies should include migration of adhesive and dye components through the container wall. In evaluating multiple-dose vials, containers must be stored in both the upright and side positions so that the liquid is in contact with the closure.

Temperature, of course, is a parameter of prime importance. The range of controlled laboratory storage conditions that have been used and reported is quite broad, beginning at below freezing and rising to above 50°C. Intervals normally studied are 5–10°C, controlled room temperature, ambient room temperature, 35–40°C and 41–50°C. It must be realized that several weeks at 50°C is strictly speaking not an accelerated test, especially when we consider prevailing summer temperatures in various parts of the country and what might be expected in the holds of ships, military depots or in trucks or trains.

Accelerated stability studies combined with basic stability information on the new drug substance, dosage forms, and containers are useful and acceptable to support expiration dates both in the original submission and for changes in the synthesis of the new drug substances, changes in the formulation of the dosage form and changes in the containers and packaging of the commercial product.

To facilitate the evaluation of a stability profile in the determination of the anticipated shelf life of a product, the applicant should include a full description of the analytical method used, its accuracy and precision, and other considerations. The method of best fit, copies of the graphs plotted, as well as a complete explanation of techniques employed in the extrapolation and curve fitting, should be included. If the Arrhenius equation is used to estimate shelf life, a minimum of four points should be used, unless full justification is provided for the use of a lesser number. In the presentation of accelerated stability data, ambient temperature data on the particular dosage form in the same container should be included.

Stability studies should consider not only the active ingredient(s) in a dosage form, but also consider other substances such as preservatives. Antimicrobial substances, in a parenteral, for example, should be tested by assay and microbial challenge. In this regard, the "Antimicrobial Preservatives—Effectiveness Test" described in the U.S.P. should be utilized.

In describing the method of testing, full information should be given regarding the method of sampling for the dosage form. For example, full information on how samples are obtained from a cream in a tube or a suspension in an aerosol is necessary for adequate evaluation. In regard to suspensions, a more vigorous treatment to assure uniformity of the sample than that recommended in the labelling should not be used.

After the original NDA has been approved, supplemental applications may be submitted providing for changes. If the change is one that requires prior approval, adequate data should be submitted to assure the continued integrity of the drug. The amount and type of data depend on the proposed change and background data available in the NDA.

In conclusion let me say that the Food and Drug Administration's role in assuring drug quality for all our people is a very important and necessary one. But in order for the FDA to carry out this function in accordance with the mandate of Congress it needs the cooperation and help of many groups, such as the medical, pharmaceutical, pharmacological and chemical societies, the pharmaceutical industry, academia and interested citizenry. With such cooperation there is no doubt that a safe, effective, and economical drug supply can and will be assured.

DISCUSSION

Dr. R. Phares (*Alza Corporation, Palo Alto, CA*): Do the new device regulations also relate to devices used in animals or are they restricted to humans?

Dr. A. Casola: The new Device Regulations relate solely to human use.

Dr. R. Phares: In considering controlled delivery of drugs and drug delivery systems, we are trying to reduce the total amount of drug the patient must receive to get a desired pharmacological response, but we are also looking to provide enough drug to last over a long period of time so as to reduce patient intervention, i.e., free the patient from frequent readministration of medication. Do you have any idea what the general feeling or attitude is concerning the fact that the patient may have in his body, at any one time in the form of a controlled drug delivery system, more than the acute toxic amount of a particular drug substance?

Dr. A. Casola: As I mentioned in my presentation, the FDA is always necessarily concerned that a patient being administered a controlled drug release system may be subjected to an unprecedented overdose of a powerful or potent drug.

Dr. R. Phares: Do you know the general philosophy or attitude of the FDA toward this question?

Dr. A. Casola: The FDA permits the marketing of such drugs provided the firms can demonstrate that they are capable of manufacturing them and have developed adequate testing procedures (e.g., *in vitro* tests correlated to *in vivo* tests) that will assure the safe use of such potent drugs.

Dr. D. Bennett (*Midland Hospital, Midland, MI*): Does the FDA feel that polymers are unique compared to other types of drug delivery systems?

Dr. A. Casola: No, I don't think so. They may be different from other forms of drug delivery systems but the FDA applies the same principles in evaluating their safety and efficacy.

Dr. Walles (*Dow Chemical Company, Midland, MI*): I wonder what safeguards there are if animals are treated with a drug when they are sick, then the meat is sold. What kind of protection is there for the public in general?

Dr. A. Casola: Before drugs may be used in food animals, data must be submitted to the Bureau of Veterinary Medicine to show its safety and efficacy. In addition, data are generated to demonstrate the time necessary to allow complete depletion of the drug in the animals. Owners of such animals cannot sell or market them until the completion of such time periods.

Dr. Walles: Well, then in general how is the consumer protected from food additives used for chickens and other animals?

Dr. A. Casola: The Bureau of Foods has the responsibility to assure the safety of food additives. When food additives are approved for use through the mechanism of food additive petitions, they become the subject of food additive regulations which are published in the Code of Federal Regulations, Title 21, Part 121.

Subject Index